Advances in Intelligent Systems and Computing

Volume 836

Series editor

Janusz Kacprzyk, Polish Academy of Sciences, Warsaw, Poland
e-mail: kacprzyk@ibspan.waw.pl

The series "Advances in Intelligent Systems and Computing" contains publications on theory, applications, and design methods of Intelligent Systems and Intelligent Computing. Virtually all disciplines such as engineering, natural sciences, computer and information science, ICT, economics, business, e-commerce, environment, healthcare, life science are covered. The list of topics spans all the areas of modern intelligent systems and computing such as: computational intelligence, soft computing including neural networks, fuzzy systems, evolutionary computing and the fusion of these paradigms, social intelligence, ambient intelligence, computational neuroscience, artificial life, virtual worlds and society, cognitive science and systems, Perception and Vision, DNA and immune based systems, self-organizing and adaptive systems, e-Learning and teaching, human-centered and human-centric computing, recommender systems, intelligent control, robotics and mechatronics including human-machine teaming, knowledge-based paradigms, learning paradigms, machine ethics, intelligent data analysis, knowledge management, intelligent agents, intelligent decision making and support, intelligent network security, trust management, interactive entertainment, Web intelligence and multimedia.

The publications within "Advances in Intelligent Systems and Computing" are primarily proceedings of important conferences, symposia and congresses. They cover significant recent developments in the field, both of a foundational and applicable character. An important characteristic feature of the series is the short publication time and world-wide distribution. This permits a rapid and broad dissemination of research results.

More information about this series at http://www.springer.com/series/11156

Oleg Chertov · Tymofiy Mylovanov
Yuriy Kondratenko · Janusz Kacprzyk
Vladik Kreinovich · Vadim Stefanuk
Editors

Recent Developments in Data Science and Intelligent Analysis of Information

Proceedings of the XVIII International Conference on Data Science and Intelligent Analysis of Information, June 4–7, 2018, Kyiv, Ukraine

 Springer

Editors

Oleg Chertov
Applied Mathematics Department
Igor Sikorsky Kyiv Polytechnic Institute
Kyiv, Ukraine

Janusz Kacprzyk
Systems Research Institute
Polish Academy of Sciences
Warsaw, Poland

Tymofiy Mylovanov
Department of Economics
University of Pittsburgh
Pittsburgh, PA, USA

Vladik Kreinovich
Department of Computer Science
University of Texas at El Paso
El Paso, TX, USA

Yuriy Kondratenko
Department of Intelligent Information
 Systems
Petro Mohyla Black Sea National University
Mykolaiv, Ukraine

Vadim Stefanuk
Institute for Information Transmission
 Problems
Russian Academy of Sciences
Moscow, Russia

ISSN 2194-5357   ISSN 2194-5365   (electronic)
Advances in Intelligent Systems and Computing
ISBN 978-3-319-97884-0   ISBN 978-3-319-97885-7   (eBook)
https://doi.org/10.1007/978-3-319-97885-7

Library of Congress Control Number: 2018950188

This Springer imprint is published by the registered company Springer Nature Switzerland AG
The registered company address is: Gewerbestrasse 11, 6330 Cham, Switzerland

This volume is dedicated to the memory of the conference founder, Prof. Dr. Tetiana Taran, an outstanding scientist in the field of artificial intelligence, a fellow of the Ukrainian Association of Developers and Users of Intelligent Systems and of Russian Association for Artificial Intelligence.

Prof. Taran was born in 1946 in Alexandrovsk-Sakhalinsky (Sakhalin region, Russia) in the family of a military serviceman. Soon, the family moved to Sevastopol (Crimea, Ukraine). In 1964, Tetiana graduated from secondary school with a gold medal, and entered Sevastopol Instrument Engineering Institute. She graduated in 1969 with honors, majoring in engineering and mathematics.

Upon graduation, T. Taran moved to Kyiv, where she worked in the Institute for Superhard Materials and the Kyiv Institute of Automatics. In 1971–1973, she was a PhD student at the Kyiv Polytechnic Institute (KPI). After defending her dissertation, Prof. Taran worked at the Applied Mathematics Department of KPI until she untimely passed away in 2007.

During her fruitful life, T. Taran had over 130 publications, including 4 books, 12 textbooks, and 113 papers in English, Ukrainian, and Russian. In particular, she is the co-author of the first textbook on artificial intelligence in Russian: T. Taran, D. Zubov, Artificial Intelligence. Theory and Applications, Luhansk (2006). Research areas of Prof. Taran spanned various fields, in particular, reflexive processes, conceptual structures, cognitive analysis, decision making, and many others.

Preface

This volume of the *Advances in Intelligent Systems and Computing* series is the proceedings of the *XVIII International Conference on Data Science and Intelligent Analysis of Information* (ICDSIAI'2018), held on June 4–7, 2018, in Kyiv (Ukraine).

The history of the conference goes back to 2001 when the workshop titled *Intelligent Analysis of Information* took place, organized by Applied Mathematics Department of Igor Sikorsky Kyiv Polytechnic Institute and Russian Association for Artificial Intelligence. In 2005, the workshop was transformed into an international conference. In 2018, the conference changed its title to International Conference on Data Science and Intelligent Analysis of Information, to reflect the broadening of its scope and composition of organizers and participants.

The conference brought together researchers working in the field of data science in all of its areas. Novel theoretical developments in methods and algorithms for big data mining and intelligent analysis of information were supplemented with applications, especially in the fields of economics, education, ecology, law, and others.

This year, sixty-one papers were submitted for review. Thirty-seven of them (61%) were selected for publication. Authors of these papers represent ten countries (Belarus, Georgia, India, Kazakhstan, Malaysia, Poland, Russia, Serbia, Ukraine, and USA). Five keynote talks were delivered:

1. Prof. Vadim Stefanuk, Institute for Information Transmission Problems, Russian Academy of Sciences, Russia; Peoples' Friendship University of Russia, Russia, "Fuzzy Set Theory is not Fuzzy."
2. Prof. Vladik Kreinovich, University of Texas at El Paso, USA, "Bounded Rationality in Decision Making Under Uncertainty: Towards Optimal Granularity."
3. Prof. Janusz Kacprzyk, Systems Research Institute, Polish Academy of Sciences, Warsaw, Poland, "Towards More Realistic, Human-Centric Decision Making Via Some Multiagent Paradigms."

4. Prof. Ashok Deshpande, College of Engineering, Pune, India, "Application of Zadeh Deshpande Formalism in Educational Data Mining."
5. Prof. Nikhil R. Pal, Indian Statistical Institute, Kolkata, India, "Making Neural Networks Smart Enough to Say Don't Know—The Open World Classification Problem."

In terms of structure, the chapters of the book are grouped into six sections:

1. Machine Learning: Novel Methods and Applications.
2. Data Analysis Using Fuzzy Mathematics, Soft Computing, Computing With Words.
3. Applications of Data Science to Economics and Applied Data Science Systems.
4. Knowledge Engineering Methods, Ontology Engineering, and Intelligent Educational Systems.
5. Intelligent Search and Information Analysis in Local and Global Networks.
6. Novel Theoretical and Practical Applications of Data Science.

We would like to express the highest gratitude to the Publication Chair, to all the members of the Program and Organization Committees, and to the reviewers, without whom the conference would not take place. Our special thanks go to Dr. Thomas Ditzinger, Executive Editor at Springer-Verlag, for his openness for collaboration and invaluable assistance in the preparation of this volume.

P. S.: Five editors of this volume would like to thank Prof. Janusz Kacprzyk, Springer-Verlag *Advances in Intelligent Systems and Computing* Series Editor, for his help in bringing this book to life.

June 2018

Oleg Chertov
Tymofiy Mylovanov
Yuriy Kondratenko
Janusz Kacprzyk
Vladik Kreinovich
Vadim Stefanuk

Organization

ICDSIAI'2018 is organized by Applied Mathematics Department of Igor Sikorsky Kyiv Polytechnic Institute and Kyiv School of Economics.

Executive Committee

Honorary Co-chairs

Mykhailo Zgurovsky	Igor Sikorsky Kyiv Polytechnic Institute, Ukraine
Oleg Kuznetsov	V. A. Trapeznikov Institute of Control Sciences, Russian Academy of Sciences, Russia

General Co-chairs

Oleg Chertov	Igor Sikorsky Kyiv Polytechnic Institute, Ukraine
Tymofiy Mylovanov	Kyiv School of Economics, Ukraine; University of Pittsburgh, USA

Program Chairs

Janusz Kacprzyk	Systems Research Institute, Polish Academy of Sciences, Poland
Yuriy Kondratenko	Petro Mohyla Black Sea National University, Ukraine
Vladik Kreinovich	University of Texas at El Paso, USA
Vadim Stefanuk	Institute for Information Transmission Problems, Russian Academy of Sciences, Russia; Peoples' Friendship University of Russia, Russia

Publication Chair

Dan Tavrov   Kyiv School of Economics, Ukraine; Igor
                                    Sikorsky Kyiv Polytechnic Institute, Ukraine

Local Organizing Committee

Chair

Ivan Dychka   Igor Sikorsky Kyiv Polytechnic Institute

Members

Olesia Verchenko   Kyiv School of Economics
Iryna Sobetska   Kyiv School of Economics
Olena Temnikova   Igor Sikorsky Kyiv Polytechnic Institute
Sergiy Kopychko   Igor Sikorsky Kyiv Polytechnic Institute

Program Committee

Oleg Chertov   Igor Sikorsky Kyiv Polytechnic Institute, Ukraine
Svetlana Cojocaru   Institute of Mathematics and Computer Science,
                                    Academy of Sciences of Moldova, Moldova
Tom Coupe   University of Canterbury, New Zealand
Ivan Dychka   Igor Sikorsky Kyiv Polytechnic Institute, Ukraine
Vladimir Golenkov   Belarusian State University of Informatics
                                    and Radioelectronics, Belarus
Eugen Ivohin   Taras Shevchenko National University of Kyiv,
                                    Ukraine
Janusz Kacprzyk   Systems Research Institute, Polish Academy
                                    of Sciences, Poland
Anton Kobrynets   Gesellberg Cryptofinance, Ukraine
Yuriy Kondratenko   Petro Mohyla Black Sea National University,
                                    Ukraine
Bart Kosko   University of Southern California, USA
Vladik Kreinovich   University of Texas at El Paso, USA
Dmytro Lande   Institute for Information Recording, National
                                    Academy of Sciences of Ukraine, Ukraine
Agassi Z. Melikov   Azerbaijan National Academy of Sciences,
                                    Azerbaijan

Jerry Mendel	University of Southern California, USA
Tymofiy Mylovanov	Kyiv School of Economics, Ukraine; University of Pittsburgh, USA
Binh P. Nguyen	Institute of High Performance Computing, Singapore
Vilém Novák	University of Ostrava, Czech Republic
Maksym Obrizan	Kyiv School of Economics, Ukraine
Irina Perfilieva	University of Ostrava, Czech Republic
Viktor Smorodin	Francisk Skorina Gomel State University, Belarus
Andrei Snarskii	Igor Sikorsky Kyiv Polytechnic Institute, Ukraine
Vadim Stefanuk	Institute for Information Transmission Problems, Russian Academy of Sciences, Russia; Peoples' Friendship University of Russia, Russia
Sergiy Syrota	Igor Sikorsky Kyiv Polytechnic Institute, Ukraine
Oleksandr Talavera	Swansea University, UK
Dan Tavrov	Kyiv School of Economics, Ukraine; Igor Sikorsky Kyiv Polytechnic Institute, Ukraine
Dongrui Wu	DataNova LLS, USA
Ronald R. Yager	Iona College, USA
Ludmila Zhilyakova	V. A. Trapeznikov Institute of Control Sciences, Russian Academy of Sciences, Russia

Subreviewers

| Iulian Secrieru | Institute of Mathematics and Computer Science, Academy of Sciences of Moldova, Moldova |
| Inga Titchiev | Institute of Mathematics and Computer Science, Academy of Sciences of Moldova, Moldova |

Contents

Novel Theoretical and Practical Applications of Data Science

Machine Learning: Novel Methods and Applications

Use of Symmetric Kernels
for Convolutional Neural Networks

Viacheslav Dudar and Vladimir Semenov[(✉)]

Faculty of Computer Science and Cybernetics,
Taras Shevchenko National University of Kyiv, Kyiv, Ukraine
slavko123@ukr.net, semenov.volodya@gmail.com

Abstract. At this work we introduce horizontally symmetric convolutional kernels for CNNs which make the network output invariant to horizontal flips of the image. We also study other types of symmetric kernels which lead to vertical flip invariance, and approximate rotational invariance. We show that usage of such kernels acts as regularizer, and improves generalization of the convolutional neural networks at the cost of more complicated training process.

Keywords: Convolutional neural network
Symmetric convolutional kernel · Horizontal flip invariance
Vertical flip invariance

1 Introduction

Convolutional neural networks (CNNs) had become one of the most powerful tools in machine learning and computer vision for last several years [1]. CNNs show state-of-art accuracies for most state-of-art benchmark datasets, such as ImageNet [2]. CNN has a set of parameters (convolutional kernels, biases, and weights of the last fully connected layers) that are adjusted during the training process. Number of such parameters is typically very large (order of millions or tens of millions). Models with so many parameters do not overfit the data much because of the following reasons:

- Data augmentation. Training set is augmented during training in different ways: affine transformations, random subimage selections, random color distortions for each pixel [2].
- Efficient regularization techniques. Dropout is one of the most powerful regularization techniques, that corresponds to approximate ensembling over exponential number of subnetworks [3].
- Inner structure of the CNN. Weight sharing is used to enforce approximate invariance of the network output to translations of the input image [1].

At this work, we focus on CNNs for classification. We propose to make the network output invariant to horizontal image flips via introduction of horizontally symmetric convolutional kernels. Thus we are modifying inner structure of the CNN to enforce additional invariance to improve generalization to the new data.

© Springer Nature Switzerland AG 2019
O. Chertov et al. (Eds.): ICDSIAI 2018, AISC 836, pp. 3–10, 2019.
https://doi.org/10.1007/978-3-319-97885-7_1

2   Symmetric Image Kernels

Let's consider typical CNN architecture that consists of several convolutional layers, followed by elementwise nonlinear function (in most cases it's RELU nonlinearity) alternating with pooling layers (it could be average or max pooling layers) followed by one or several fully connected layers with softmax activation function and trained with categorical cross-entropy loss.

Consider the first convolutional layer of the net. This layer is translation equivariant, so output of the layer is changed in the same way as the input for translations. But it's not equivariant to the horizontal image flip in case of arbitrary convolution kernel.

We will focus on kernels of size 3×3 that are the most widely used [1]. General 3×3 convolution kernel:

$$k = \begin{bmatrix} a & b & c \\ d & e & f \\ g & h & i \end{bmatrix}$$

We propose to use horizontally symmetric kernels of the form:

$$k = \begin{bmatrix} a & b & a \\ d & e & d \\ g & h & g \end{bmatrix}$$

We show that in this case convolution layer becomes equivariant to horizontal image flips, and the whole network, under certain structure, becomes invariant to horizontal flips.

It is enough to show equivariance in one-dimensional case (for each row of the image). Consider arbitrary vector:

$$x = (x_1 \ldots x_n)$$

and one-dimensional symmetric kernel:

$$k = (a, b, a)$$

At the moment we consider convolution with stride 1 and no padding.

$$x * k = (a(x_1 + x_3) + bx_2 \ldots a(x_{n-2} + x_n) + bx_{n-1})$$

Convolution with flipped vector $\hat{x}$:

$$\hat{x} = (x_n \ldots x_1)$$

$$\hat{x} * k = (a(x_{n-2} + x_n) + bx_{n-1} \ldots a(x_1 + x_3) + bx_2) = \widehat{x * k}$$

Thus convolution with symmetric kernel of the flipped image is equal to the flip of convolution with initial image. Thus symmetric kernel makes convolution equivariant. Clearly, this result generalizes for 3D convolutions used in CNNs.

Consider now other types of operations performed in CNN. Elementwise application of non-linear function, max and average pooling layers are also clearly flip equivariant. Thus superposition of 3D convolutions, non-linear functions and poolings is also flip equivariant.

The only transformation used in CNNs that does not have this property is the flatten layer that maps tensor to vector before fully connected layers. That's why we consider only cases when the last layer is global pooling (max or average). This condition is not restrictive, as the newest architectures (as DenseNet [4]) use global pooling before Fully Connected layers.

Since global pooling (pools tensor to vector of the same depth) is invariant to horizontal flips, the whole network output becomes invariant to horizontal flips. Thus if symmetric kernels are used then posterior probabilities $p\left(C_i|x\right)$ produced by the CNN are exactly the same for the flipped image $\widehat{x}$:

$$p\left(C_i|x\right) = p\left(C_i|\widehat{x}\right)$$

3   Levels of Symmetry

We experimented with several levels of symmetry of convolutional kernels. They are summarized in the Table 1. Note, that the third column contains induced equivariances for convolutional layers that in turn correspond to induced invariances of the network output (it happens in case global pooling is used before fully connected layer).

Table 1. Symmetry levels of convolutional kernels

Symmetry Level	Kernel form	Induced network invariances
0	$k = \begin{bmatrix} a & b & c \\ d & e & f \\ g & h & i \end{bmatrix}$	No induced invariances
1	$k = \begin{bmatrix} a & b & a \\ d & e & d \\ g & h & g \end{bmatrix}$	Horizontal flip
2	$k = \begin{bmatrix} a & b & a \\ d & e & d \\ a & b & a \end{bmatrix}$	Horizontal flip Vertical flip
3	$k = \begin{bmatrix} a & b & a \\ b & e & b \\ a & b & a \end{bmatrix}$	Horizontal flip Vertical flip 90° rotations
4	$k = \begin{bmatrix} a & a & a \\ a & e & a \\ a & a & a \end{bmatrix}$	Horizontal flip Vertical flip 90° rotations Approximate arbitrary rotations

Different symmetry levels are applicable to different datasets. For example, for the MNIST dataset levels 2 and higher are not applicable, since one can obtain digit 9 from the digit 6 with consecutive horizontal and vertical flip, so the network trained with such kernels will not distinguish between 6 and 9. But for datasets that contain photos of real world images high symmetry levels are applicable. On the other hand experiments show that training of a network with high symmetry level is a complicated problem, so in practice levels higher that 2 should not be used.

4   Backpropagation Equations

At this section we describe the modification of the backpropagation procedure that is used to find gradients of the error function with respect to the network weights. For simplicity, we show forward and backward pass of the network only for 1-dimensional convolution for symmetry levels 0 and 1, as extension to 2D convolution and other symmetry levels is straightforward.

Let us denote elements of the convolutional layer in such a way: input vector: x, output vector: y, general convolutional kernel: (a, b, c), symmetric convolutional kernel: (a, b, a). We denote by δx and δy derivatives of the error function with respect to vectors x and y, and by δa, δb, δc derivatives of error function with respect to convolutional kernel elements. Equations for forward and backward passes then become (Table 2):

Table 2. Forward and backward pass for symmetric 1D convolution

Level, pass	Operation
Level 0, Forward	$y_i \mathrel{+}= a x_{i-1} + b x_i + c x_{i+1}$
Level 1, Forward	$y_i \mathrel{+}= a \left(x_{i-1} + x_{i+1} \right) + b x_i$
Level 0, Backward	$\delta x_{i-1} \mathrel{+}= a \cdot \delta y_i; \delta x_i \mathrel{+}= b \cdot \delta y_i; \delta x_{i+1} \mathrel{+}= c \cdot \delta y_i$ $\delta a \mathrel{+}= \delta y_i \cdot x_{i-1}; \delta b \mathrel{+}= \delta y_i \cdot x_i; \delta c \mathrel{+}= \delta y_i \cdot x_{i+1}$
Level 1, Backward	$\delta x_{i-1} \mathrel{+}= a \cdot \delta y_i; \delta x_i \mathrel{+}= b \cdot \delta y_i; \delta x_{i+1} \mathrel{+}= a \cdot \delta y_i$ $\delta a \mathrel{+}= \delta y_i \cdot \left(x_{i-1} + x_{i+1} \right); \delta b \mathrel{+}= \delta y_i \cdot x_i$

Note, that distributive law makes forward and backward pass for level 1 slightly faster than for level 0. The same holds for higher symmetry levels.

5   Experiments

To test the given approach, we use CIFAR-10 dataset, which consists of photos of size $3 \times 32 \times 32$ (3 color channels) distributed among 10 classes which include animals, cars, ships and other categories. Training and test sample sizes are 50000 and 10000 correspondingly. As a basic network we chose a variant of DenseNet

Table 3. CNN configuration

Block	Description
Dense block 1	Number of layers: 1; Convolutional depth: 30
	Input: $3 \times 32 \times 32$; Output: $33 \times 32 \times 32$
Pooling 1	Average pooling 2×2
	Output: $33 \times 16 \times 16$
Dense block 2	Number of layers: 1; Convolutional depth: 30
	Output: $63 \times 16 \times 16$
Pooling 2	Average pooling 2×2
	Output: $63 \times 8 \times 8$
Dense block 3	Number of layers: 1; Convolutional depth: 30
	Output: $93 \times 8 \times 8$
Pooling 3	Average pooling 2×2
	Output: $93 \times 4 \times 4$
Dense block 4	Number of layers: 1; Convolutional depth: 30
	Output: $123 \times 4 \times 4$
Pooling 4	Full average pooling 4×4
	Output: $123 \times 1 \times 1$
Fully connected + Softmax	Input length: 123
	Output length: 10

[4] - one of the most efficient recent architectures. Exact configuration of the net we use is given in the Table 3.

Note, that we are using RELU nonlinearity for each layer of dense block.

We use this network architecture with each symmetry level for convolutional kernels. Since symmetry levels induce parameter sharing, total number of parameters for next levels is decreased.

We train all the networks with stochastic optimization method ADAM [5] with initial learning rate 0.02, multiplying it by 0.97 after every 5 epochs. We use minibatch size of 1280 in all cases.

Final results for different symmetry levels are given in the Table 4.

Table 4. Loss functions and accuracies for different symmetry levels

Level	Model coefficients	Train error	Train accuracy	Test error	Test accuracy
0	95520	0.19	93.15%	1.17	68.92%
1	62280	0.28	89.92%	1.33	69.72%
2	42120	0.73	74.38%	1.07	65.54%
3	32040	1.02	63.75%	1.20	58.72%
4	21960	1.16	58.96%	1.30	54.25%

To see if usage of symmeric kernels improves regularization, we recorded train and test error function values and accuracies after every 5-th epoch during training. Scatterplots based on these tables are shown on Figs. 1 and 2.

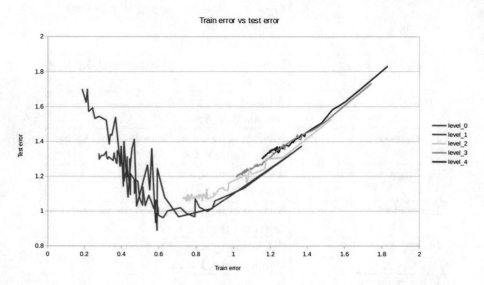

Fig. 1. Relation between train and test error function values for different symmetry levels

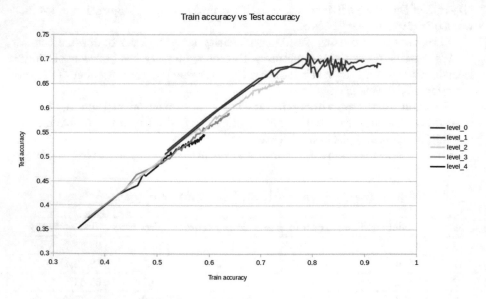

Fig. 2. Relation between train and test accuracies for different symmetry levels

6 Conclusion

At this work we presented symmetric kernels for convolutional neural networks. Use of such kernels guarantees the network will be invariant under certain transformations, such as horizontal flips for the lowest symmetry level, and approximate rotational symmetry for the highest symmetry level.

We tested this approach by training convolutional neural net with the same DenseNet architecture on CIFAR-10 dataset under different symmetry levels. Since most of the parameters in such network are convolutional kernels (all except biases and $123 * 10$ matrix for the last fully connected layer) so total number of coefficients adjusted during training varies a lot: from 21960 (highest symmetry level) to 95520 (no symmetry).

Experiments suggest that CNN training is more complicated for higher symmetry levels (as expected) and that only level 1 symmetry shows improvement in generalization. This can be seen on Fig. 1 where net without symmetries has higher test error values than net with horizontally symmetric kernels for low train error levels (0.2–0.4). The same effect is observed on Fig. 2 where the network with horizontally symmetric kernels stabilizes at the highest test accuracy level. This shows networks with horizontally symmetric kernels tend to overfit less.

Why networks with higher symmetry levels (2, 3 and 4) do not show improvement in generalization despite providing additional output invariances? From our point of view the reason is as follows. From a common point of view trained convolutional neural network extracts low level image features such as edges and corners at first convolutional layers and combines them into more complex shapes in subsequent layers. With the introduction of DenseNets this interpretation became not so clear since deeper layers have direct dependency on input, but convolutional kernel subtensors acting on input still extract these low level features. The problem with convolutional kernels of high symmetry levels is that they cannot extract image edges or corners of certain orientation (in fact units of convolutional layer respond to edges of different orientations in the same way). Thus such units cannot find joint orientation of edges within the image, besides the general network output is invariant under these transformations. From our point of view this is the reason networks with high symmetry levels do not show improvement in generalization.

Thus we suggest to use convolutional neural networks with horizontally symmetric kernels (symmetry level 1) in practice, since they show lower test error function values and higher test set accuracies as the same network with general convolutional kernels. At the same time such networks have lesser total number of parameters (approximately 2/3) and their output is guaranteed to be invariant under horizontal image flips.

References

1. Goodfellow, I., Bengio, Y., Courville, A.: Deep Learning. MIT Press, Cambridge (2016). http://www.deeplearningbook.org
2. Krizhevsky, A., Sutskever, I., Hinton, G.E.: ImageNet classification with deep convolutional neural networks. In: Proceedings of Advances Conference in Neural Information Processing System (NIPS 2012), Lake Tahoe, NE (2012)
3. Srivastava, N., Hinton, G., Krizhevsky, A., Sutskever, I., Salakhutdinov, R.: Dropout: a simple way to prevent neural networks from overfitting. J. Mach. Learn. Res. **15**, 1929–1958 (2014)
4. Huang, G., Liu, Z., van der Maaten, L., Weinberger, K.Q.: Densely connected convolutional networks. In: Proceedings of the IEEE Conference on Computer Vision and Pattern Recognition (2017)
5. Kingma, D.P., Ba, J.: Adam: a method for stochastic optimization. In: International Conference for Learning Representations, Banff, Canada (2014)

Neural Network User Authentication by Geometry of the Auricle

Berik Akhmetov[1], Ihor Tereikovskyi[2(✉)], Liudmyla Tereikovska[3], and Asselkhan Adranova[4]

[1] Yessenov University, Aktau, Kazakhstan
berik.akhmetov@kguti.kz
[2] Igor Sikorsky Kyiv Polytechnic Institute, Kyiv, Ukraine
terejkowski@ukr.net
[3] Kyiv National University of Construction and Architecture, Kyiv, Ukraine
tereikovskal@ukr.net
[4] Kazakh National Pedagogical University Named after Abay, Almaty, Kazakhstan
assel.adranova@gmail.com

Abstract. The article is devoted to the development of a neural network model intended for use in the system of biometric user authentication based on the analysis of the geometry of the auricle. It is determined that from the point of view of using neural network methods the main features of the recognition task are the number of recognizable users, the size and quality of the auricle images, and the number and parameters of the characteristic features of the auricle. There was shown the expediency of using a convolutional neural network, the parameters of which must be adapted to the peculiarities of the recognition problem. There are proposed the principles of adaptation of structural parameters. The number of convolutional layers should correspond to the amount of image recognition levels of the auricle by an expert. The amount of feature maps in the n convolutional layer should be equal to the amount of features at the n recognition level. The map of n layer features, corresponding to the j recognition feature, is associated only with those maps of features of the previous layer, which are used for the construction of this figure. The size of the convolution kernel for the n convolutional layer must be equal to the size of the recognizable features on the n hierarchical level. The use of convolutional layers should not distort the geometric parameters of the features used for the image recognition of the auricle. Based on the proposed principles and revealed features of the problem of image recognition of the auricle, there was developed an appropriate method of adapting the structural parameters of the convolutional neural network. Conducted computer experiments showed satisfactory recognition accuracy, which confirms the prospects of the proposed solutions. It is shown that further research should be connected with the formation of the methodological base for adapting the main components of the mathematical support to the features of image recognition of the user's auricle in a biometric authentication system.

Keywords: Biometric authentication · Auricle · Adaptation
Convolution layer · Sub-sample layer · Convolutional neural network

© Springer Nature Switzerland AG 2019
O. Chertov et al. (Eds.): ICDSIAI 2018, AISC 836, pp. 11–19, 2019.
https://doi.org/10.1007/978-3-319-97885-7_2

1 Introduction

Nowadays, one of the main trends in the development of information security systems is the introduction of neural networks of biometric users authentication, based on the analysis of the blood vessel image of the eyeground, voice, fingerprints, and the geometry of the human face. The use of these biometric characteristics in the authentication systems is explained by the possibility of implementing tacit control, good testing and low cost of reading devices. At the same time, the constantly increasing requirements for the effectiveness of these tools predetermine the need for their improvement due to the expansion of the spectrum of analyzed biometric characteristics [1–3]. One of the ways of such an extension is the use of an additional module in the means of authentication that implements user recognition based on the geometry of the auricle. This predetermines the relevance of the research in the direction of neural network models development designed to recognize users by the geometry of the auricle.

2 Features of the Biometric Authentication Task on the Geometry of the Auricle

According to the data from [5–8], the external human auricle can be characterized by a number of anatomical components. As shown on Fig. 1, these components include: 1 – triangular fossa; 2 – tubercle of the auricle; 3 – scaphoid fossa; 4 – crus of helix; 5 – cavity of the auricle; 6– anthelix; 7 – helix; 8– antitragus; 9 – ear lobe; 10– intertragic notch; 11– tragus; 12– tuberculum supratragicum; 13 – sulcus of crus of helix; 14 – crura of antihelix. Therefore, the geometry of the auricle provides a significant set of different structures that can be used to obtain a whole set of measurements that are exclusive for each person.

The procedure of users' recognition by the geometry of the auricle can be viewed as a typical procedure for biometric authentication based on the recognition of graphic biometric features. The analyzed image of the auricle is presented as a vector of predetermined parameters. After that, in order to establish identity there is realized a comparison of this vector with vectors stored in the database. Recognition can be realized for a flat or three-dimensional cloud of dots, depicting surface of the auricle.

The analysis allows to state that from the point of view of using neural network methods, the main features of the recognition problem are the size and quality of the auricle images, the amount and parameters of the characteristic features of the auricle, and the amount of characteristic vectors stored in the database, which correspond to known users [9].

On the basis of theoretical works devoted to neural network methods of biometric authentication, we can state that an important direction at increasing the efficiency of such systems is the adaptation of the structure of the neural network model to the conditions of use [4, 10]. Since the source of information is a two-dimensional image of the geometry of the auricle, then due to the proven effectiveness for recognition a convolutional neural network (CNN) should be used. At the same time, in the available literature there is no description of the mechanism of adaptation of CNN structural parameters to the task of biometric users' authentication according to the geometry of the auricle. Insufficient adaptation negatively affects the accuracy of users' recognition by the auricle, which in turn can lead to unauthorized access to the information of the protected system and/or to blocking of information for legitimate users.

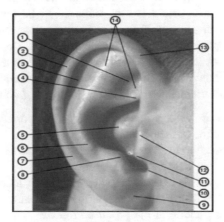

Fig. 1. Anatomic components of the auricle

The main purpose of the study is to develop a method for determining the structural parameters of a convolutional neural network used for biometric users' authentication in the geometry of the auricle. In order to achieve this purpose, the following tasks should be solved: to determine the principles of adaptation of the structural parameters of a convolutional neural network; to detail the components of the method; to carry out experimental studies aimed at verifying the proposed solutions.

3 Principles of Adaptation of the Structural Parameters of a Convolutional Neural Network

On the basis of theoretical works devoted to the CNN [11–15] we can state that their main structural parameters are the size of the input field (a_0), the amount of input neurons (L_{in}), the amount of output neurons (L_{out}), the amount of neurons in a fully connected layer (L_f), the amount of convolution layers (K_{ls}), the amount of feature maps in each convolution layer ($L_{h,k}$), the amount of sub-sample layers (K_{ld}), the scale coefficient for each sub-sample layer (m_l), the convolution kernel size for each k convolutional layer ($b \times b)_k$, displacement of the receptive field for each k convolution procedure conduction d_k, the size of feature maps for each k convolution layer $(a \times a)_k$, the structure of connections between adjacent layers of the convolution/sub-sample.

In this case, the size of the sub-sample layer and the size of the feature map are calculated as follows:

$$c_l = a_k/m_l, \qquad (1)$$

$$a_k = (a_{k-1} - b_k + 2r_k)/d_k + 1, \qquad (2)$$

where a_k is the size of the convolution layer that precedes the l layer of the sub-sample, r_k is amount of additional zeros for the k convolutional layer.

The adaptation of these parameters is proposed to be done on the fact that in systems of biometric authentication the recognition process of the CNN of the auricle should be as close as possible to its biological prototype. Under the term biological prototype there is meant the process of recognition of geometric parameters of the auricle by the expert. Integration of the proposed approach with the concept of CNN functioning allowed to formulate the following group of principles of adaptation:

Principle 1. The amount of convolution layers should correspond to the amount of recognition levels of the auricle by an expert.

Principle 2. The amount of feature map in the n convolutional layer should be equal to the amount of features at the n recognition level.

Principle 3. The map of features of the n layer, corresponding to the j recognition feature, is associated only with those maps of features of the previous layer, which are used for the construction of the proposed figure.

Principle 4. The size of the convolution kernel for the n convolutional layer should be equal to the size of recognizable features on the n hierarchical level.

Principle 5. The use of convolutional layers should not distort the geometric parameters of the features used to recognize the auricle.

4 Method for Determining Structural Parameters of a Convolutional Neural Network

In accordance with [3, 4] the starting point for the development of a neural network model is the determination of the amount of input and output neurons. Typically, the amount of output neurons is equal to the amount of recognized users, and the amount of input neurons is equal to the amount of pixels of the recognized image (the image of the auricle). Using modern Full-HD webcams as a data source the full size of the captured image is 1920×1080 pixels. At the first approximation we can assume that the captured image corresponds to the profile of the user's head. Through full-scale experiments it was established that the linear dimensions of the projection of the auricle are about 25% of the linear dimensions of the profile of the user's head. Therefore, the maximum dimensions of the square describing the outer contour of the auricle are 480×270 pixels. Since the input field of the classical CNN must have a square shape it can be argued that the side of the specified square should have a size of 270 to 480 pixels. Wherein $L_{in} = 72900\ldots 230400$.

Further, the development of the method was based on the formulated principles of adaptation. Therefore, by expert estimation it is determined that the minimum amount of auricle recognition levels is 3. At the first level, there are recognized the signs of the anatomical components of the auricle, on the second one - separate individual anatomical components and on the third level - realized the classification of the whole auricle image by feeding into the fully connected layer of the anatomical components. Thus, the minimum amount of convolution layers is $K_{ls,min} = 2$. According to the analysis of the images of different types of anatomical components shown in Fig. 1 there was determined that for their recognition it is enough to have 6 elementary figures shown on Fig. 2. The size of the figures is determined by analogy with [12, 15] and is 5×5 pixels.

Fig. 2. Recognizable features of anatomical components of the auricle.

Consequently the size of the convolution kernel is $(b \times b)_1 = (5 \times 5)$ and the amount of feature maps in the first convolutional layer $L_{h,1} = 6$. On the basis of data [11–14] and taking into account the need to minimize the computing resources of the network it is determined that the displacement of the receptive field should be:

$$d_k = Round(b_k/2), \tag{3}$$

Thus, $d_1 = 3$. Substituting the obtained results into expressions (2) it is determined that the size of the feature map of the first convolutional layer is equal to

$$a_1 = (a_0 - 5 + 2r_1)/3 + 1, \tag{4}$$

where a_0 is the size of the figure of the auricle.

There was accepted the precondition on compliance of the feature maps of the second convolution layer with the anatomical components of the auricle. Because the implementation of the convolution procedure can negatively affect the detailing of the anatomical components, it is necessary to abandon the sub-sample layer located between the first and second convolutional layers. Therefore $L_{h,2} = 14$. By analogy with the first convolutional layer, it is determined that $(b \times b)_2 = (5 \times 5)$, $d_2 = 3$, and the size of the feature map of the second convolutional layer is calculated using (2).

Based on the analysis of Figs. 1 and 2 and using the third principle of adaptation there is determined the need for a fully connected structure of links between adjacent convolution/sub-sample layers. Note that, in accordance with the fifth principle, the final calculation of the amount and coefficient factors of the sub-sample layers there should be taken into account the need to convolve the two-dimensional input data set to the characteristic vector.

Calculation of the minimum number of neurons of a fully connected layer can be made from positions of minimum sufficiency determined by Hecht-Nielsen theorem:

$$L_{f,\min} = 2(L_{h,K} + 1),\tag{5}$$

where $L_{h,K}$ is the number of feature maps in the last convolutional layer.

Finally, the amount of neurons in a fully connected layer should be determined by numerical experiments.

On the basis of the proposed principles there was developed a method for determining the structural parameters of CNN, the implementation of which provides the implementation of 7 stages.

Stage 1. To identify the set of recognizable users whose elements will correspond to the output neurons of the network.

Stage 2. To determine the amount of input parameters of the network. In order to do this there should be calculated the sizes of images used for recognition.

Stage 3. On the basis of the first and second principles, using the geometric characteristics of the standards of recognized images, to determine the amount of convolution layers and the amount of feature maps in each convolutional layer.

Stage 4. On the basis of the third principle to determine the structure of the links between adjacent feature maps.

Stage 5. On the basis of the fifth principle to determine the presence and parameters of sub-sample layers.

Stage 6. On the basis of the fourth principle, taking into account the need to convolve the image to the feature vector, to determine the kernel size and convolution displacement step for each recognition level.

Stage 7. On the basis of the method of determining the number of hidden neurons in a multilayer perceptron [4], to determine the number of neurons in a fully connected layer.

5 Experimental Results

For verification of the developed method of determining the structural parameters of CNN there were conducted experiments for the auricles recognition of five users. The size of the analyzed images of the auricles was 300-300 pixels. Consequently, $a_0 \times a_0 = 300 \times 300$, $L_{in} = 89401$, $L_{out} = 5$. Using the developed method, by means of trivial substitutions in (2) and (3), the following CNN parameters were calculated: $a_1 = 98$, $a_2 = 30$. At calculating there is taken into account that the following parameters have already been determined in the developed procedure: $(b \times b)_1 = (b \times b)_2 = (5 \times 5)$, $d_1 = d_2 = 3$, $K_{ls,\min} = 2$, $L_{h,1} = 5$, $L_{h,2} = 48$, $r_1 = 1$, $r_2 = 0$. In this case, the calculation of the parameters r_1 and r_2 was based on the fact that the amount of feature maps defined in expression (2) in the convolutional layer must be an integer. The subsequent construction of CNN is implemented due to the need to reduce the two-dimensional feature map of the size to the characteristic vector. As a result, the network was supplemented by two convolutional layers and two sub-sample layers. The selected layer parameters are: $(b \times b)_3 = (5 \times 5)$, $(b \times b)_4 = (4 \times 4)$, $d_3 = d_4 = 1$, $r_3 = r_4 = 0$, $m_1 = m_2 = 2$, $L_{h,3} = 48$, $L_{h,4} = 32$. In this case, in accordance with the expression (2): $a_3 = 26$, $c_1 = 13$, $a_4 = 10$, $c_2 = 5$. There should be noted that the overlay of the fifth convolution kernel of the size $(b \times b)_5 = (5 \times 5)$ on the fourth sub-sample layer map allows to make the final convolution of the input field to the feature vector. Using (5), we obtained $L_{f,min} = 66$. Finally, the amount of neurons in a fully connected layer $L_f = 300$ is chosen by analogy with [4, 14] on the basis of the balance of the required computing resources and the network memory size.

The constructed CNN became the main for the development of the software complex, EarDetection, designed for auricle recognition. The program is written in the Python programming language using the TensorFlow library. The recognition system was trained and tested on a sample collected by the authors. The experiments calculated the accuracy of recognition of test samples at different amounts of learning epochs (t_e) and at different amounts of feature map $(L_{h,1}, L_{h,2}, L_{h,3}, L_{h,4})$. The experiments were performed using a computer with an IntelCorei7-7700 (4.2 GHz) processor and 8 GB RAM that was running the Windows 10 operating system. The results of the experiments are shown in Figs. 3 and 4.

The maximum recognition accuracy, equal to about 0.72, is achieved at the described CNN configuration parameters. The obtained results make it possible to state about the promising use of the additional module for user recognition according to the geometry of the auricle is in biometric authentication systems.

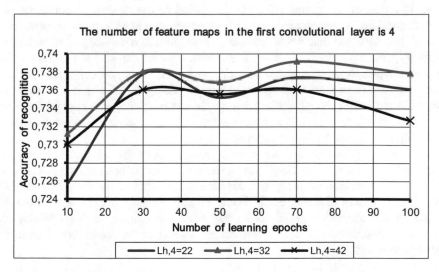

Fig. 3. Dependence of recognition accuracy on the number of learning epoch with $L_{h,1} = 4$, $L_{h,2} = 38$, $L_{h,3} = 38$ and different values of $L_{h,4}$

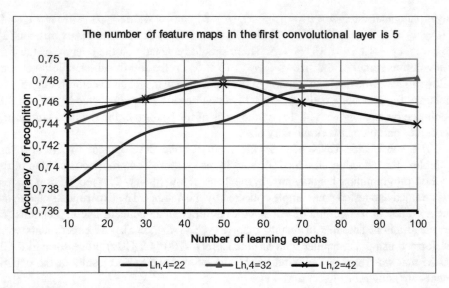

Fig. 4. Dependence of recognition accuracy on the number of learning epoch with $L_{h,1} = 5$, $L_{h,2} = 48$, $L_{h,3} = 48$ and different values of $L_{h,4}$

6   Conclusions

There is proposed an approach for the adaptation of the structural parameters of a convolutional neural network for user auricles recognition in a biometric authentication system. It is assumed that the adaptation is based on the conditions of its maximum similarity with the process of auricles recognition by an expert. On the basis of the proposed approach and using the generally accepted concept of the convolutional neural network there were formulated additional principles for adapting its structure. Using the proposed principles of adaptation there was developed a method for determining the structural parameters of a convolutional neural network designed for faces recognition in a biometric authentication system.

Computer experiments for user auricle recognition of the adapted convolutional neural network in conditions close to the biometric authentication system demonstrated satisfactory recognition accuracy and confirmed the prospects of the proposed solutions. The next step of the research should be appropriately correlated with the formation of the adaptation principles of the main components of the mathematical support to the peculiarities of user auricle recognition in the conditions of biometric authentication system.

References

1. Lakhno, V., Kazmirchuk, S., Kovalenko, Y., Myrutenko, L., Zhmurko, T.: Design of adaptive system of detection of cyber-attacks, based on the model of logical procedures and the coverage matrices of features. East.-Eur. J. Enterp. Technol. 3(9(81)), 30–38 (2016)
2. Bapiyev, I.M., Aitchanov, B.H., Tereikovskyi, I.A., Tereikovska, L.A., Korchenko, A.A.: Deep neural networks in cyber attack detection systems. Int. J. Civil Eng. Technol. (IJCIET) **8**, 1086–1092 (2017)
3. Aitchanov, B.: Perspectives for using classical neural network models and methods of counteracting attacks on network resources of information systems. In: Aitchanov, B., Korchenko, A., Tereykovskiy, I., Bapiyev, I. (eds.) News of the national academy of sciences of the Republic of Kazakhstan. Series of Geology and Technical Sciences. vol. 5, pp. 202–212 (2017)
4. Korchenko, A., Tereykovsky, I., Karpinsky, N., Tynymbayev, S.: Neural network models, methods and means of safety parameters assessment of the Internet focused information systems. Our Format. p. 275 (2016). (in Russian)
5. Abdel-Mottaleb, M., Zhou, J.: Human ear recognition from face profile images. In: ICB, pp. 786–792 (2006)
6. Akkermans, H.M., Kevenaar, T.A.M., Schobben, D.W.E.: Acoustic ear recognition for person identification. In: Fourth IEEE Workshop on Automatic Identification Advanced Technologies, pp. 219–223 (2005)
7. Burge, M., Burger, W.: Ear biometrics In: Jain, A.K., Bolle, R., Pankanti, S. (eds.) Biometrics: Personal Identification in Networked Society, pp. 273–286. Kluwer Academic, Dordrecht (1998)
8. Burge, M., Burger, W.: Ear biometrics in computer vision. In: Proceedings ICPR 2000, pp. 822–826 (2002)
9. Chang, K., Bowyer, K.W., Sarkar, S., Victor, B.: Comparison and combination of ear and face images in appearance-based biometrics. IEEE Trans. PAMI. **25**, 1160–1165 (2003)
10. Arsentyev, D.A., Biryukova, T.S.: Method of flexible comparison on graphs as algorithm of images recognition. Bulletin of Ivan Fedorov MGUP, No. 6. pp. 74–75 (2015). (in Russian)
11. Bryliuk, D., Starovoitov, V.: Application of recirculation neural network and principal component analysis for face recognition. In: The 2nd International Conference on Neural Networks and Artificial Intelligence. BSUIR, Minsk, pp. 136–142 (2001)
12. Chirchi, V.R.E., Waghmare, L.M.: Iris biometric authentication used for security systems. Int. J. Image, Graph. Signal Process. **6**(9), 54–60 (2014)
13. Connaughton, R., Bowyer, K.W., Flynn, P.J.: Fusion of face and iris biometrics. In: Handbook of Iris Recognition, pp. 219–237. Springer, London (2013)
14. Fedotov, D.V., Popov, V.A.: Optimisation of convolutional neural network structure with self-configuring evolutionary algorithm in one identification problem. VestnikSibGAU, **16** (4), 857–862 (2013). (in Russian)
15. Zoubida, L., Adjoudj, R.: Integrating face and the both irises for personal authentication. Int. J. Intell. Syst. Appl. (IJISA) **9**(3), 8–17 (2017)

Race from Pixels: Evolving Neural Network Controller for Vision-Based Car Driving

Borys Tymchenko[✉] and Svitlana Antoshchuk

Odessa National Polytechnical University, Odesa, Ukraine
tim4bor@gmail.com

Abstract. Modern robotics uses many advanced precise algorithms to control autonomous agents. Now arises tendency to apply machine learning in niches, where precise algorithms are hard to design or implement. With machine learning, for continuous control tasks, evolution strategies are used. We propose an enhancement to crossover operator, which diminishes probability of degraded offsprings compared to conventional crossover operators. Our experiments in TORCS environment show, that presented algorithm can evolve robust neural networks for non-trivial continuous control tasks such as driving a racing car in various tracks.

Keywords: Neuroevolution · Genetic algorithms · Neural networks
Deep learning · Continuous control · Crossover · Autonomous vehicles
TORCS

1 Introduction

Nowadays, deep learning (DL) shows great success in various tasks of different complexity. Researches show, that both supervised and reinforcement learning can be applied to such control tasks as autonomous vehicle locomotion [2], games [3], robotics [4] and navigation [5]. Within these successes, often lies deep neural networks (DNN), which are used to control agents. Usually, these DNNs are used as end-to-end action policies, processing inputs from sensors, extracting features and providing action decisions. So, in such control tasks, deep neural network policies take state observations directly as input and outputs a decision, which action to take.

In order to optimize this kind of policies, different gradient-based methods are used. For supervised learning, variations of stochastic gradient descent (SGD) are used [1]. In reinforcement learning, policy gradient methods have been proposed to estimate gradients approximately.

Apart from gradient based methods, stand other alternative solutions for stochastic optimization. Recent discoveries show, that they can be applied to direct policy search problem [6].

O. Chertov et al. (Eds.): ICDSIAI 2018, AISC 836, pp. 20–29, 2019.
https://doi.org/10.1007/978-3-319-97885-7_3

Evolution strategies (ES) is a class of black box optimization techniques that can search in the policy space without relying on the gradients. At each iteration, ES samples a candidate population of genotypes from a distribution over the parameter space, evaluates the fitness function on these candidates (parameter vectors), and constructs a new probability distribution over the parameter space using the candidates with the high value of the fitness function. This process is repeated until fitness function is maximized or limit of iterations is reached. Evolution strategies applied to neural networks is usually called *neuroevolution* in literature.

ES showed surprising success in evolving deep neural networks in complex environments, such as Atari games [6], MuJoCo and different mazes [7]. However, their sample efficiency in reinforcement learning tasks is lower, than policy-gradient [6] because they cannot fully exploit temporal properties in RL problems or network structure advancements.

Operating the same principle of black box optimization, very similar to ES stand genetic algorithms. GA introduces mandatory crossover operator, which is optional in ES. It is used to provide more diversity of good policies in the population. Generally, when used with neural networks, this operator is disruptive. Performed in parameters space, straightforward crossover tends to break existing good hierarchical "data paths" by exchanging parameters between two policies, causing degeneration of an individual agent and drop in performance of generation in general.

There is a challenge to design such crossover operator, that will effectively combine parent neural networks and generate non-degenerate children. In this paper, we propose a method of crossover, that can effectively combine parent neural networks in informed way, that leads to better performance, than existing naive crossover operators.

2   Background and Related Work

Researchers show interest in applying neuroevolution approach to RL tasks. Early applications of neuroevolution to control tasks started with ESP [8]. NEAT [9] was also used for black box policy optimization. Unlike other methods, NEAT evolves not only neural network parameters but also topology. It provides distinct codification of genotypes and tracks history of every agent in order to perform crossover of neural networks with different topologies. HyperNEAT was used to play ATARI games at human level [10].

In recent discovery, Salimans et al. [6] provide a method of black box policy optimizations. At each iteration, algorithms samples candidate genotypes from normal distribution with mean m. The distribution vector is then updated in the direction of weighted average of candidate genotypes, where weights are proportional to each candidate's fitness. This showed a great success in complex environments, however, the method is very sample inefficient.

Zhang et al. [11] prove that Salimans' [6] neuroevolution can be an approximation of SGD methods. Additionally, unlike SGD methods, neuroevolution can work when the neural network is not differentiable.

Such et al. [12] provide simple genetic algorithm, that works in deep-RL domains including Atari and MuJoCo. It can evolve DNNs of a decent size with outstanding performance. Their "Deep GA" successfully evolves networks with over four million free parameters, the largest neural networks ever evolved with a traditional evolutionary algorithm. They use neural networks of fixed structure, special mutation operator and no crossover operator. In addition, in the same paper, Such et al. show that simple random search can outperform sophisticated algorithms in several domains.

Lehman et al. [7] show, that mutation operator in GAs can be done in an informed way, using gradient information from neural network. This greatly improves overall performance of GA because prevents drastically degenerate genotypes from spawning. In addition, they provide a method of approximation when the gradient information is unavailable.

3   Genetic Algorithm

Genome Representation. In our work, we assume neural network of fixed topology. Every weight becomes a single gene. Every gene is represented by a 32-bit floating-point number. It is stored as a real value.

This encoding was prevailed over binary strings representation for compatibility with used crossover and mutation operators.

With binary string, mutation operator can change only a part of an encoded floating point number (especially, exponent), which will lead to severe change in neural network weight, encoded with this gene. As we are minimizing drastic changes in parameters, we cannot use binary strings genome representation.

3.1   Overall Algorithm

Our procedure of evolving policies is based on a conventional genetic algorithm [21] with selection, crossover and mutation operators. Policies are represented by convolutional neural networks, detailed structure is described in (4.1). Initial population consists of neural networks, that are initialized at random with He weights initialization [13].

We use *elitism* to preserve best agents, so best solutions are never lost during the evolution process.

3.2   Environment and Rewards

We use TORCS racing environment [14], which states a control task of steering simulated racing car over various tracks. Although, this environment can provide shaped rewards on every tick, we focused on sparse rewards in the end of every episode.

We use high-dimensional input space, namely an array of RGB pixels from front-facing camera on agent. As output, we consider steering angle. This environment was previously solved using deep deterministic policy gradient (DDPG) [15]. In their experiments, researchers used the following shaped reward function:

$$r_t = v_x cos(\theta) - v_x sin(\theta) \tag{1}$$

where r_t is an immediate reward, v_x is longitudinal velocity of the car and θ is an angle of heading deviation from track center line. This reward is designed to increase forward velocity and diminish traverse velocity.

We introduce more penalizing reward function, which is designed to discourage neural network from oscillating too much by minimizing traverse velocity and excessive steering:

$$r_t = v_x cos(\theta) - v_x sin(\theta) - v_x |\alpha| \tag{2}$$

where α is an agent's steering angle normalized to $[-1, 1]$ range.

Episode terminates after collision with track edge, going in wrong direction on the track, or when desired number of laps is finished successfully.

We assume that fitness of each agent is equal to it's sparse reward in the end of episode and is defined as:

$$R = \sum_{t=1}^{T} r_t \tag{3}$$

where r_t is an immediate reward from (2) and T is a number of timesteps in the episode.

This setting allows future transfer to environments with true sparse rewards, e.g. real-life racing vehicles with only time score as reward.

3.3   Selection

We used deterministic tournament selection [16]. Given a generation of k policies, it returns one, which then will be used in crossover step. We perform selection operator twice, in order to get two possibly best parents independently. With tournament selection, it is easy to control selective pressure by varying tournament pool size.

In our experiments we used fixed tournament pool size.

3.4   Safe Mutation

Mutation operator is a way to induce exploration. However, generic mutations (like single point or Gaussian noise) are extremely disruptive in context of neural networks.

To stick to concept of non-differentiable policies, we used safe mutation through rescaling (SM-R) from [7]. As measure of divergence, we used Euclidian distance between parent and child outputs run on the data batches recorded by parent.

The main idea of this method is to add Gaussian noise of proper scale to the weights in order to make neural network behave close enough to its parents' behavior. To implement this, parents' experiences that consist of observations and actions are written in a buffer, and then are used to evaluate the mutated offspring.

3.5   Layer Blend Crossover

In this paper, we introduce Layer Blend Crossover (LBC). This operator mixes two policies and produces a child policy. All policies have the same network architectures and differ only by weights.

As a neural network can be represented as a highly non-linear function with weights as parameters, weights cannot be changed significantly without severe change in the output result. Also, each layer of a neural network can be viewed as a "building block" with limited functionality within neural network. Previous researches show [17], that simple swapping of the whole layers can be beneficial with relatively small networks, however, inefficient with larger ones.

To decrease disruption, while maintaining the production of diverse off-springs, we apply alpha blending for each corresponding layer in parents' neural networks:

$$l_i^{child} = \alpha_i \cdot l_i^a + (1 - \alpha_i) \cdot l_i^b \tag{4}$$

where l_i represents parameters of i-th layer, and indices a and b denote parents.

Different blending parameter α_i is chosen for each layer in the following manner:

$$\mu = \frac{f_a}{f_a + f_b} \tag{5}$$

$$\alpha_i = N(\mu, \sigma^2, 0, 1) \tag{6}$$

where f_a and f_b are parents' fitnesses and $f_a > f_b$, N is a truncated normal distribution with mean μ, variance σ^2, lower bound 0 and upper bound 1. σ becomes an additional parameter, which defines the "randomness" of crossover process. Closer to 0, all layers are blended with the same ratio (same as BLX-a algorithm [18]). In addition, with high selective pressure, μ tends to 0.5 because selected parents have nearly the same fitness.

With $\sigma \to +\infty$, the distribution tends to uniform, which can be also disruptive. We consider using σ in range $[0.3, 0.6]$, in our experiments we used fixed value of 0.45, which was found empirically.

To supply parent's experience to safe mutations after crossover, we merge experiences of parents at crossover phase.

4   Experiments

In this section, we perform experiments to measure robustness and efficiency of the proposed algorithm. We begin by describing setup of experiments, followed by learning curves and comparison with existing methods.

4.1   Setup

We performed all our experiments within TORCS racing environment. We benchmark 7 continuous control racing car steering tasks. All these tasks have different complexity level based on road conditions. All our control policies are represented by 3-layer CNN with LReLU [13] activations between convolutional layers and hyperbolic tangent between fully connected layers. Structure of neural network is depicted at Fig. 1.

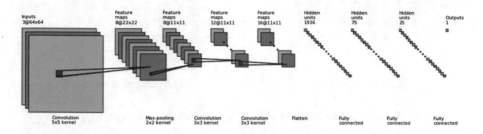

Fig. 1. CNN used in experiments [20]

In all our experiments, we used population size of 50, tournament pool size is set to 10% of population, elite number is set to 10% of population.

For safe mutations, target divergence is set to 0.2, crossover is performed every time new offspring is produced. For LBC, σ was set to 0.45.

We show results of LBC against naive n-point crossover. In experiments, we used n set to number of layers in neural network. Crossover points were set between layers.

Also, we compare total number of episodes to the same fitness of genetic algorithm equipped with LBC against DDPG from [15].

4.2 Performance

To measure crossover performance, we run GA to learn policy to operate in TORCS. During the training, we picked 10 bins of parent-child at random and plotted their fitness (Fig. 2).

The left subplot depicts LBC. From plot it's clearly seen the advantage of LBC over naive n-point crossover. We observe, that in many cases, LBC produces more evenly distributed (based on fitness) offsprings; their fitness tends to be close to mean of parents' fitness. Contrary, n-point crossover tends to produce degenerate offsprings with fitness close to or less than least fit parent.

To make more demonstrative comparison, we evolve the same network with LBC and n-point crossover (Fig. 3). Plots are averaged over 10 runs and normalized for demonstrative purposes.

While LBC converges to target fitness in 67 generations, usage of n-point crossover took above 200 generations to evolve. In addition, LBC has more smooth change in fitness between generations, which is the sign of higher robustness.

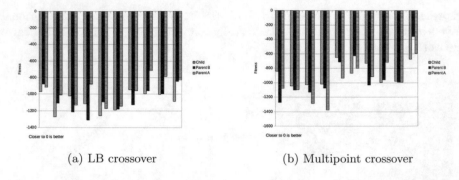

(a) LB crossover   (b) Multipoint crossover

Fig. 2. 10 random parents-child pairs crossover comparison

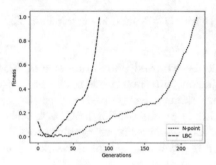

Fig. 3. Fitness of the top-fit agent over generations

4.3   Comparison with DDPG

To compare our algorithm with existing state-of-the-art technologies, we chose
to compare it with existing implementation of the DDPG algorithm from [19].
In order to keep results correct, we trained our own policies using [19] and our
proposed reward function (2). We also modified actor and critic neural networks
in DDPG approach to match Fig. 1.

As we cannot directly compare generations, we compared total number of
episodes played by all agents. Agents were trained to reproducibly complete a
full lap on tracks with different complexity. Results are shown in Table 1. Track
outlines are shown in Appendix A.

Here, GA with LBC can learn a good continuous control policy from pixels
on a par with conventional DDPG algorithm. Considering track outlines, we
state, that GA outperforms DDPG method on tracks with smaller amount of
hard turns. This result is consistent to [22], because cornering can be considered
as a narrowing path fitness landscape, where tiny deviation in control, especially
at high speed, leads to avalanche-like error windup and subsequent agent crash,
while at straight parts, agent has more space to recover.

Table 1. Train episodes to complete the track

TORCS track name	Length (m)	GA-LBC	DDPG
evo-4-r	714.16	1200	947
evo-3-l	979.82	1450	933
Aalborg	2587.54	**2550**	2980
CG track 2	3185.83	**2300**	2365
Ruudskogen	3274.20	**2500**	2542
E-Track 2	5380.50	4100	3634
Spring	22129.80	**4650**	8762

4.4 Agents Robustness

To check the robustness of trained agents and verify, that learned policies are not overfit to a particular track, we tested policies on different tracks. Results are shown in Table 2. Numbers in cells represent fraction of track length that agent can pass without collision with track wall.

Table 2. Cross-testing of evolved policies

Trained	Tested						
	evo-4-r	evo-3-l	Aalborg	CG track 2	Ruudskogen	E-Track 2	Spring
evo-4-r	1.0	1.0	0.73	0.22	0.6	0.40	0.1
evo-3-l	1.0	1.0	0.91	0.1	0.53	0.41	0.11
Aalborg	1.0	1.0	1.0	1.0	0.87	0.2	0.19
CG track 2	0.85	1.0	0.01	1.0	0.93	0.76	0.64
Ruudskogen	1.0	1.0	1.0	0.63	1.0	0.73	0.74
E-Track 2	1.0	1.0	1.0	0.44	1.0	1.0	0.63
Spring	1.0	1.0	1.0	1.0	0.79	0.92	1.0

61% of agents can finish tracks, different from the one, on which they were trained. 84% of agents can finish more than 50% of an unknown track. Low performance on several tracks is caused by sufficient visual difference. Obviously, agents that were trained at more complex track can complete also easier tracks if they have similar appearance.

5 Conclusion

We present Layer Blend Crossover (LBC), a new crossover operator for genetic algorithms. LBC does more efficient crossover, by lesser disrupting existing parents' behavior. Our experiments show the benefits of it in the context of continuous control tasks. It has on par performance with state-of-the-art methods, with possibility of being applied to domains wits sparse rewards and non-differentiable policies. LBC shows stability and robustness of agents. Future advances in evolution strategies will also benefit from usage of such technique in RL tasks.

Appendix A   Track Outlines

See Appendix Fig. 4.

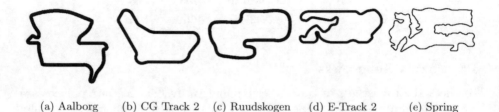

(a) Aalborg   (b) CG Track 2   (c) Ruudskogen   (d) E-Track 2   (e) Spring

Fig. 4. Different track outlines

References

1. Kingma, D.P., Ba, J.: Adam: A Method for Stochastic Optimization. CoRR 1412.6980 (2014). http://arxiv.org/abs/1412.6980
2. Bojarski, M., Testa, D.D., Dworakowski, D., Firner, B., Flepp, B., Goyal, P., Jackel, L.D., Monfort, M., Muller, U., Zhang, J., Zhang, X., Zhao, J., Zieba, K.: End to End Learning for Self-Driving Cars. CoRR 1604.07316 (2016). http://arxiv.org/abs/1604.07316
3. Mnih, V., Kavukcuoglu, K., Silver, D., Graves, A., Antonoglou, I., Wierstra, D., Riedmiller, M.A.: Playing atari with deep reinforcement learning. CoRR 1312.5602 (2013). http://arxiv.org/abs/1312.5602
4. Nair, A., McGrew, B., Andrychowicz, M., Zaremba, W., Abbeel, P.: Overcoming exploration in reinforcement learning with demonstrations. CoRR 1709.10089 (2017). http://arxiv.org/abs/1709.10089
5. Bruce, J., Sünderhauf, N., Mirowski, P., Hadsell, R., Milford, M.: One-shot reinforcement learning for robot navigation with interactive replay. CoRR 1711.10137 (2017). http://arxiv.org/abs/1711.10137
6. Salimans, T., Ho, J., Chen, X., Sutskever, I.: Evolution strategies as a scalable alternative to reinforcement learning. CoRR 1703.03864 (2017). http://arxiv.org/abs/1703.03864
7. Lehman, J., Chen, J., Clune, J., Stanley, K.O.: Safe mutations for deep and recurrent neural networks through output gradients. CoRR 1712.06563 (2017). http://arxiv.org/abs/1712.06563
8. Gomez, F.J., Miikkulainen, R.: Solving non-Markovian control tasks with neuroevolution. In: Proceedings of the 16th International Joint Conference on Artificial Intelligence, IJCAI 1999, San Francisco, CA, USA, vol. 2, pp. 1356–1361. Morgan Kaufmann Publishers Inc. (1999). http://dl.acm.org/citation.cfm?id=1624312.1624411

9. Stanley, K.O., Miikkulainen, R.: Evolving neural networks through augmenting topologies. Evol. Comput. **10**(2), 99–127 (2002). https://doi.org/10.1162/106365602320169811

10. Stanley, K.O., D'Ambrosio, D.B., Gauci, J.: A hypercube-based encoding for evolving large-scale neural networks. Artif. Life **15**(2), 185–212 (2009). https://doi.org/10.1162/artl.2009.15.2.15202

11. Zhang, X., Clune, J., Stanley, K.O.: On the relationship between the openai evolution strategy and stochastic gradient descent. CoRR 1712.06564 (2017). http://arxiv.org/abs/1712.06564

12. Such, F.P., Madhavan, V., Conti, E., Lehman, J., Stanley, K.O., Clune, J.: Deep neuroevolution: Genetic algorithms are a competitive alternative for training deep neural networks for reinforcement learning. CoRR 1712.06567 (2017). http://arxiv.org/abs/1712.06567

13. He, K., Zhang, X., Ren, S., Sun, J.: Delving deep into rectifiers: surpassing human-level performance on imagenet classification. In: Computing Research Repository, 1502.01852 (2015). http://arxiv.org/abs/1502.01852

14. Loiacono, D., Cardamone, L., Lanzi, P.L.: Simulated car racing championship: competition software manual. CoRR 1304.1672 (2013). http://arxiv.org/abs/1304.1672

15. Lillicrap, T.P., Hunt, J.J., Pritzel, A., Heess, N., Erez, T., Tassa, Y., Silver, D., Wierstra, D.: Continuous control with deep reinforcement learning. CoRR 1509.02971 (2015). http://arxiv.org/abs/1509.02971

16. Blickle, T., Thiele, L.: A comparison of selection schemes used in evolutionary algorithms. Evol. Comput. **4**(4), 361–394 (1996). https://doi.org/10.1162/evco.1996.4.4.361

17. Koehn, P.: Combining genetic algorithms and neural networks: the encoding problem. The University of Tennessee, Knoxville (1994). http://homepages.inf.ed.ac.uk/pkoehn/publications/gann94.pdf

18. Gwiazda, T.: Genetic Algorithms Reference Volume 2 Mutation Operator for Numerical Optimization Problems. In: Genetic Algorithms Reference. Simon and Schuster (2007). https://books.google.com.ua/books?id=O1FVGQAACAAJ

19. Lau, B.: Using keras and deep deterministic policy gradient to play TORCS (2016). https://yanpanlau.github.io/2016/10/11/Torcs-Keras.html

20. Ding, W.G.: Python script for illustrating convolutional neural network (convnet) (2018). https://github.com/gwding/drawconvnet

21. Sastry, K., Goldberg, D., Kendall, G.: Search Methodologies: Introductory Tutorials in Optimization and Decision Support Techniques. Springer, Boston (2005). https://doi.org/10.1007/0-387-28356-0_4

22. Lehman, J., Chen, J., Clune, J., Stanley, K.O.: ES is more than just a traditional finite-difference approximator. CoRR 1712.06568 (2017). http://arxiv.org/abs/1712.06568

Application of Neuro-Controller Models for Adaptive Control

Viktor Smorodin$^{(\boxtimes)}$ ⓘ and Vladislav Prokhorenko ⓘ

Francisk Skorina Gomel State University, Gomel, Belarus
smorodin@gsu.by

Abstract. In this paper, a method for constructing a model of a controller based on recurrent neural network architecture for implementation of control for the optimal trajectory finding problem is considered. A type of a neuro-controller based on recurrent neural network architecture with long short-term memory blocks as a knowledge base on the external environment and previous states of the controller is proposed. The formalization of the technological cycle of a special type for adaptive control of a production process using the model of the neuro-controller is given.

Keywords: Artificial neural networks · Recurrent neural networks
Neuro-controller · Knowledge base · Artificial intelligence
Mathematical models · Control system · Adaptive control

1 Introduction

Recently, artificial neural networks are increasingly being applied to solving new classes of practical problems, different from the typical problems solved effectively by neural networks (such as classification problems, data compression, prediction, approximation, etc.). Despite the complexity of the practical problems that can be solved using the construction of mathematical models [1], neural networks are a fairly effective tool for modeling complex systems as a set of models of the components of the object under study when solving important tasks of complex process control in the areas of activity, which are difficult to formalize.

Such an approach may be of interest when solving problems of rapid response in the process of operation of complex technological objects when it is necessary to take human factor into account in the control process.

Since neural networks are able to extract complex dependencies between input and output data during their training and have the ability to generalize, application of mathematical models of neural networks is also relevant to ensure reliability of operation and safety of control of the technological production cycle.

A recurrent neural network based on a multilayer perceptron is a well-known architecture that is characterized by the presence of feedback connections between layers in the form of time delay elements, which allows the network to accumulate memory of its previous states and to reproduce sequences of reactions. This feature

© Springer Nature Switzerland AG 2019
O. Chertov et al. (Eds.): ICDSIAI 2018, AISC 836, pp. 30–38, 2019.
https://doi.org/10.1007/978-3-319-97885-7_4

makes it possible to apply recurrent neural networks to various real-world problems [2, 3], including control tasks [4–6].

There are interesting potential capabilities of recurrent neural networks [2, 7] that require further study due to the large number of variants of possible architectures [8–11] and the complexity of their analysis, as well as the fact that their training by gradient descent methods is difficult in tasks where long-term time dependencies need to be taken into account [12].

To solve this problem, the architecture of the long short-term memory (LSTM) block was introduced, which is capable of learning long-term dependencies [13]. Currently, neural network architectures, which include LSTM-blocks, are successfully used for solving such problems as time series analysis [14], speech recognition [15], and handwritten text recognition [2].

It should be noted that the neural network controller modeling is effective when a high-quality controller of the controlled system is available [16]. In this case, the neural network acts as an approximator of its function and is trained to simulate the effects of the controller on the controlled system. In some situations, it may be more practical to use the neuro-controller constructed in that way than the original controller. For example, this can be the case because of the common properties of neural networks, which include the ability to generalize, the ability to process noisy data, and fault tolerance due to structural parallelism.

Considering the importance of adaptive control problems, a method for constructing a model of a controller based on recurrent neural network architecture for implementation of control for the optimal trajectory finding problem is being proposed in this paper. A task of finding a trajectory to the designated target for the controlled object in a phase plane of system states is being considered. The phase plane contains passable and impassable regions. The neural network is used as a controller for the object in this task and determines the direction of its next move at each time point.

2   Formalization of the Problem and Architectures of Neural Networks

In the considered task of trajectory finding in the phase plane of system states, the controlled object moves across a two-dimensional region divided into cells that may be passable or impassable. A passable cell is assigned a value of 0, while the impassable one is assigned a value of 1. In this region, a target cell is designated. It is guaranteed that a path from the starting position of the controlled object to the target cell exists in the region.

At each moment of time, the controller receives a vector of seven elements: data on four cells adjacent to the current position of the controlled object, the distance to the target cell, and the direction to the target cell.

The result produced by the neuro-controller at a given moment of time is a four-element vector that determines the direction of the next move of the controlled object in the region. The controlled object continues to move until the target cell is reached.

In this paper, we consider two types of controllers based on neural networks: a neuro-controller that uses a recurrent architecture based on a multilayer perceptron and a neuro-controller that uses a recurrent architecture that contains LSTM blocks.

The recurrent architecture based on a multilayer perceptron includes three fully connected layers consisting of eight, twenty, and four neurons, respectively. There is also a feedback connection through a time delay elements between the second layer and the first layer. A hyperbolic tangent is used as an activation function in the hidden layers of the network.

In Fig. 1, the scheme of the proposed neural network architecture is shown.

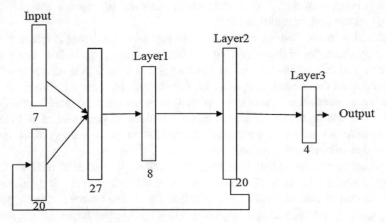

Fig. 1. Scheme of the recurrent neural network architecture based on a multilayer perceptron

When implementing complex pathfinding strategies, information from the past moments in time that may be significantly distant from the current moment in time is required to make a decision. In order to solve the task of selecting a direction of the next movement for the controlled object in the region at each moment of time, the neuro-controller has to take into account information about external environment and decisions made at previous moments in time.

Therefore, it makes sense to use LSTM blocks that will be responsible for long-term storage of information in the considered task. LSTM blocks will serve as a neural network equivalent of a knowledge base, which contains information about the external environment and is accumulated by the controller. The recurrent neural network architecture with LSTM block is a modification of the previous architecture.

The module has a state of size 16 and is connected to the second layer of the neural network through the elements of a time delay. Its current state is passed to the input of the third layer.

In Fig. 2, the scheme of this architecture is shown.

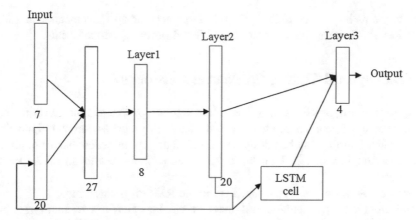

Fig. 2. Scheme of the recurrent neural network architecture with the LSTM block

3 Generating Data for Training and Testing of the Neuro-Controllers

The neural network training procedure requires a large amount of diverse data samples to be generated. 75,000 regions of 30 × 30 cells with different amounts and placements of impassable areas were generated to be used in the training of neuro-controllers.

In Fig. 3, examples of the generated areas are shown. It is guaranteed that a path from the starting position to the target cell exists.

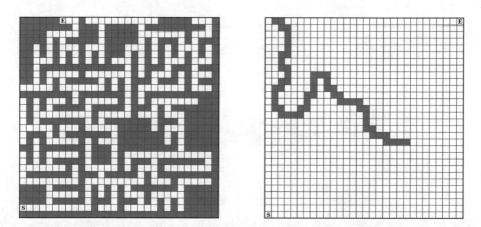

Fig. 3. Examples of the generated regions used for training and testing of the neuro-controllers

Recurrent neural networks are trained on sequences of input and output data.

In the considered pathfinding problem, data sequences used in training are parts of a path from the starting position to the target cell. The best-first search was used to generate the paths in each region. 155,000 sequences of 40 movements and 250,000 sequences of 25 movements were obtained based on the generated paths.

4   Training and Testing the Neural Networks

During the training process such values of the network parameters (connection weights and bias values of neurons) are found that the network produces correct outputs for the given inputs. Training can be considered a non-linear optimization task of minimizing some loss function specified on the training set with respect to all of the network parameters.

The neuro-controllers were trained using the RMSProp algorithm.

The cross entropy function was used as the loss function to minimize during training. The parameters of the neural networks were corrected after presenting each batch of 50 sequences.

In Fig. 4, the loss function values during 30 epochs of training are shown.

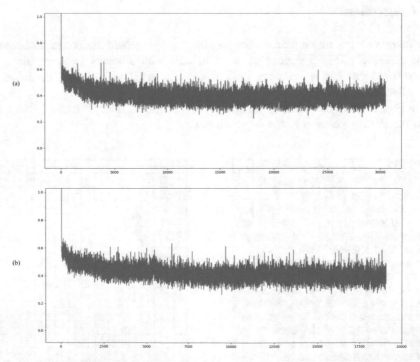

Fig. 4. The loss function values during 30 epochs of training of the recurrent neural network based on multilayer perceptron: (a) training on sequences of 25 elements; (b) training on sequences of 40 elements

A recurrent neural network based on a multilayer perceptron demonstrated inability to learn to reproduce sequences of 25 and 40 movements. The neuro-controller based on this neural network architecture demonstrated poor performance, especially in the areas with large amount of impassable cells.

In Fig. 5, the loss function values during 30 epochs of training are shown.

Figure 6 shows examples of pathfinding by the neuro-controller.

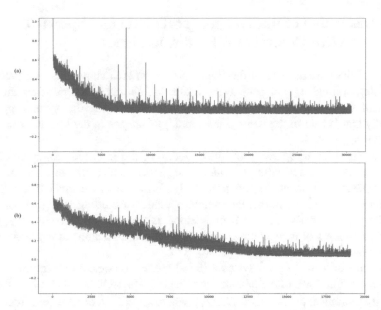

Fig. 5. The loss function values during 30 epochs of training of the recurrent neural network with the LSTM block: (a) training on sequences of 25 elements; (b) training on sequences of 40 elements

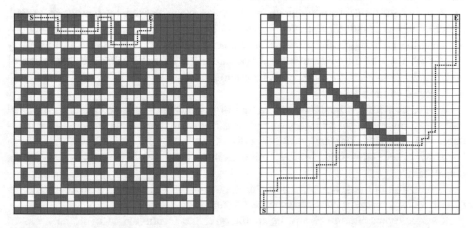

Fig. 6. Examples of successful pathfinding by the neuro-controller based on the recurrent network with the LSTM block after training.

The recurrent neural network with the LSTM block was successfully trained on training sets consisting of sequences of 25 and 40 elements. The network demonstrated a relatively high efficiency in pathfinding and was able to generate a path to the target cell in 73% of the test regions.

5   Formalization of the Technological Cycle of a Special Type for Adaptive Control of a Production Process

Analysis of the current state of developments in the area shows that the problem of establishing optimal control parameters within the chosen criterion arises due to the insufficient effectiveness of existing analysis methods for the system operation. Especially that is true in the cases when topology changes in the operation process of complex system.

Let us consider an approach of formalization of the technological cycle of a production process of a special type, which is operating under control of an automated control system of a technological process (ACSTP). It is assumed that a physical controller exists, which controls the system operation according to a given program.

In the paper, adaptive control is understood as the ability of the control system to alter the parameters of its operation, depending on control actions of the controller and random (external) disturbances (impacts).

Probabilistic technological systems (PTS) are technological objects the operation parameters of which are of probabilistic nature [17]. The technological cycle consists of a set of technological operations $\{MTXO_{ij}\}$, where $i, j = \overline{1, N}$, and the resources that are being consumed by the technological operations during their implementation, possibly with competition for the required resources.

The steelmaking process can be an example of such a technological system, where the melting operation, the consumption of the resources by the open-hearth furnaces, and the casting operation are being controlled.

Another group of technological processes is characterized by having the graph structure of the technological cycle, which is implemented as a result of interactions between the technological operations in the set (TXO_i). The operations, in turn, may consist of a set of mircotechnological operations $\{MTXO_{ij}\}$, where i is the index of TXO_i and j is the index of the mircotechnological operation $MTXO_{ij}$ inside the technological operation TXO_i.

Let us construct a mathematical model of the PTS in the state space. The model includes a complete set of input, output, and state variables of the finite set of interconnected mathematical component-models.

During the operation $MTXO_{ij}$, the technological cycle resources are being consumed.

Simulation of the given types of technological processes is being realized on a basis of critical or average values of resource consumption rates.

Such a representation allows us to interpret the technological cycle as a closed technological system. In order to study that system, neural network technologies can be applied that are based on constructing mathematical models of artificial neural networks.

In this paper the technological processes of continuous nature that operate in real time are considered. The control over the operation of the technological cycle is being realized using the appropriate automation equipment.

A mathematical model of the neuro-controller based on existing prototype is used. The procedure of synthesis of the adaptive controller is based on the training of a recurrent neural network with LSTM blocks (Fig. 2). The LSTM blocks have the ability to store information for long time, thus allowing the possibility of creating a neural network analog of the knowledge base of the environment (random disturbances) and the previous controller states.

The choice of this method of adaptation to external influences satisfies both the adaptation process performance requirements and the quality requirements for the processes in the control system when there is a lack of current information about the nature of the possible random disturbances.

The model of the neuro-controller was realized in Python programming language; the models of recurrent neural network architectures with LSTM blocks were built using the TensorFlow technology, which simplifies the process of integration of self-learning elements and functions of artificial intelligence. The joint operation of the mathematical model and the control system of the technological cycle is implemented on the basis of the interface between the computer technology and the control units of the ACSTP.

6 Conclusion

The experiments carried out on the models have shown that the neuro-controller based on the recurrent neural network with LSTM blocks can be successfully used to solve adaptive control tasks. Such blocks allow the neural network to store information about the states of the system from the past moments in time that may be significantly distant from the current moment in time, which allows the neural network to learn long-term dependencies and to reproduce long sequences of reactions to random disturbances and external influences. The possibility of increasing the efficiency of the existing architecture by adding additional memory modules and training on longer data sequences depends on the specific parameters of the modeling object operation.

Application of recurrent neural networks with additional LSTM blocks as a basis for the development of intelligent decision support systems provides new opportunities for implementation of control actions based on the current situation during the operation of the object under study, and determining the optimal control parameters for the object.

The proposed approach can provide an opportunity for the development of adaptive control segmentation algorithms for complex technological systems. Algorithms based on neural networks will be able to take into account the changes in the parameters of the control object and the real-time external disturbances using the appropriate knowledge bases.

References

1. Maximey, I.V., Smorodin, V.S., Demidenko, O.M.: Development of simulation models of complex technical systems. F. Skorina State University, Gomel, Belarus (2014)
2. Multiple Object Recognition with Visual Attention. https://arxiv.org/abs/1412.7755. Accessed 25 Mar 2018
3. Karpathy, A.: Deep visual-semantic alignments for generating image descriptions. IEEE Trans. Pattern Anal. Mach. Intell. **39**(4), 664–676 (2017)
4. Mayer, H., Gomez, F., Wierstra, D., Nagy, I., Knoll, A., Schmidhuber, J.: A system for robotic heart surgery that learns to tie knots using recurrent neural networks. In: 2006 IEEE/RSJ International Conference on Intelligent Robots and Systems, Beijing, China, pp. 543–548. IEEE (2006)
5. Huh, D., Todorov, E.: Real-time motor control using recurrent neural networks. In: Proceedings of IEEE Symposium on Adaptive Dynamic Programming and Reinforcement Learning (IEEE ADPRL), Nashville, TN, USA, pp. 42–49. IEEE (2009)
6. Nouri, K., Dhaouadi, R.: Adaptive control of a nonlinear DC motor drive using recurrent neural networks. Appl. Soft Comput. **8**(1), 371–382 (2008)
7. Siegelmann, H.T.: Computation beyond the turing limit. Science **268**(5210), 545–548 (1995)
8. Elman, J.L.: Finding structure in time. Cognit. Sci. **14**(2), 179–211 (1990)
9. Jordan, M.I.: Serial order: a parallel distributed processing approach. Adv. Psychol. **121**, 471–495 (1997). Neural-Network Models of Cognition
10. Goller, C., Kuchler, A.: Learning task-dependent distributed structure-representations by backpropagation through structure. In: IEEE International Conference on Neural Networks, Washington, DC, USA, pp. 347–352. IEEE (1996)
11. Neural Turing Machines. https://arxiv.org/abs/1410.5401. Accessed 25 Mar 2018
12. Bengio, Y., Simard, P., Frasconi, P.: Learning long-term dependencies with gradient descent is difficult. IEEE Trans. Neural Netw. **5**(2), 157–166 (1994)
13. Hochreiter, S., Schmidhuber, J.: Long short-term memory. Neural Comput. **9**(8), 1735–1780 (1997)
14. Malhotra, P., Vig, L., Shroff, G., Agarwal, P.: Long short term memory networks for anomaly detection in time series. In: ESANN 2015 Proceedings, European Symposium on Artificial Neural Networks, Computational Intelligence and Machine Learning, pp. 89–95. Presses universitaires de Louva, Bruges, Belgium (2015)
15. Graves, A., Mohamed A., Hinton G.: Speech recognition with deep recurrent neural networks. In: 2013 IEEE International Conference on Acoustics Speech and Signal Processing (ICASSP), Vancouver, BC, Canada, pp. 6645–6649. IEEE (2013)
16. Omidvar, O., Elliott, D.L.: Neural Systems for Control. Academic Press, New York (1997)
17. Maximey, I.V., Demidenko, O.M., Smorodin, V.S.: Problems of theory and practice of modeling complex systems. F. Skorina State University, Gomel, Belarus (2015)

Forecasting of Forest Fires in Portugal Using Parallel Calculations and Machine Learning

Yaroslav Vyklyuk[1]([✉]), Milan M. Radovanović[2],
Volodymyr Pasichnyk[3], Nataliia Kunanets[3], and Sydor Petro[1]

[1] Bukovinian University, Chernivtsi, Ukraine
vyklyuk@ukr.net
[2] Geographical Institute "Jovan Cvijić", Serbian Academy of Sciences and Arts,
Belgrade, Serbia
[3] Information Systems and Networks Department,
Lviv Polytechnic National University, Lviv, Ukraine

Abstract. Forest fires that occurred in Portugal on June 18, 2017 caused several dozens of human casualties. The cause of their emergence, as well as many others that occurred in Western Europe at the same time, remained unknown. The heliocentric hypothesis has indirectly been tested, according to which charged particles are a possible cause of forest fires. We must point out that it was not possible to verify whether in this specific case the particles by reaching the ground and burning the plant mass create initial phase of the formation of flame. Therefore, we have tried to determine whether during the critical period, i.e. on June 15–19, there was a certain statistical connection between certain parameters of the solar wind (SW) and meteorological elements. Based on the hourly values of the charged particles flow, a correlation analysis was performed with hourly values of individual meteorological elements including time lag at Monte Real station. The application of the adaptive neuro-fuzzy inference systems has shown that there is a high degree of connection between the flow of protons and the analyzed meteorological elements in Portugal. However, further verification of this hypothesis requires further laboratory testing.

Keywords: Forest fires · Heliocentric hypothesis · ANFIS models
Portugal

1 Introduction

Forest fires that occurred on June 18, 2017 in the central part of Portugal are among the most endangered in the history of this country. About 60 forest fires were reported during the night of June 17/18, 2017. The number of victims was 64, including six firefighters. Many died in their cars or near the vehicles when they tried to escape from fire. About 200 people were injured. On June 20, a plane with two crew members who participated in the extinguishing of the fire collapsed. On that day, the fires were localized. More than 45,000 hectares of forest burnt. The question arose whether it was possible to predict these natural disasters, and therefore take measures to eliminate them and minimize human and material casualties.

© Springer Nature Switzerland AG 2019
O. Chertov et al. (Eds.): ICDSIAI 2018, AISC 836, pp. 39–49, 2019.
https://doi.org/10.1007/978-3-319-97885-7_5

Severe weather conditions were mentioned as a possible reason. It is possible that the fire was caused by a thunderstorm, as the investigators found a tree that was hit by a "thunderstorm without rain," the Portuguese media reported, referring to police sources. In addition, some officials expressed the view that fires were deliberately set.

Gomes and Radovanović [2] presented a number of critical views on the so-called generally accepted attitudes regarding explanation of the initial phase of the flame. In short, there was a suspicion of anthropogenic activity being possibly responsible in 95% of cases, as it was accepted not only in the media, but also in many scientific circles. And first of all, because it is hard to believe that, for example, over 30,000 fires have been reported over the years, and the area of the burnt vegetation is over 350,000 ha.

Milenković et al. [5] found the connection between the number of forest fires in Portugal and the Atlantic multidecadal oscillation. Considering absolute values, there some other so-called potential candidates look suspicious, such as lightning or high air temperature. Thunderstorms are followed by rain almost as a rule, and when it comes to lightning, they strike without rain. As for air temperature, it has been determined that a minimum of 300 °C is required to show the initial phase of the flame [12]. This temperature has never been measured on the ground by far (including desert areas), not to mention the air temperature.

As far as we know, Gomes and Radovanović [1] argued for the first time that the processes on the Sun could represent a potential explanation for the occurrence of forest fires. In the meantime, more research was published on this subject: Radovanović et al. [10], Radovanović, Vyklyuk et al. [8], Radovanović, Pavlović et al. [9], Radovanović, Vyklyuk et al. [7], etc. In these works, the effectiveness of adaptive neuro-fuzzy inference systems (ANFIS) and artificial neural networks (ANN) was proved. In this paper, therefore, we tried to re-examine whether there is justification for this kind of approach. According to the mentioned hypothesis, it is necessary to have a coronary hole and/or energy region in the geoeffective position on the Sun before the occurrence of a fire. On June 15, coronary holes CH807 and CH808, as well as energy region 12663 were in the geoeffective position. The characteristic of this region was beta-gamma. The maximum speed of the SW at La Grange Point was 594 km/s[1]. On that day, on June 16, according to the same source, the Ap index was around 3–48.

2 Data and Methods

Bearing in mind that it was not possible to directly record the possible propagation of particles to the ground, as a potential reason that causes initial phase of the flame in the burning plant mass, we decided to test the heliocentric hypothesis indirectly. The following hourly averaged real-time data were used as input parameters: differential electron (energy ranges 38–53 and 175–315 keV)—$E1$, $E2$, proton flux (energy ranges 47–68, 115–195, 310–580, 795–1193 and 1060–1900 keV)—$P1$–$P4$, and integral proton Flux—$I1$, $I2$[2]; hourly averaged real-time bulk parameters of the solar wind

---

[1] http://www.solen.info/solar/indices.html.

[2] ftp://ftp.swpc.noaa.gov/pub/lists/ace2/201706_ace_epam_1h.txt.

plasma proton density (p/cc)—**W1**, bulk speed (km/s)—**W2**, and ion temperature (degrees K) **W3**[3]. ACE Satellite—Solar Wind Electron Proton Alpha Monitor is located at La Grange point so that it measures the data in real time that come from the Sun to our planet. Hourly meteorological data relating to Monte Real Station have been used as an output (Latitude: 39° 49' 52" N, Longitude: 8° 53' 14" W). This station is located in the military air-base near Leiria. The data include air temperature (°C), humidity (%), and air pressure (hPa). All the data used in the paper refer to June 15–19, 2017. This station was selected because it is located near the fire-affected area and the data are available on the Internet (Table 1).

The goal of calculations was to investigate the functional dependencies between the characteristics of the SW and air temperature T, humidity H, and pressure P. The measurement step was 1 h.

The solution of this problem consists of several stages.

Table 1. Tested input fields and output fields and correlation between them

Input fields		Correlation (R)		
		T	H	P
Differential Flux particles/cm2-s-ster-MeV, electrons				
E1	38–53	−0.11	0.14	0.16
E2	175–315	0.06	−0.02	−0.17
Differential Flux particles/cm2-s-ster-MeV, protons				
P1	47–68	0.01	0.03	0.11
P2	115–195	0.14	−0.12	−0.07
P3	310–580	0.15	−0.13	−0.17
P4	795–1193	0.11	−0.08	−0.30
P5	1060–1900	−0.10	0.06	0.22
Integral Proton Flux				
I1	> 10MeV	−0.57	0.50	0.85
I2	> 30 MeV	−0.55	0.48	0.85
Solar Wind				
W1	Proton Density (p/cc)	−0.11	0.16	0.41
W2	Bulk Speed (km/s)	0.34	−0.34	−0.47
W3	Ion Temperature (degrees K)	0.14	−0.12	−0.06

2.1 Preliminary Analysis

The feature of this dataset is the presence of missed data (gaps) with a maximum duration of 3 h. The spline interpolation using not-a-knot end conditions was used to fill in these gaps. The interpolated value at a query point is based on a cubic interpolation of the values at neighboring grid points in each respective dimension [3].

Correlation analysis was performed to establish the presence of a linear connection between input and output fields. Calculations show that Pearson correlation coefficients (R, Table 1) are sufficiently small in all cases except $I1$ and $I2$. It means that any linear

---

[3] ftp://ftp.swpc.noaa.gov/pub/lists/ace2/201706_ace_swepam_1h.txt.

dependencies of these data are not observed. High values of R for $I1$ and $I2$ indicate the presence of strongly expressed nonlinear relationships. The presence of lagging (time) between input and output fields may be another reason for small correlation coefficients.

To establish the lag dependence, the transformation of the dataset was conducted. Output fields were fixed, after that the time series of each input field was shifted vertically downward by the number of rows equal to the lag studied. After that, the correlation coefficient between input and output fields was calculated. We investigated the lag from 0 to 5 h.

Calculations show that the smallest R is observed for electron and proton flux (E and P). It means that these input fields do not impact output fields for all lags. The largest R is observed for $I1$ and $I2$ relating to air pressure. R grows weakly with increasing lag. It means that there are nonlinear inertial dependencies between these fields and output fields. These lags mean that it is possible to predict output fields for few hours forward. Similar situation was observed for fields $W1$, $W2$, $W3$.

For further research, an autocorrelation analysis should be carried out to reconcile the interconnection between the input fields. Calculations show strong linear relationship between $I1$ and $I2$ fields. It means that only one of them should be used in calculations.

2.2   Search of Best Models

As can be seen from lag correlation and autocorrelation analysis, the best models for all output fields must be dependent on integral proton flux and solar wind with lag 5:

$$T(H, P) = F((I1 \text{ or } I2)_5, W1_5, W2_5, W3_5), \qquad (1)$$

where subscribe index 5 means lag 5.

We must know which of them ($I1$ or $I2$) is better. Therefore we tested models $T(H, P) = F(I1_5, W1_5, W2_5, W3_5)$ and $T(H, P) = F(I2_5, W1_5, W2_5, W3_5)$.

For checking this decision, models with all possible combinations of lags from 0 to 5 were tested ($6^4 = 1{,}296$ models). Models with $I1$ or $I2$ input fields were tested separately. Theoretical investigations [6] showed that electrons must have nonlinear impact on output fields. Therefore, similar calculations for models containing one of the fields $E1$ or $E2$ were carried out ($6^5 = 7{,}776$ models). In addition, models that take into account only differential flux of electrons and protons were tested ($6^7 = 279{,}936$ models). Linear regression analysis and ANFIS were used as models in this investigation. Two Gauss membership functions were created for each input field in ANFIS.

Since ANFIS is a Sugeno-type system, the output membership function type was selected as constant. Each ANFIS system was trained during 100 epochs, initial step size was 0.01, step size decrease rate was 0.9, step size increase rate was 1.1. Hybrid method was selected as optimization method for membership function parameter training. For learning process, the dataset was split into training and test sets in proportion 90/10. This method is a combination of least-squares estimation and back-propagation.

Taking into account that 894,240 models have been investigated, and all of them are independent of each other, the parallel calculation was used to solve this problem. It decreased time of calculation about 3.5 times. The longest calculation lasted about

60 h. The total time consisted of about 200 h ($\sim$ 8 days). Results of these calculations are presented in Table 2.

Table 2. Comparison of correlation coefficients of the best models with models with lag 5 input fields

Models	Linear		ANFIS		Number of models	Position of (1) model (linear/ANFIS)
	Lag 5	Best	Lag 5	Best		
Temperature						
F(I1, W1, W2, W3)	0.8199	0.8251	0.8529	0.8621	1296	19 16
F(I2, W1, W2, W3)	0.8113	0.8178	0.8483	0.8540	1296	11 11
F(I1, W1, W2, W3, E1)	0.8337	0.8431	0.8698	0.8839	7776	54 41
F(I1, W1, W2, W3, E2)	0.8239	0.8287	0.8843	0.9051	7776	65 359
F(E1, E2, P1, P2, P3, P4, P5)	0.3372	0.4241	0.3567	0.4733	279936	95, 927 196, 615
Humidity						
F(I1, W1, W2, W3)	0.7559	0.7680	0.8026	0.8247	1296	20 24
F(I2, W1, W2, W3)	0.7437	0.7597	0.7953	0.8115	1296	26 24
F(I1, W1, W2, W3, E1)	0.7705	0.7861	0.8216	0.8455	7776	46 123
F(I1, W1, W2, W3, E2)	0.7672	0.7831	0.8538	0.8910	7776	50 733
F(E1, E2, P1, P2, P3, P4, P5)	0.3553	0.4363	0.3689	0.4610	279936	56, 329 130, 930
Pressure						
F(I1, W1, W2, W3)	0.8985	0.9037	0.9483	0.9637	1296	178 247
F(I2, W1, W2, W3)	0.8767	0.8988	0.9338	0.9506	1296	932 512
F(I1, W1, W2, W3, E1)	0.8994	0.9055	0.9521	0.9679	7776	917 929
F(I1, W1, W2, W3, E2)	0.8991	0.9061	0.9576	0.9673	7776	1280 309
F(E1, E2, P1, P2, P3, P4, P5)	0.4090	0.5879	0.4338	0.6153	279936	256, 184 276, 420

As it can be seen from Table 2, correlation coefficient between real data and models' data was selected as a criterion of accuracy. First of all, it should be noted that all ANFIS models have higher correlation coefficient than linear ones. It is clearly seen that models based on differential flux of electrons and protons have the smallest R. This means that they are not the main factors of influence on output fields.

Comparing models containing *I1* and *I2* factors allows us to conclude that factor *I1* makes it possible to better describe output fields. It is true for linear and ANFIS models.

As the calculations have shown, taking into account the factor describing differential flux of electrons allowed slight increase of the correlation coefficients for linear and ANFIS models. It should be noted that the influence of factors *E1* and *E2* is approximately the same. Therefore, factor *E2* was chosen for further calculations.

So, in the next stage we investigated the most accurate models:

$$T(H, P) = F(I1_1, W1_1, W2_1, W3_1, E2_1),$$ (2)

where subscript 1 stands for lag 1.

As it can be clearly seen from Table 2, there is a large number of models that are more accurate than (1) (columns 2–3 and 4–5). As the last column of the table shows, models (2) occupy by far not the first places among exact models. Therefore, the classical approach to the definition of exact models (1), described in the lag analysis stage, is not suitable for this class of problems. Figure 1 represents the distribution of correlation coefficients for all models (2). As it can be seen from Fig. 1, there are lots of models that can make forecasts of output fields with high level of accuracy. It can be the base for the creation of multimodels expert system for forecasting crisis events.

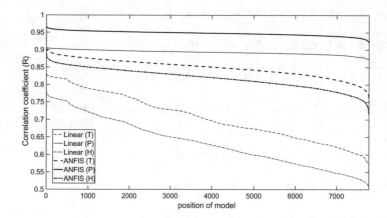

Fig. 1. Distribution of correlation coefficients of all possible models (2)

2.3 Accuracy Analysis

To confirm this conclusion, the adequacy of the three most accurate models was tested:

- linear with higher correlation coefficient;
- ANFIS with higher correlation coefficient;
- model with higher total ($R_{Linear} + R_{ANFIS}$) correlation coefficient.

Information about input fields of these models is presented in Table 3.

Table 3. Lags (2) of the best models for forecasting T, H, P

Model	Best for:	II_1	WI_1	$W2_1$	$W3_1$	$E2_1$	R
Temperature							
T_1	Linear	1	5	0	0	5	0.8287
T_2	ANFIS	0	5	5	3	5	0.9051
T_3	Linear + ANFIS	5	5	2	3	5	1.7215
Humidity							
H_1	Linear	1	5	0	0	5	0.7831
H_2	ANFIS	5	4	4	2	5	0.8910
H_3	Linear + ANFIS	4	5	5	3	4	1.6392
Pressure							
P_1	Linear	4	3	3	3	4	0.9061
P_2	ANFIS	5	0	5	3	1	0.9673
P_3	Linear + ANFIS	5	0	5	3	0	1.8684

As it can be seen from Table 3, the best linear model for temperature is:

$$T_1 = F(II_1, WI_5, W2_0, W3_0, E2_5).$$

The best ANFIS model is as follows:

$$T_2 = F(II_0, WI_5, W2_5, W3_3, E2_5).$$

The model with higher total correlation coefficient is as follows:

$$T_3 = F(II_5, WI_5, W2_2, W3_3, E2_5).$$

As it can be seen, lags of input fields of these models are different.

In addition, lags of the input fields of these models are sometimes much smaller than in (1). On the other hand, it confirms that multimodels approach can make forecasting on different time periods from 0–5 h.

According to the data for the lags from Table 3, nine linear and nine ANFIS models were constructed.

After training each ANFIS model, we obtained a set of membership functions, rules, fuzzification, and defuzzification methods etc. [4]. The attributes of obtained membership functions of input factors are presented in Table 4. As we mentioned above, each input factor consists of 2 membership functions.

As can be seen from Table 4, during training, only parameters of II membership functions were changed. It confirms that this field is most significant in these models.

For testing accuracy of these models, results of model forecasting were compared with real data and correlation coefficient was calculated (Table 5).

As can be seen from Table 5, all models have high R. All ANFIS models have higher correlation coefficient than linear ones. It confirms that ANFIS models are more accurate and take into account nonlinear effects. For visual comparison of results, predicted values obtained by the models in comparison with real data are presented in Fig. 2.

2.4   Adequacy Analysis

As can be seen from Fig. 2, ANFIS models better describe and predict fluctuation in amplitude. It is clearly seen for humidity models. Despite the high coefficient of ANFIS models, linear models also accurately describe explored output fields. Therefore, an adequacy analysis and sensitivity analysis is required to select the correct type of model.

Table 4. Parameters of membership functions for ANFIS models

[σ c]

Model	II_1	WI_1	$W2_1$	$W3_1$	$E2_1$
Temperature					
T_1	[0.05 1.97] [0.03 2.10]	[11.0 1.0] [11.0 27.0]	[111 362] [111624]	[16706130599] [167061423999]	[63953–99999] [6395350600]
T_2	[0.02 1.95] [0.04 2.09]				
T_3	[0.041.95] [0.05 2.09]				
Humidity					
H_1	[0.06 1.98] [0.04 2.11]	[11.0 1.0] [11.0 27.0]	[111 362] [111624]	[16706130599] [167061423999]	[63953–99999] [6395350600]
H_2	[0.04 1.96] [0.05 2.09]				
H_3	[0.04 1.96] [0.053 2.097]				
Pressure					
P_1	[0.03 1.96] [0.03 2.11]	[11.0 1.0] [11.0 27.0]	[111 362] [111624]	[16706130599] [167061423999]	[63953–99999] [6395350600]
P_2	[0.03 1.97] [0.04 2.10]				
P_3	[0.04 1.96] [0.04 2.10]				

Table 5. Pierson correlation coefficients of models from Table 4

Model	Linear	ANFIS
T_1	0.8287	0.8714
T_2	0.7697	0.9051
T_3	0.8204	0.9012
H_1	0.7831	0.8374
H_2	0.7076	0.8910
H_3	0.7629	0.8763
P_1	0.9061	0.9581
P_2	0.9010	0.9673
P_3	0.9032	0.9652

For this purpose, for each model, the following calculations were carried out:

1. For each row of the training set, each value of the input parameter of turning was changed by 10%.
2. A relative change in the output field on change of separate input field was calculated.
3. Data was averaged on all records.

Results of these calculations are presented in Table 6.

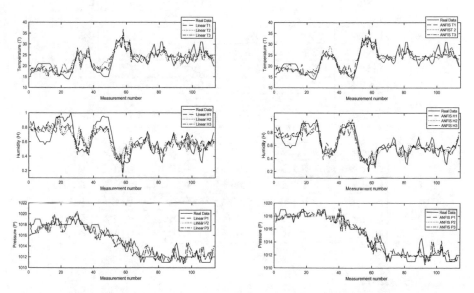

Fig. 2. Comparison of predicted values obtained by the linear and ANFIS models from Table 5 in comparison with the real data

As can be seen from Table 6, −82% for linear *T1* means that on average if factor *I1* increases by 10%, the temperature will decrease by 82% after 1 h. Similarly, the same increase in *I1* will lead to decreasing temperature after 5 h by 57% according to ANFIS model. As can be seen, the most significant factors are *I1* and *W2*. These results have also confirmed that electrons do not impact investigated output fields. The results show that increasing *I1* will lead to decreasing temperature and increasing humidity. In contrast, increasing *W2* will lead to increasing temperature and decreasing humidity. It is clearly seen that input factors have weak impact on pressure despite the highest correlation coefficient of models.

Table 6. Sensitivity analysis of models from Table 2

Model	Best for:	II_1	$W1_1$	$W2_1$	$W3_1$	$E2_1$
Temperature						
T_1	Linear	−82%	2%	3%	−1%	0%
	ANFIS	−39%	0%	3%	0%	0%
$T2$	Linear	−80%	1%	2%	−1%	0%
	ANFIS	−30%	3%	31%	−1%	0%
T_3	Linear	−91%	1%	0%	0%	0%
	ANFIS	−57%	1%	11%	1%	0%
Humidity						
H_1	Linear	96%	−3%	−7%	3%	0%
	ANFIS	72%	−1%	−15%	1%	0%
H_2	Linear	116%	−2%	3%	0%	0%
	ANFIS	91%	1%	−16%	−1%	0%
H_3	Linear	121%	−2%	1%	0%	0%
	ANFIS	58%	−2%	−18%	2%	0%
Pressure						
P_1	Linear	1.29%	0.00%	−0.05%	0.01%	0.00%
	ANFIS	0.36%	0.00%	−0.07%	0.01%	0.00%
P_2	Linear	1.30%	0.00%	−0.04%	0.01%	0.00%
	ANFIS	0.39%	0.00%	−0.05%	0.02%	0.00%
P_3	Linear	1.30%	0.00%	−0.04%	0.01%	0.00%
	ANFIS	0.43%	−0.01%	−0.07%	0.01%	0.00%

3 Conclusions

Forest fires that occurred on June 18 and 19, 2017 in Portugal are among the most catastrophic ones in the history of the country. As in many other cases, the cause of their emergence has remained unknown. Relying on recent results, we have tried to test the heliocentric hypothesis of the occurrence of forest fires in this case. ACE satellite registered a sudden inflow of temperature, speed, and density of the SW particles a couple of days before the formation of fires. The basic starting point was that if there is any connection between the process on the Sun and forest fires, then during critical days the meteorological parameters would have to "react" to some extent to certain parameters of the SW. In that sense, we have tried to determine whether there is any statistical connection between the flow of protons and electrons in some energy ranges on the one hand and the air temperature, relative humidity, and air pressure in Monte Real, on the other one. Calculations included hourly values, but with a time lag shift from 0 to 5 h in the period June 15–19, 2017. The largest R is observed for (II) proton flux > 10 meV and ($I2$) proton flux > 30 meV (0.89 and 0.86 respectively) relative to air pressure.

Linear regression analysis and ANFIS were used as models in this investigation. Taking into account that 894,240 models have been investigated, and all of them are independent of each other, parallel calculations were used to solve this problem. As it can be seen from Fig. 1, there are lots of models that can make forecasts of output fields with high level of accuracy. According to the data for the lags from Table 3, nine linear and nine ANFIS models were constructed. As it can be seen from Table 5, ANFIS models are more accurate and take into account nonlinear effects.

Obtained results indicate the need of further improvement of the presented methods for the purpose of creation of scientifically based Web-oriented multimodels expert system for making forecasting of crisis events in different time periods of 0–5 h. Especially if we keep in mind that, depending on the repeatability of certain processes in the Sun, we can expect more or less similar weather and environmental conditions in certain locations on Earth [11].

References

1. Gomes, J.F.P., Radovanovic, M.: Solar activity as a possible cause of large forest fires a case study: analysis of the Portuguese forest fires. Sci. Total Environ. **394**(1), 197–205 (2008)
2. Gomes, J.F.P., Radovanovic, M.: Solar Activity and Forest Fires, pp. 1–139. Nova Science Publishers, New York (2009). ISBN 978-1-60741-002-7
3. Hazewinkel, M.: Encyclopaedia of Mathematics: Monge - Ampère Equation - Rings and Algebras, 929 p. Springer, Dordrecht (2013). ISBN 978-0-7923-2976-3
4. Jang, J.-S.R.: ANFIS: adaptive-network-based fuzzy inference system. IEEE Trans. Syst. Man Cybern. **23**(3), 665–685 (1993)
5. Milenković, M., Yamashkin, A.A., Ducić, V., Babić, V., Govedar, Z.: Forest fires in Portugal - the connection with the atlantic multidecadal oscillation (AMO). J. Geogr. Inst. Jovan Cvijic SASA **67**(1), 27–35 (2017)
6. Radovanović, M.: Forest fires in Europe from July 22nd to 25th 2009. Arch. Biolog. Sci. **62** (2), 419–424 (2010)
7. Radovanović, M., Vyklyuk, Y., Jovanović, A., Vuković, D., Milenković, M., Stevančević, M., Matsiuk, N.: Examination of the correlations between forest fires and solar activity using Hurst index. J. Geogr. Inst. Jovan Cvijic SASA **63**(3), 23–32 (2013)
8. Radovanović, M.M., Vyklyuk, Y., Malinović Milićević, B.S., Jakovljević, M.D., Pecelj, R. M.: Modelling of forest fires time evolution in the USA on the basis of long term variations and dynamics of the temperature of the solar wind protons. Therm. Sci. **19**(Suppl. 2), S437–S444 (2014)
9. Radovanović, M.M., Pavlović, T.M., Stanojević, G.B., Milanović, M.M., Pavlović, M.A., Radivojević, A.R.: The influence of solar activities an occurrence of the forest fires in south Europe. Therm. Sci. **19**(2), 435–446 (2014)
10. Radovanović, M.M., Vyklyuk, Y., Milenković, M., Vuković, B.D., Matsiuk, N.: Application of ANFIS models for prediction of forest fires in the USA on the basis of solar activity. Therm. Sci. **19**(5), 1649–1661 (2015)
11. Todorović, N., Vujović, D.: Effect of solar activity on the repetitiveness of some meteorological phenomena. Adv. Space Res. **54**(11), 2430–2440 (2014)
12. Viegas, D.X.: A mathematical model for forest fires blowup. Combust. Sci. Technol. **177**, 27–51 (2005)

A New Non-Euclidean Proximal Method for Equilibrium Problems

Lyubov Chabak[1], Vladimir Semenov[2($\boxtimes$)], and Yana Vedel[2]

[1] Infrastructure and Technologies State University, Kyiv, Ukraine
chabaklm@ukr.net
[2] Faculty of Computer Sciences and Cybernetics,
Taras Shevchenko National University of Kyiv, Kyiv, Ukraine
semenov.volodya@gmail.com, yana.vedel@gmail.com

Abstract. The paper analyzes the convergence of a new iterative algorithm for approximating solutions of equilibrium problems in finite-dimensional real vector space. Using the Bregman distance instead of the Euclidean, we modified the recently proposed two-stage proximal algorithm. The Bregman distance allows us to take into account the geometry of an admissible set effectively in some important cases. Namely, with the suitable choice of distance, we obtain a method with explicitly solvable auxiliary problems on the iterative steps. The convergence of the algorithm is proved under the assumption that the solution exists and the bifunction is pseudo-monotone and Lipschitz-type.

Keywords: Equilibrium problem · Two-stage proximal method
Bregman distance · Pseudo-monotonicity · Lipschitz property
Convergence

1 Introduction

Consider the equilibrium problem for nonempty convex closed set $C \subseteq R^d$ and bifunction $F : C \times C \to R$:

$$\text{find } x \in C \text{ such that} \quad F(x,y) \geq 0 \quad \forall y \in C, \tag{1}$$

where $F(y,y) = 0$ for all $y \in C$.

The equilibrium problem (1) is very general in the sense that it includes, as special cases, many applied mathematical models such as: variational inequalities, optimization problems, saddle point problems, and Nash equilibrium point problems. Algorithms for solving the equilibrium problems were the subject of many papers, for example, [1–9,11,12].

In this paper, we propose and analyze a new iterative method for solving the equilibrium problem. Using the Bregman distance (Bregman divergence) instead of the Euclidean one we modify the two-stage proximal algorithm from [6]. The Bregman distance allows us to take into account the geometry of an admissible

© Springer Nature Switzerland AG 2019
O. Chertov et al. (Eds.): ICDSAI 2018, AISC 836, pp. 50–58, 2019.
https://doi.org/10.1007/978-3-319-97885-7_6

set effectively in some cases. Namely, with the suitable choice of distance, we obtain a method with explicitly solvable auxiliary problems on the iterative steps.

Remark 1. The Bregman distance, however, is not a metric, because it is, in general, not symmetric, and does not satisfy the triangle inequality.

We note that, in the particular case of a variational inequality, the obtained algorithm coincides with the version of the method of mirror descent recently studied by one of the authors [10].

2   Assumptions

We assume that the bifunction F satisfies the following conditions:

(A1) For all x, $y \in C$ from $F(x,y) \geq 0$, it follows that $F(y,x) \leq 0$ (pseudo-monotonicity).
(A2) $F : C \times C \to R$ is lower semicontinuous on $C \times C$.
(A3) For all $x \in C$ the function $F(x, \cdot)$ in convex on C.
(A4) For all $y \in C$ the function $F(\cdot, y)$ is upper semicontinuous on C.
(A5) For all x, y, $z \in C$, the following inequality holds

$$F(x,y) \leq F(x,z) + F(z,y) + a\,\|x - z\|^2 + b\,\|z - y\|^2,$$

where a, b are positive constants (the Lipschitz-type property).

Let us consider the following dual equilibrium problem:

$$\text{find } y \in C \text{ such that} \quad F(x,y) \leq 0 \quad \forall x \in C. \tag{2}$$

We denote the sets of solutions for problems (1) and (2) as S and S^*, respectively. In the considered case, the sets S and S^* are equal, convex, and closed. Further, we assume that the solution set S is nonempty.

3   The Algorithm

Let's recall some facts about Bregman distance. Let $\varphi : R^d \to R$ be the continuously differentiable convex function on C. Assume that φ is strongly convex with the parameter $\sigma > 0$ in the norm $\| \cdot \|$, i.e.

$$\varphi(a) - \varphi(b) \geq (\nabla\varphi(b), a - b) + 2^{-1}\sigma\|a - b\|^2 \quad \forall a \in C, \ b \in C.$$

The Bregman distance (generated by function φ) on the set C is defined by

$$D_\varphi(a,b) = \varphi(a) - \varphi(b) - (\nabla\varphi(b), a - b) \quad \forall a \in C, \ b \in C.$$

Remark 2. We consider three classical examples. If $\varphi(x) = \frac{1}{2}\|x\|_2^2$, where $\|\cdot\|_2$ the Euclidean norm, we have $D_\varphi(x,y) = \frac{1}{2}\|x-y\|_2^2$. For a standard probability simplex

$$\Delta_d = \left\{ x \in R^d : \ x_i \geq 0, \ \sum_{i=1}^d x_i = 1 \right\}$$

and negative entropy

$$\varphi(x) = \sum_{i=1}^d x_i \ln x_i$$

(it is strongly convex with parameter 1 in ℓ_1-norm on Δ_d), we obtain the well-known Kullback-Leibler divergence (KL-divergence) on Δ_d:

$$D_\varphi(x,y) = \sum_{i=1}^d x_i \ln \left(\frac{x_i}{y_i} \right) \quad \forall x \in \Delta_d, \ y \in \mathrm{ri}\, \Delta_d.$$

In the case of a product of simplexes $\Pi = \Delta_{d_1} \times \ldots \times \Delta_{d_k}$, and choosing

$$\varphi(x) = \sum_{l=1}^k \sum_{i=1}^{d_l} x_i^l \ln x_i^l,$$

where $x = (x_1^1, \ldots, x_{d_1}^1, x_1^2, \ldots, x_{d_k}^k) \in \Pi$, we obtain the corresponding Bregman divergence in the form of sum of Kullback-Leibler divergences:

$$D_\varphi(x,y) = \sum_{l=1}^k \sum_{i=1}^{d_l} x_i^l \ln \left(\frac{x_i^l}{y_i^l} \right).$$

The following useful three-point identity holds:

$$D_\varphi(a,c) = D_\varphi(a,b) + D_\varphi(b,c) + (\nabla\varphi(b) - \nabla\varphi(c), a - b).$$

The strong convexity of φ implies the estimate

$$D_\varphi(a,b) \geq \frac{\sigma}{2}\|a-b\|^2. \tag{3}$$

The minimization problem

$$F(a,y) + \lambda^{-1} D_\varphi(y,b) \to \min_{y \in C} \quad (a, b \in C, \ \lambda > 0)$$

always has only one solution.

Suppose that we are able to solve efficiently this problem. For example, it is possible in the case of probability simplex, linearity F for the second argument, and the Kullback-Liebler divergence. Indeed, the solution of the problem

$$\sum_{i=1}^d a_i y_i + \frac{1}{\lambda} \sum_{i=1}^d y_i \ln \left(\frac{y_i}{b_i} \right) \to \min_{y \in \Delta_d} \quad (a \in R^d, \ b \in \mathrm{ri}\, \Delta_d, \ \lambda > 0)$$

has the form

$$z_i = \frac{b_i e^{-\lambda a_i}}{\sum_{j=1}^{d} b_j e^{-\lambda a_j}}, \quad i = 1, \ldots, d.$$

Now, we introduce the following iterative algorithm for solving the equilibrium problem (1).

Algorithm 1. Two-Stage Proximal Algorithm

For $x_1, y_1 \in C$, generate sequences of elements $x_n, y_n \in C$ with the iterative scheme

$$\begin{cases} x_{n+1} = \underset{y \in C}{\operatorname{argmin}} \left(F(y_n, y) + \frac{1}{\lambda} D_\varphi(y, x_n) \right), \\ y_{n+1} = \underset{y \in C}{\operatorname{argmin}} \left(F(y_n, y) + \frac{1}{\lambda} D_\varphi(y, x_{n+1}) \right), \end{cases}$$

where $\lambda > 0$.

In Algorithm 1, at each iterative step, we must solve two optimization programs onto C with strongly convex functions.

If $\varphi(\cdot) = \frac{1}{2} \| \cdot \|_2^2$, then Algorithm 1 takes the following form:

$$\begin{cases} x_{n+1} = \operatorname{prox}_{\lambda F(y_n, \cdot)} x_n, \\ y_{n+1} = \operatorname{prox}_{\lambda F(y_n, \cdot)} x_{n+1}, \end{cases} \tag{4}$$

where prox_g is the proximal operator, associated with convex lower semicontinuous proper function g:

$$R^d \ni x \mapsto \operatorname{prox}_g x = \underset{y \in \operatorname{dom} g}{\operatorname{argmin}} \left(g(y) + \frac{1}{2} \|y - x\|_2^2 \right) \in \operatorname{dom} g.$$

Two-step proximal algorithm (4) was introduced in [6]. In the special case of variational inequality problem, i.e., if $F(x, y) = (Ax, y - x)$, it takes the form

$$\begin{cases} x_{n+1} = P_C(x_n - \lambda A y_n), \\ y_{n+1} = P_C(x_{n+1} - \lambda A y_n), \end{cases}$$

where P_C is the operator of metric projection onto the set C.

For variational inequalities, Algorithm 1 has the form

$$\begin{cases} x_{n+1} = \Pi_C \left((\nabla \varphi)^{-1} (\nabla \varphi(x_n) - \lambda A y_n) \right), \\ y_{n+1} = \Pi_C \left((\nabla \varphi)^{-1} (\nabla \varphi(x_{n+1}) - \lambda A y_n) \right), \end{cases}$$

where Π_C is the Bregman projection operator onto the set closed convex C defined by the following rule:

$$\Pi_C x = \underset{y \in C}{\operatorname{argmin}} D_\varphi(y, x).$$

This method was introduced in [10].

4   Convergence Results

We note first that if, for some number $n \in N$, the following equalities are satisfied:

$$x_{n+1} = x_n = y_n \tag{5}$$

then $y_n \in S$ and the following stationarity condition holds:

$$y_k = x_k = y_n \quad \forall k \geq n.$$

Further, we assume that for all numbers $n \in N$, condition (5) doesn't hold. Finally, we have the following technical lemmas.

Lemma 1. *Let us have the non-negative sequences $(a_n), (b_n)$ such that*

$$a_{n+1} \leq a_n - b_n.$$

Then, there exists the limit $\lim_{n \to \infty} a_n \in R$, and $\sum_{n=1}^{\infty} b_n < +\infty$.

Lemma 2. *The element $x \in C$ is the solution of the equilibrium problem (1) if and only if*
$$F(x,y) + \lambda^{-1} D_\varphi(y,x) \geq 0 \quad \forall y \in C,$$
where $\lambda > 0$.

We start the analysis of the convergence with the proof of important inequality for sequences (x_n) and (y_n), generated by Algorithm 1.

Lemma 3. *Let sequences $(x_n), (y_n)$ be generated by Algorithm 1, and let $z \in S^*$. Then, we have:*

$$D_\varphi(z, x_{n+1}) \leq D_\varphi(z, x_n) - \left(1 - \frac{2\lambda b}{\sigma}\right) D_\varphi(x_{n+1}, y_n)$$
$$- \left(1 - \frac{4\lambda a}{\sigma}\right) D_\varphi(y_n, x_n) + \frac{4\lambda a}{\sigma} D_\varphi(x_n, y_{n-1}). \tag{6}$$

Proof. We have

$$D_\varphi(z, x_{n+1}) = D_\varphi(z, x_n) - D_\varphi(x_{n+1}, x_n)$$
$$+ (\nabla\varphi(x_{n+1}) - \nabla\varphi(x_n), x_{n+1} - z)$$
$$= D_\varphi(z, x_n) - D_\varphi(x_{n+1}, y_n) - D_\varphi(y_n, x_n)$$
$$- (\nabla\varphi(y_n) - \nabla\varphi(x_n), x_{n+1} - y_n) + (\nabla\varphi(x_{n+1}) - \nabla\varphi(x_n), x_{n+1} - z). \tag{7}$$

From the definition of points x_{n+1} and y_n, it follows that

$$\lambda F(y_n, z) - \lambda F(y_n, x_{n+1}) \geq (\nabla\varphi(x_{n+1}) - \nabla\varphi(x_n), x_{n+1} - z), \tag{8}$$

$$\lambda F(y_{n-1}, x_{n+1}) - \lambda F(y_{n-1}, y_n) \geq -(\nabla\varphi(x_n) - \nabla\varphi(y_n), y_n - x_{n+1}). \tag{9}$$

Using inequalities (8), (9) for the estimation of inner products in (7), we get

$$D_\varphi(z, x_{n+1}) \leq D_\varphi(z, x_n) - D_\varphi(x_{n+1}, y_n) - D_\varphi(y_n, x_n)$$
$$+ \lambda \left(F(y_n, z) - F(y_n, x_{n+1}) + F(y_{n-1}, x_{n+1}) - F(y_{n-1}, y_n) \right). \tag{10}$$

From the inclusion $z \in S^*$, it follows that $F(y_n, z) \leq 0$, and Lipschitz-type continuity property of F guarantees satisfying of the inequality

$$-F(y_n, x_{n+1}) + F(y_{n-1}, x_{n+1}) - F(y_{n-1}, y_n)$$
$$\leq a \|y_{n-1} - y_n\|^2 + b \|y_n - x_{n+1}\|^2.$$

Using the above estimates for (10), we get:

$$D_\varphi(z, x_{n+1}) \leq D_\varphi(z, x_n) - D_\varphi(x_{n+1}, y_n) - D_\varphi(y_n, x_n)$$
$$+ \lambda a \|y_{n-1} - y_n\|^2 + \lambda b \|y_n - x_{n+1}\|^2. \tag{11}$$

The term $\|y_{n-1} - y_n\|^2$ can be estimated in the following way:

$$\|y_{n-1} - y_n\|^2 \leq 2 \|y_{n-1} - x_n\|^2 + 2 \|y_n - x_n\|^2.$$

Substituting this inequality in (11), we obtain inequality

$$D_\varphi(z, x_{n+1}) \leq D_\varphi(z, x_n) - D_\varphi(x_{n+1}, y_n) - D_\varphi(y_n, x_n)$$
$$+ 2\lambda a \|y_{n-1} - x_n\|^2 + 2\lambda a \|y_n - x_n\|^2 + \lambda b \|y_n - x_{n+1}\|^2. \tag{12}$$

After estimation the norms in (12) using inequality (3), we obtain:

$$D_\varphi(z, x_{n+1}) \leq D_\varphi(z, x_n) - D_\varphi(x_{n+1}, y_n) - D_\varphi(y_n, x_n)$$
$$+ \frac{4\lambda a}{\sigma} D_\varphi(x_n, y_{n-1}) + \frac{4\lambda a}{\sigma} D_\varphi(y_n, x_n) + \frac{2\lambda b}{\sigma} D_\varphi(x_{n+1}, y_n),$$

i.e. the inequality (6).

Lemma 4. Let $\lambda \in \left(0, \frac{\sigma}{2(2a+b)}\right)$. Then all limit points of the sequence (x_n) belong to the set S.

Proof. Let $z \in S^*$. Assume

$$a_n = D_\varphi(z, x_n) + \frac{4\lambda a}{\sigma} D_\varphi(x_n, y_{n-1}),$$
$$b_n = \left(1 - \frac{4\lambda a}{\sigma}\right) D_\varphi(y_n, x_n) + \left(1 - \frac{4\lambda a}{\sigma} - \frac{2\lambda b}{\sigma}\right) D_\varphi(x_{n+1}, y_n).$$

Inequality (6) takes the form $a_{n+1} \leq a_n - b_n$. Then from Lemma 1, we can conclude that the finite limit exists

$$\lim_{n\to\infty} \left(D_\varphi(z, x_n) + \frac{4\lambda a}{\sigma} D_\varphi(x_n, y_{n-1}) \right),$$

and

$$\sum_{n=1}^{\infty} \left(\left(1 - \frac{4\lambda a}{\sigma}\right) D_\varphi(y_n, x_n) + \left(1 - \frac{4\lambda a}{\sigma} - \frac{2\lambda b}{\sigma}\right) D_\varphi(x_{n+1}, y_n)\right) < +\infty.$$

Hence, we get:

$$\lim_{n\to\infty} D_\varphi(y_n, x_n) = \lim_{n\to\infty} D_\varphi(x_{n+1}, y_n) = 0, \qquad (13)$$

and convergence of sequence $(D_\varphi(z, x_n))$ for all $z \in S^*$. From (13), it follows that

$$\lim_{n\to\infty} \|y_n - x_n\| = \lim_{n\to\infty} \|y_n - x_{n+1}\| = 0, \qquad (14)$$

and naturally $\lim_{n\to\infty} \|x_n - x_{n+1}\| = 0$. From inequality

$$D_\varphi(z, x_n) \geq \frac{\sigma}{2}\|z - x_n\|^2$$

and (14), it follows that sequences (x_n), (y_n) are bounded. Now we consider the subsequence (x_{n_k}), which converges to some feasible point $\bar{z} \in C$. Then, from (14) it follows that $y_{n_k} \to \bar{z}$ and $x_{n_k+1} \to \bar{z}$. Let us show that $\bar{z} \in S$. We have:

$$F(y_{n_k}, y) + \frac{1}{\lambda}D_\varphi(y, x_{n_k}) \geq F(y_{n_k}, x_{n_k+1}) + \frac{1}{\lambda}D_\varphi(x_{n_k+1}, x_{n_k}) \quad \forall y \in C. \quad (15)$$

Passing to the limit in (15), taking into account conditions (A2), (A4), we get:

$$F(\bar{z}, y) + \frac{1}{\lambda}D_\varphi(y, \bar{z}) \geq \overline{\lim}_{k\to\infty} \left\{ F(y_{n_k}, y) + \frac{1}{\lambda}D_\varphi(y, x_{n_k})\right\}$$

$$\geq \underline{\lim}_{k\to\infty} \left\{ F(y_{n_k}, x_{n_k+1}) + \frac{1}{\lambda}D_\varphi(x_{n_k+1}, x_{n_k})\right\} = F(\bar{z}, \bar{z}) = 0 \quad \forall y \in C.$$

Lemma 2 implies the inclusion $\bar{z} \in S$.

Remark 3. Lemma 4 is valid without assumption (A1) about the pseudomonotonicity of bifunction F.

Lemma 5. *Sequences $(x_n), (y_n)$ generated by Algorithm 1 converge to the solution $\bar{z} \in S$ of the problem (1).*

Proof. Let $\bar{z} \in S = S^*$ and $x_{n_k} \to \bar{z}$, $y_{n_k} \to \bar{z}$. The limit

$$\lim_{n\to\infty} D_\varphi(\bar{z}, x_n) = \lim_{n\to\infty} (\varphi(\bar{z}) - \varphi(x_n) - (\nabla\varphi(x_n), \bar{z} - x_n))$$

exists, and

$$\lim_{n\to\infty} D_\varphi(y_n, x_n) = \lim_{n\to\infty} (\varphi(y_n) - \varphi(x_n) - (\nabla\varphi(x_n), y_n - x_n)) = 0.$$

Since $D_\varphi(\bar{z}, x_{n_k}) \to 0$, then $D_\varphi(\bar{z}, x_n) \to 0$. Hence $x_n \to \bar{z}$ and $y_n \to \bar{z}$.

Summing up, we formulate the main result.

Theorem 1. *Let $C \subseteq R^d$ be a nonempty convex closed set, for bifunction F : $C \times C \to R$, conditions (A1)–(A5) are satisfied, and $S \neq \emptyset$. Assume that $\lambda \in \left(0, \frac{\sigma}{2(2a+b)}\right)$. Then, sequences $(x_n), (y_n)$ generated by Algorithm 1 converge to the solution $\bar{z} \in S$ of the equilibrium problem (1).*

5   Conclusions

In this paper, a novel iterative method for solving the equilibrium programming problem is proposed and analyzed. The method is the extension (using the Bregman divergence instead of the Euclidean distance) of a two-stage proximal algorithm [6]. The Bregman divergence allows to take into account the geometry of an admissible set effectively in some cases. For example, for a variational inequality on a probability simplex

$$\text{find } x \in \Delta_d \text{ such that } \quad (Ax, y - x) \geq 0 \; \forall y \in \Delta_d, \tag{16}$$

where $A : \Delta_d \to R^d$, choosing KL-divergence we get Algorithm 1 in the form [10]:

$$\begin{cases} x_i^{n+1} = \dfrac{x_i^n \exp\left(-\lambda(Ay_n)_i\right)}{\sum_{j=1}^m x_j^n \exp\left(-\lambda(Ay_n)_j\right)}, & i = 1, \dots, d, \\ y_i^{n+1} = \dfrac{x_i^{n+1} \exp\left(-\lambda(Ay_n)_i\right)}{\sum_{j=1}^m x_j^{n+1} \exp\left(-\lambda(Ay_n)_j\right)}, & i = 1, \dots, d, \end{cases} \tag{17}$$

where $(Ay_n)_i \in R$ is the i-th coordinate of the vector $Ay_n \in R^d$, $\lambda > 0$.

Analysis of the convergence of the method was carried out under assumption of the existence of a solution under the pseudo-monotonicity and Lipschitz-type continuity conditions of the bifunction.

The paper also helps us in the design and analysis of more practical algorithms. In one of the future work we plan to consider a randomized version of Algorithm 1 and carry out the corresponding convergence analysis. It will help to have a progress in using this method for solving equilibrium problems of huge size.

References

1. Anh, P.N.: Strong convergence theorems for nonexpansive mappings and Ky Fan inequalities. J. Optim. Theory Appl. **154**, 303–320 (2012)
2. Anh, P.N., Hai, T.N., Tuan, P.M.: On ergodic algorithms for equilibrium problems. J. Glob. Optim. **64**, 179–195 (2016)
3. Antipin, A.S.: Equilibrium programming: proximal methods. Comput. Math. Math. Phys. **37**, 1285–1296 (1997)
4. Combettes, P.L., Hirstoaga, S.A.: Equilibrium programming in Hilbert spaces. J. Nonlinear Convex Anal. **6**, 117–136 (2005)

5. Iusem, A.N., Sosa, W.: On the proximal point method for equilibrium problems in Hilbert spaces. Optimization **59**, 1259–1274 (2010)
6. Lyashko, S.I., Semenov, V.V.: A new two-step proximal algorithm of solving the problem of equilibrium programming. In: Goldengorin, B. (ed.) Optimization and Its Applications in Control and Data Sciences. SOIA, vol. 115, pp. 315–325. Springer, Cham (2016)
7. Moudafi, A.: Proximal point algorithm extended to equilibrium problem. J. Nat. Geom. **15**, 91–100 (1999)
8. Quoc, T.D., Muu, L.D., Hien, N.V.: Extragradient algorithms extended to equilibrium problems. Optimization **57**, 749–776 (2008)
9. Semenov, V.V.: Hybrid splitting methods for the system of operator inclusions with monotone operators. Cybern. Syst. Anal. **50**, 741–749 (2014)
10. Semenov, V.V.: A version of the mirror descent method to solve variational inequalities. Cybern. Syst. Anal. **53**, 234–243 (2017)
11. Takahashi, S., Takahashi, W.: Viscosity approximation methods for equilibrium problems and fixed point problems in Hilbert spaces. J. Math. Anal. Appl. **331**, 506–515 (2007)
12. Vuong, P.T., Strodiot, J.J., Nguyen, V.H.: Extragradient methods and linesearch algorithms for solving Ky Fan inequalities and fixed point problems. J. Optim. Theory Appl. **155**, 605–627 (2012)

Data Analysis Using Fuzzy Mathematics, Soft Computing, Computing with Words

Perceptual Computing Based Method for Assessing Functional State of Aviation Enterprise Employees

Dan Tavrov[1,2(✉)], Olena Temnikova[1], and Volodymyr Temnikov[3]

[1] Igor Sikorsky Kyiv Polytechnic Institute, Kyiv, Ukraine
[2] Kyiv School of Economics, Kyiv, Ukraine
dtavrov@kse.org.ua
[3] National Aviation University, Kyiv, Ukraine

Abstract. Safety of aircraft flights and aviation security in general highly depend on the correctness of actions performed by ground traffic control officers and on how well informed are the decisions made by the supervising officers. That said, the task of developing methods and tools for increasing the level of functional security of aviation enterprises is pressing. Such methods and tools must be aimed at determining factors causing negative influence of the human factor on aviation security, and mitigating their negative impact, including increasing awareness of decision makers.

In this paper, we present methods for increasing efficiency of control over aviation enterprise employees functional state during recruitment and afterwards. The proposed approach allows the decision maker to rank employees according to their functional state and thereby choose the most fit for the job.

Special features of the domain in question are the qualitative nature of assessments by experts of the components of the functional state of a person (physiological, physical, psychological, etc.), the presence of a wide range of uncertainties, and its ill-defined nature. To account for these features, we propose to apply results from the theory of perceptual computing, thereby making expert assessment of the functional state of employees of airports in a linguistic form.

The proposed approach is evaluated using examples involving expert assessment of the functional state of airport employees of two different work profiles.

Keywords: Perceptual computing · Aviation security
Type-2 fuzzy set · Psycho-physiological state

1 Introduction

Analysis of recent reports [1,2] on aviation security (AS) and flight safety (FS) shows that their state all over the world, including Ukraine, stays at a rather

© Springer Nature Switzerland AG 2019
O. Chertov et al. (Eds.): ICDSIAI 2018, AISC 836, pp. 61–70, 2019.
https://doi.org/10.1007/978-3-319-97885-7_7

unsatisfactory level. According to [3–6], currently, to enforce AS and FS, it is expedient to apply a wide variety of technical, organizational measures, and measures aimed at reducing the human factor impact.

Main threats to AS and FS are:

- intentional acts of unlawful intrusion (AUI) (e.g., unauthorized access to the craft or restricted zones of the airport);
- unintentional acts (mistakes) committed by airport employees and air traffic controllers, caused by constant emotional and nervous tension (in other words, human factor related threats).

As mentioned above, a wide variety of measures must be employed to effectively mitigate these threats. Measures aimed at reducing human factor impact are arguably the most important and at the same time the hardest to implement in practice.

Quality of airport protection from AUI and the overall level of AS significantly depend on the correctness of actions carried out by security service (SS) officers in order to protect aircrafts, baggage, and information from intentional malicious acts. In its own turn, efficiency of SS officers' actions depends on the quality of organization of their workspace, technical equipment they use, and their functional state (FuncS), i.e. health, stress tolerance, and emotional stability. The latter one is especially important as unintentional mistakes leading to accidents are often caused by being in inadequate physical, physiological, or mental state. Moreover, quality of air traffic control, and therefore FS, highly depend on the FuncS of air traffic controllers.

That said, early detection of a person's inability to serve as an air traffic controller or SS officer is an important and pressing problem, which has the following features making it especially hard to solve:

- significant dependence of correctness of air traffic controllers' and SS officers' actions on their FuncS;
- incompleteness and uncertainty of information a decision maker possesses.

One of the key factors determining ability of a person to serve as an air traffic controller or an SS officer (alongside her background and skills) is her FuncS. To increase control over the FuncS of aviation enterprise employees (or applicants) during periodic monitoring, we propose to improve the test methodology during recruitment process and afterwards by introducing additional elements of psychophysiological monitoring.

In the paper, we use the term FuncS to include the following indicators, each consisting of a number of factors:

- physiological state: hearing, sight, internal organs, skin, etc.;
- psychological state: memory, response time, emotional stability, communicability, etc.;
- mental state: states of the central and autonomic nervous systems, human mental activity, etc.;
- physical state: limbs, stamina, etc.

Efficient control over a person's FuncS consists of:

- monitoring of FuncS;
- forecasting changes in FuncS during the specified time period;
- correction of FuncS after monitoring and predicting its change.

Control over FuncS can be split into three parts:

1. Proactive control over FuncS, i.e. determining the level of a person's ability to perform given tasks. Such control is done during job application process, periodic monitoring, and after testing on simulators.
2. Checking the FuncS during pre-shift examination, whose main goal is to assess a person's readiness to perform work responsibilities during the work shift.
3. Adaptive control over FuncS during the work shift.

In this paper, we focus on the first step, which is conducted as follows. Experts assess FuncS of employees (applicants) (e.g., physicians conduct medical examination), after which these assessments are aggregated (using different relative weights) to produce the overall ranking of candidates, thereby enabling the decision maker to select the most highly fit individuals to perform the job.

Since the very definition of FuncS and the nature of measurements involved are very subjective and imprecise, in this paper, we propose to conduct such expert assessments using words rather than numbers (including relative weights). Then, aggregation of such linguistic assessments can be carried out using computing with words (CWW) methodology [7], and in particular, perceptual computing [8]. Benefits of this approach are threefold:

- the task faced by the experts becomes more understandable;
- subjectivity of assessments is reduced by making use of word models constructed by surveying many experts;
- the resulting overall assessments of the FuncS enable the decision maker to rank job candidates, whereas in conventional approaches, each individual is typically marked as "fit" or "not fit" for the job.

The rest of the paper is structured as follows. In Sect. 2, we briefly describe the basics of perceptual computing and explain how it can be applied to the task set above. In Sect. 3, we discuss application of the proposed approach to model data corresponding to 10 air traffic controllers and 10 SS officers. Conclusions and further re-search ideas are outlined in Sect. 4.

2 Perceptual Computing Basics

2.1 Perceptual Computer Structure

Perceptual computing is a type of CWW used for making subjective judgments [8]. As such, any perceptual computer (Per-C) works with words rather than numbers, both as inputs and outputs. Words in the Per-C are modeled using type-2 fuzzy sets. In practice, the most widely used type-2 fuzzy sets are interval type-2 fuzzy sets (IT2FS), which, unlike type-1 fuzzy sets characterized by a membership function $\mu_A(x) : X \rightarrow [0,1]$, are characterized by two membership functions $\underline{\mu}_{\tilde{A}}(x)$ and $\overline{\mu}_{\tilde{A}}(x)$, called the lower membership function (LMF) and the upper membership function (UMF) of $\tilde{A}$, respectively. Each element x belongs to $\tilde{A}$ with the interval of degrees J_x, and thus LMF and UMF bound the footprint of uncertainty of an IT2FS FOU $\left(\tilde{A}\right) = \bigcup_{\forall x \in X} J_x$. The FOU shows the uncertainty associated with membership grades of each x in the IT2FS.

IT2FSs are superior as models for words to type-1 fuzzy sets because [9] the latter ones have well-defined (certain) membership functions, whereas words mean different things to different people (are uncertain).

In this paper, words are represented using trapezoidal IT2FS, for which both LMF and UMF are trapezoidal membership functions:

$$\mu_{trap}(x; a, b, c, d, h) = \begin{cases} (x-a)/(b-a), & a \le x \le b \\ h, & b \le x \le c \\ (d-x)/(d-c), & c \le x \le d \\ 0, & \text{otherwise} \end{cases} \tag{1}$$

Then, each IT2FS $\tilde{A}$ can be uniquely represented as a vector $(a, b, c, d, e, f, g, i, h)$ such that $\underline{\mu}_{\tilde{A}}(x) = \mu_{trap}(x; e, f, g, i, h)$ and $\overline{\mu}_{\tilde{A}}(x) = \mu_{trap}(x; a, b, c, d, 1)$.

In a Per-C, input words are processed in three stages [10]:

- encoder, which transforms words into their IT2FS representation. A collection of words with associated representations is called a codebook;
- CWW engine, which processes IT2FSs and outputs one or more other IT2FSs;
- decoder, which converts the output of the CWW engine to a representation suitable for the task being solved.

2.2 CWW Engine

In this paper, due to the nature of calculations involved, we use the linguistic weighted average (LWA) operator as a CWW engine. In the most general case, LWA can be expressed as

$$Y_{LWA} = \frac{\sum_{i=1}^{n} X_i W_i}{\sum_{i=1}^{n} W_i}, \tag{2}$$

where each X_i is a factor to be weighted and each W_i is its relative weight. All factors and weights involved are expressed as words (IT2FSs).

To calculate (2), proceed along the following lines. First, consider the case, in which X_i and W_i are intervals $[a_i, b_i]$ and $[c_i, d_i]$, respectively, $i = 1, \ldots, n$. Such LWA is called interval weighted average (IWA), and can be expressed as $Y_{IWA} = [y_l, y_r]$:

$$y_l = \frac{\sum_{i=1}^{L} a_i d_i + \sum_{i=L+1}^{n} a_i c_i}{\sum_{i=1}^{L} d_i + \sum_{i=L+1}^{n} c_i}, \qquad y_r = \frac{\sum_{i=1}^{R} b_i c_i + \sum_{i=R+1}^{n} b_i d_i}{\sum_{i=1}^{R} c_i + \sum_{i=R+1}^{n} d_i}, \tag{3}$$

where L and R are switch points found, e.g., using the enhanced Karnik-Mendel algorithms [11].

Consider now the case when X_i and W_i in (2) are type-1 fuzzy sets. Such LWA is called fuzzy weighted average (FWA). We can compute FWA using IWA by noting the fact that, according to the well-known α-cut decomposition theorem [12], each type-1 fuzzy set A can be represented as $A = \bigcup_{\alpha \in [0,1]} \alpha A$, where $\mu_{\alpha A}(x) = \alpha \cdot \mu_{\alpha A}(x)$, and $^{\alpha}A = \{x | \mu_A(x) \geq \alpha\}$ is called an α-cut of A.

Using this insight, we can compute FWA as follows:

1. For each $\alpha \in \{\alpha_1, \ldots, \alpha_m\}$, compute α-cuts of X_i and W_i.
2. For each $\alpha \in \{\alpha_1, \ldots, \alpha_m\}$, compute $^{\alpha}Y_{FWA} = [^{\alpha}y_l, ^{\alpha}y_r]$ of α-cuts of $^{\alpha}X_i$ and $^{\alpha}W_i$.
3. Connect all left coordinates $(^{\alpha}y_l, \alpha)$ and right coordinates $(^{\alpha}y_r, \alpha)$ to get Y_{FWA}.

Then, LWA (2) is an IT2FS Y_{LWA}, whose UMF is an FWA of UMFs of IT2FSs X_i using UMFs W_i as weights, and whose LMF is an FWA of LMFs of IT2FSs X_i using LMFs W_i as weights.

2.3 Decoder

In general, the output of the Per-C can be [6] a word most similar to the output of the CWW engine; a class, into which the output of the CWW engine is mapped; or a rank of competing alternatives. Taking into account discussion in Sect. 1, in this paper, we use the latter approach. Since each output of the CWW engine is an IT2FS, to rank them, we use the ranking method based on IT2FS average centroids [13].

Average centroid of an IT2FS is defined as

$$c\left(\tilde{A}\right) = \frac{c_l\left(\tilde{A}\right) + c_r\left(\tilde{A}\right)}{2}, \tag{4}$$

where $c_l\left(\tilde{A}\right)$ and $c_r\left(\tilde{A}\right)$ are IWA:

$$c_l\left(\tilde{A}\right) = \frac{\sum_{i=1}^{L} x_i \overline{\mu}_{\tilde{A}}(x_i) + \sum_{i=L+1}^{n} x_i \underline{\mu}_{\tilde{A}}(x_i)}{\sum_{i=1}^{L} \overline{\mu}_{\tilde{A}}(x_i) + \sum_{i=L+1}^{n} \underline{\mu}_{\tilde{A}}(x_i)},$$

$$c_r\left(\tilde{A}\right) = \frac{\sum_{i=1}^{R} x_i \underline{\mu}_{\tilde{A}}(x_i) + \sum_{i=R+1}^{n} x_i \overline{\mu}_{\tilde{A}}(x_i)}{\sum_{i=1}^{R} \underline{\mu}_{\tilde{A}}(x_i) + \sum_{i=R+1}^{n} \overline{\mu}_{\tilde{A}}(x_i)}.$$

Then, the ranking method is simply to compute (4) for every IT2FS, and then rank them according to obtained average centroids.

2.4 Application of Perceptual Computer to Proactive Control over the Functional State of Aviation Enterprise Employees

The decision making process for assessing FuncS of aviation enterprise employees is shown in Fig. 1. This process is hierarchical and distributed [10], since it is carried out by aggregating independent expert assessments of a number of factors.

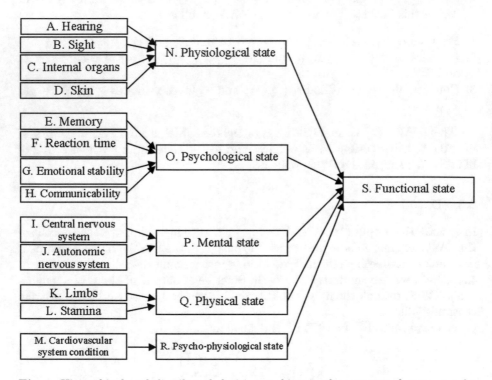

Fig. 1. Hierarchical and distributed decision making in the perceptual computer for determining the functional state of aviation enterprise employees

In the Per-C for assessing FuncS, there are two levels of hierarchy:

- level of indicators: assessment for each indicator (N–R) is done by computing LWA of assessment for individual factors (A–D for indicator N, E–H for indicator O, and so forth);
- level of the functional state: assessment is done by computing LWA of assessments for indicators.

It is important to note that in Fig. 1, we also propose to include an additional factor "M. Cardiovascular system condition," due to the fact that it enables to determine the state of regulatory systems of a human body and its adaptive capabilities [14, 15].

Expert assessments of factors A–M, as well as weights for LWAs, are expressed as words from the codebook of the Per-C.

3 Application of the Perceptual Computing to Model Data

3.1 Constructing the Codebook

Each word in the codebook was defined on the interval $[0, 1]$, which doesn't affect the result because rescaling the weights doesn't influence the LWA. Words both for assessment and for weights were chosen so that they collectively cover the interval $[0, 1]$ and enable the experts to perform the assessment. The following words were chosen:

- for assessment: good (G), normal (N), satisfactory (S), acceptable (A), unsatisfactory (U);
- for weights: essential effect (Es), significant effect (Sg), average effect (Av), insignificant effect (In), little or no effect (Lt).

IT2FS models for each word were created using the interval approach [16], whose detailed description we omit here due to space limitations. The main idea of this approach is to survey a number of domain experts and elicit from them a set of intervals that, in their mind, best correspond to each word. Then, each set of intervals corresponding to a given word is converted to an IT2FS.

For this paper, we surveyed 20 experts and obtained the results presented in Table 1 up to two significant digits. We report means (m) and standard deviations (σ) for left (l) and right (r) ends of the intervals obtained. Also, we report IT2FS models in accordance with the format given in Sect. 2.1.

Table 1. Statistics for words from the codebook and their IT2FS models

Word	Interval statistics				Parameters of trapezoidal T2FS								
	m_l	σ_l	m_r	σ_r	a	b	c	d	e	f	g	i	h
G	0.88	0.04	1.00	0.00	0.74	0.98	1.00	1.00	0.93	1.00	1.00	1.00	1.00
N	0.74	0.03	0.87	0.04	0.66	0.77	0.83	0.93	0.74	0.80	0.80	0.86	0.68
S	0.63	0.03	0.73	0.03	0.56	0.65	0.71	0.83	0.65	0.68	0.68	0.71	0.45
A	0.54	0.04	0.62	0.03	0.48	0.55	0.61	0.68	0.55	0.58	0.58	0.61	0.49
U	0.00	0.00	0.53	0.04	0.00	0.00	0.05	0.78	0.00	0.00	0.04	0.65	1.00
Es	0.88	0.04	1.00	0.00	0.74	0.98	1.00	1.00	0.93	1.00	1.00	1.00	1.00
Sg	0.70	0.07	0.87	0.04	0.57	0.72	0.83	0.93	0.73	0.78	0.78	0.86	0.54
Av	0.46	0.10	0.69	0.07	0.24	0.48	0.68	0.84	0.52	0.58	0.58	0.64	0.37
In	0.29	0.11	0.45	0.10	0.16	0.31	0.41	0.55	0.34	0.37	0.37	0.44	0.49
Lt	0.00	0.00	0.28	0.11	0.00	0.00	0.04	0.53	0.00	0.00	0.01	0.13	1.00

3.2   Assignment of Weights

In Table 2, weights are shown corresponding to arcs in Fig. 1 that connect factors and indicators. In Table 3, weights are shown that correspond to arcs in Fig. 1 that connect indicators and FuncS. All weights were assigned by the third author of this paper based on his expertise in the domain. Weights in Tables 2 and 3 differ for air traffic controllers and SS officers, to emphasize the fact that for different professions, the model can differ in order to capture features of each profession.

Table 2. Weights of arcs in the Per-C model connecting factors and indicators for air traffic controller and SS officer

Factor	Weights of arcs for air traffic controller	Weights of arcs for SS officer
A	Sg	Av
B	Es	Sg
C	Av	Av
D	Lt	Av
E	Sg	Av
F	Es	Sg
G	Es	Es
H	Av	Es
I	Es	Sg
J	Sg	Sg
K	In	Sg
L	Av	Es
M	Es	Av

Table 3. Weights of arcs in the Per-C model connecting indicators and overall functional state for air traffic controller and SS officer

Profession	Indicators				
	N	O	P	Q	R
Air traffic controller	Sg	Es	Sg	Av	Es
SS officer	Av	Sg	Av	Es	Sg

3.3   Discussion of the Obtained Results

Table 4 shows the result of applying the Per-C with proposed specification to 20 model data sets: expert assessments for each factor for 10 air traffic controllers, and expert assessments for each factor for 10 SS officers. The data were modeled by the third author of the paper based on his domain expertise. The last column gives the rank of each employee, where 1 stands for the most highly fit one.

As we can see, application of perceptual computing to the task of assessing the FuncS of an aviation enterprise employee supports the claim that it is important to take into account the whole variety of factors, in particular, physiological and psychological state factors, along with the cardiovascular system condition.

Another important conclusion is that perceptual computing enables the decision maker to rank employees by FuncS, and thereby select the most fit for the job. This approach is superior to the conventional one, when each candidate is labeled "fit" or "not fit," without any viable means to compare them.

Table 4. Linguistic assessments for factors for ten air traffic controllers and ten SS officers from the example, centroids of the output IT2FSs, and overall ranks

#	Linguistic assessment for each factor													Centroid			Rank
	A	B	C	D	E	F	G	H	I	J	K	L	M	c_l	c_r	c	
Air traffic controllers																	
1	G	G	N	N	G	G	G	G	G	G	N	G	G	0.86	0.97	0.91	1
2	N	N	A	A	N	N	S	S	S	S	N	N	N	0.70	0.80	0.75	8
3	N	G	N	S	G	G	G	N	N	N	S	S	N	0.77	0.87	0.82	5
4	G	G	N	A	S	S	N	N	S	A	N	N	N	0.72	0.82	0.77	7
5	S	S	A	S	N	N	S	A	A	A	S	S	A	0.59	0.70	0.65	10
6	G	G	N	S	N	G	N	N	N	N	N	N	N	0.78	0.87	0.83	4
7	N	G	S	S	G	G	G	S	N	N	S	S	G	0.78	0.89	0.84	3
8	S	S	S	N	N	N	N	N	A	A	A	A	S	0.63	0.74	0.68	9
9	N	N	N	G	G	G	G	G	N	N	S	S	N	0.77	0.86	0.81	6
10	N	N	N	G	G	G	G	N	G	G	N	N	G	0.83	0.93	0.88	2
SS officers																	
1	G	N	G	G	G	G	N	N	G	G	G	G	G	0.85	0.96	0.91	1
2	S	N	N	N	N	N	A	A	N	N	S	S	S	0.66	0.77	0.71	8
3	N	S	S	N	N	G	N	S	G	G	G	N	N	0.77	0.88	0.82	6
4	S	N	N	N	G	G	G	N	S	S	N	N	S	0.72	0.83	0.77	7
5	A	S	S	A	S	S	A	S	N	N	S	A	A	0.60	0.71	0.65	10
6	N	N	N	N	G	G	G	G	N	G	N	N	G	0.82	0.91	0.86	3
7	N	S	S	G	N	G	S	S	G	G	G	G	N	0.78	0.90	0.84	5
8	A	A	A	S	S	S	S	N	N	N	N	N	A	0.65	0.76	0.71	9
9	N	S	S	N	N	N	N	G	G	G	G	G	N	0.80	0.90	0.85	4
10	G	N	N	G	N	N	N	G	G	G	G	G	G	0.85	0.95	0.90	2

4   Conclusions and Further Research

In this paper, we presented the perceptual computing based approach to solving one of the most important stages of proactive control over the functional state of an aviation enterprise employee, i.e. conducting its expert assessment. The task of expert assessment in our work boils down to defining the level of functional state of an employee based on linguistic evaluations of individual factors, including physiological, psychological, mental, and physical ones.

Further research directions should include expanding the set of words used in the perceptual computer and conducting field studies using the proposed approach to eval-uate its practical efficiency.

References

1. Safety Report. International Civil Aviation Organization. Published in Montréal, Canada (2017)
2. Safety Review. State Aviation Administration of Ukraine (2016)
3. Doc 9859: Safety Management Manual. International Civil Aviation Organization (2013)
4. Doc 8973/9: Aviation Security Manual. International Civil Aviation Organization (2014)
5. Doc 9808: Human Factors in Civil Aviation Security Operations. International Civil Aviation Organization (2002)
6. Annex 17: Security: Safeguarding International Civil Aviation Against Acts of Unlawful Interference. International Civil Aviation Organization (2017)
7. Zadeh, L.A.: From computing with numbers to computing with words—from manipulation of measurements to manipulation of perceptions. IEEE Trans. Circ. Syst. I: Fundam. Theory Appl. **46**(1), 105–119 (1999)
8. Mendel, J.M.: The perceptual computer: an architecture for computing with words. In: Proceedings of Modeling with Words Workshop in the Proceedings of FUZZ-IEEE 2001, pp. 35–38 (2001)
9. Mendel, J.M.: Fuzzy sets for words: a new beginning. In: Proceedings of FUZZ-IEEE 2003, St. Louis, MO, pp. 37–42 (2003)
10. Mendel, J.M., Wu, D.: Perceptual Computing. Aiding People in Making Subjective Judgments. Wiley, Hoboken (2010)
11. Wu, D., Mendel, J.M.: Enhanced Karnik-Mendel algorithms. IEEE Trans. Fuzzy Syst. **17**(4), 923–934 (2009)
12. Klir, G.J., Yuan, B.: Fuzzy Sets and Fuzzy Logic. Theory and Applications. Prentice Hall, Upper Saddle River (1995)
13. Wu, D., Mendel, J.M.: A comparative study of ranking methods, similarity measures and uncertainty measures for interval type-2 fuzzy sets. Inf. Sci. **179**(8), 1169–1192 (2009)
14. Bayevsky, R.M., Kirillov, O.I., Kletskin, S.Z.: Mathematical analysis of heart rate changes under stress. Medicine, Moscow (1984). (in Russian)
15. Bayevsky, R.M., Berseneva, A.P.: Assessment of the adaptive capabilities of the body and the risk of developing diseases. Medicine, Moscow (1997). (in Russian)
16. Liu, F., Mendel, J.M.: Encoding words into interval type-2 fuzzy sets using an interval approach. IEEE Trans. Fuzzy Syst. **16**, 1503–1521 (2008)

Multi-criteria Decision Making and Soft Computing for the Selection of Specialized IoT Platform

Yuriy Kondratenko$^{(\boxtimes)}$ ⓘ, Galyna Kondratenko ⓘ,
and Ievgen Sidenko ⓘ

Intelligent Information Systems Department,
Petro Mohyla Black Sea National University, Mykolaiv, Ukraine
{yuriy.kondratenko, halyna.kondratenko,
ievgen.sidenko}@chmnu.edu.ua

Abstract. The task of the appropriate selection of the specialized Internet of Things (IoT) platform is very relevant today. The complexity of the selection process is due to (a) the large number of IoT platforms, which are available on the IoT services market, and (b) the variety of services and features, which they offer. In this paper, the multi-criteria decision making (MCDM) and the soft computing approaches for choosing the specialized IoT platform are considered. Authors illustrate solving MCDM problem using the linear convolution method with simple ranking approach to forming weight coefficients for criteria. MCDM methods have some limitations: (a) the need to take into account weight coefficients of the criteria; (b) the composition of the Pareto-optimal set of alternative decisions; (c) the lack of ability to change the dimension of the vector of alternatives and criteria in real time; (d) significant impact of weight coefficients that the expert determines on the result. Thus, the authors propose to use the soft computing approach, in particular, Mamdani-type fuzzy logic inference engine for selection of the specialized IoT platform. Relevant factors (reliability, dependability, safety, and security of IoT platforms) are considered as the most important ones for decision making in the IoT platform selection processes. In addition, analysis and research of the influence level of various factors on the selection of specialized IoT platform have been carried out. Special cases of choosing the specialized IoT platform with confirmation of the appropriateness of the using soft computing approach are discussed.

Keywords: IoT platform · Criterion · Alternative
Multi-criteria decision making · Factor · Fuzzy logic inference engine

1 Introduction

Decision making process involves selecting one of the possible variants of decisions according to the criteria under deterministic and/or uncertain conditions. These possible variants of decisions are called alternatives. When there are many alternatives, a decision maker (DM) cannot take enough time and attention to analyze each of them, so there is a need for means to support the choice of decisions. In the modern theory of

© Springer Nature Switzerland AG 2019
O. Chertov et al. (Eds.): ICDSIAI 2018, AISC 836, pp. 71–80, 2019.
https://doi.org/10.1007/978-3-319-97885-7_8

decision making, it is considered that the variants of decisions are characterized by different indicators of their attractiveness for DM. These indicators are called features, factors, attributes, or quality measures. They all serve as criteria for selecting a decision [1–3].

Internet of Things (IoT) describes a network of interconnected smart devices, which are able to communicate with each other for a certain goal. In simple words, the purpose of any IoT device is to connect with other IoT devices and applications (mostly in a cloud-based way) to relay information using Internet transfer protocols. The gap between the device sensors and data networks is filled by an IoT platform. At the same time, the question remains about the choice of the IoT platform when designing IoT systems [4–7].

The aim of this research is (a) to formulate the multi-criteria decision making (MCDM) problem for selecting the IoT platform from a set of alternatives, and (b) to solve the MCDM for increasing efficiency of decision-making processes using linear convolution method and soft computing approach, based on Mamdani-type fuzzy logic inference engine.

2 Related Work and Problem Statement

In most cases, the choice of the IoT platform for the development of IoT systems involves comparative analysis of their capabilities provided by developers of IoT platforms. Besides, IoT developers often give preference to the well-known IoT platforms, without considering criteria (factors) that in the future may affect the development, maintenance, updating, reliability, safety, and scaling of the developed IoT systems. One of the approaches to selecting an IoT platform is based on defining a reference platform architecture that includes benefits and capabilities of the existing modern IoT platforms [8]. Later on, a comparative analysis of the selected platforms with the reference one is carried out, and the best IoT platform is determined.

Now, several MCDM methods are well tested, in particular, the analytic hierarchy process, the pair-comparison approach, and the Delphi method, etc. [1–3, 9–13].

An important problem is the choice of the IoT platform, taking into account criteria influencing the decision making processes. The necessity of using the MCDM when choosing the specialized IoT platform is concerned with the complexity of taking into account all features, possibilities, and application spheres of IoT platforms. Besides, an incorrectly selected IoT platform may lead to the reduction of reliability and safety of IoT systems [4].

Let's consider the most popular specialized (cloud services) IoT platforms, which can be used to create different IoT systems. As a matter of fact, IoT platform is not dominated by one single criteria. In most cases, IoT platforms that, for example, have the highest level of safety, do not support variability of databases and visualization tools, and vice versa [6, 14, 15]. The following IoT platforms (alternatives) can be considered most suited for such description: Kaa IoT Platform (E_1) [4]; AWS IoT Platform (E_2) [5]; Microsoft Azure IoT Platform (E_3) [14]; Bosch IoT Suite—MDM IoT Platform (E_4) [6, 16]; IBM Watson IoT Platform (E_5) [8]; Google Cloud IoT Platform (E_6) [8]; Samsung Artik Cloud IoT Platform (E_7) [6].

Each platform has advantages and limitations. Examination of these IoT platforms in detail will be held in order to determine, which one is best suited for specialized IoT networks [4–6].

Kaa IoT Platform (E_1). Main features of Kaa IoT platform [4] are: manage an unlimited number of connected devices; perform real-time device monitoring; perform remote device provisioning; collect and analyze sensor data; analyze user behavior deliver targeted notifications; create cloud services for smart products.

AWS IoT Platform (E_2). Main features of AWS IoT platform [5] are: the registry for recognizing devices; software development kit for devices; device shadows; secure device gateway; rules engine for inbound message evaluation.

Microsoft Azure IoT Platform (E_3). Main features of Azure IoT platform [14] are: device shadowing; a rules engine; identity registry; information monitoring.

Bosch IoT Suite - MDM IoT Platform (E_4). Main features of Bosch IoT platform [6, 16] are: the platform as a Service (PaaS); remote manager; analytics; cost-effective; ready to use.

IBM Watson IoT Platform (E_5). IBM Watson offers some great security possibilities based on machine learning and data science. Users of IBM Watson [8] get: device management; secure communications; real-time data exchange; recently added data sensor and weather data service.

Google Cloud IoT Platform (E_6). Some of the features of the Google Cloud platform [8] are: accelerate the business; speed up the devices; cut cost with cloud service; partner ecosystem.

Samsung Artik Cloud IoT Platform (E_7). Samsung [6] aims to provide a platform that takes care of the total security with services like Artik Module, Cloud, Security and ecosystem. Artik provides a fast and open platform that is responsible for developing of products.

Certain criteria influence the evaluation and the choice of decisions from the set of alternatives $E = (E_1, E_2, \ldots, E_i, \ldots, E_m)$, $i = 1, 2, \ldots, m$, where $m = 7$ for above-mentioned IoT platforms. These criteria are selected by developers based on their own experience and depend on the scope of the task. Such tasks are called multi-criteria tasks. The selection of the criteria is an important and rather complex task since it involves formulation of a set of factors that influence the decision making. Properly selected criteria increase efficiency of decision making while solving multi-criteria problems in various types of their applications.

Some criteria are not relevant (within the scope of a special application) in selection process for the specialized IoT platform. According to various studies and our own experience, we propose to use the following important (main) criteria when selecting IoT platform [8]: level of safety and reliability (Q_1); device management (Q_2); integration level (Q_3); level of processing and action management (Q_4); database functionality (Q_5); protocols for data collection (Q_6); usefulness of visualization (Q_7); variety of data analytics (Q_8).

Level of Safety and Reliability (Q_1) required to operate an IoT platform is much higher than general software applications and services. Millions of devices being connected with an IoT platform mean that it is necessary to anticipate a proportional number of vulnerabilities [12, 14].

Device Management (Q_2). The IoT platform should maintain a list of devices connected to it and track their operational status; it should be able to handle configuration, firmware (or any other software) updates, and provide device-level error reporting and error handling [16].

Integration Level (Q_3) is another important criterion expected from an IoT platform. The API should provide access to important operations and data that need to be exposed from the IoT platform [4, 6].

Level of Processing and Action Management (Q_4). Data obtained from IoT devices ultimately affect events in real life. Therefore, the IoT platform must be able to build processes, event-action-triggers, and other "smart actions" based on specific sensor data [5].

Database Functionality (Q_5). Requirement for this criterion is an attempt to restore order in the processing and transfer of data from, for example, different platforms or even to other information systems [6].

Another important criterion, which needs attention, is the types of **protocols for data collection** (Q_6) used for data communication between components of an IoT platform. Lightweight communication protocols should be used to enable low energy use as well as low network bandwidth functionality [5, 8, 16].

Visualization (Q_7) enables humans to see patterns and observe trends from visualization dashboards where data are vividly portrayed through line-, stacked-, or pie charts, 2D- or even 3D-models [6, 14].

Variety of Data Analytics (Q_8). Data collected from sensors connected to an IoT platform needs to be analyzed in an intelligent manner in order to obtain meaningful insights. There are four main types of analytics, which can be conducted on IoT data: real-time, batch, predictive, and interactive analytics [4, 5, 8].

On the basis of developed criteria and alternative decisions, we proceed to formulation of the multi-criteria problem statement with its subsequent solution using one of the appropriate methods of MCDM, in particular, the linear convolution method with simple ranking approach to forming weight coefficients for criteria [1, 3].

3 Multi-criteria Decision Making for IoT Platform Selection

The analysis of many real practical problems naturally led to the emergence of a class of multi-criteria problems. The solution of the corresponding problems is found through the use of such methods as the selection of the main criterion, the linear, multiplicative and max-min convolutions, the ideal point method, the sequential concessions methodology, the lexicographic optimization [1–3, 9, 11].

The task of selecting the IoT platform is formulated as an MCDM problem and has the following form (decisions matrix):

$$Q(E_i) = \begin{pmatrix} Q_1(E_1) & Q_1(E_2) & \cdots & Q_1(E_m) \\ Q_2(E_1) & Q_2(E_2) & \cdots & Q_2(E_m) \\ \cdots & \cdots & & \cdots \\ Q_n(E_1) & Q_n(E_2) & \cdots & Q_n(E_m) \end{pmatrix}; E_i \in E; (i = 1, 2, \ldots, m; j = 1, 2, \ldots, n), \quad (1)$$

where $Q(E_i)$ is a vector criterion of quality for ith alternative; $Q_j(E_i)$ is the jth component of the vector criterion of quality $Q(E_i)$.

Evaluation of the ith alternative by the jth criterion $Q_j(E_i)$ has a certain scale of assessment and is presented by experts based on their experience, knowledge, and experimental research in the field of specialized IoT platforms [1].

To solve the IoT platform selection problem, it is necessary to find the best alternative $E^* \in E$ using data (1):

$$E^* = Arg \max_{i=1\ldots m} (Q(E_i)), E_i \in E, i = 1, 2, \ldots, 7. \quad (2)$$

Let's apply one of the existing MCDM methods, for example, the linear convolution method (LCM) with simple ranking approach to forming weight coefficients for criteria to solve the corresponding task of multi-criteria selection of IoT platforms [3].

With the help of LCM, the global criterion is presented as a linear combination of the vector criterion of quality with weight coefficients $\lambda_j, (j = 1, \ldots, n)$:

$$Q(E_i) = \sum_{j=1}^{n} \lambda_j Q_j(E_i) \Rightarrow Max; \; F_i \in F; \; \lambda_j > 0; \; \sum_{j=1}^{n} \lambda_j = 1; (i = 1, 2, \ldots, m), \quad (3)$$

where λ_j is a weight coefficient, which denotes importance of criterion Q_j.

Experts are invited to evaluate alternative decisions (IoT platforms) $E_1, E_2, \ldots, E_7$ according to the specified criteria $Q_1, Q_2, \ldots, Q_8$ using the 10-point rating scale (from 0 to 9), where 9 points correspond to the largest (the best) value of the alternative decision by the criterion.

Let us consider an example with experts' evaluation of the specified criteria $Q_j(E_i)$ for IoT platform selection in the case of experts' data presented in Table 1.

Table 1. Decisions matrix for multi-criteria selection of the IoT platform

	E_1	E_2	E_3	E_4	E_5	E_6	E_7
Q_1	7	8	8	8	6	7	8
Q_2	8	7	9	6	9	8	6
Q_3	8	6	7	6	7	6	8
Q_4	9	7	8	7	8	8	9
Q_5	8	9	8	5	9	8	7
Q_6	9	6	7	9	5	6	8
Q_7	5	7	6	9	7	7	6
Q_8	8	8	7	6	7	8	6

Weights $\lambda_1 = 0.2222$, $\lambda_2 = 0.1528$, $\lambda_3 = 0.0556$, $\lambda_4 = 0.0278$, $\lambda_5 = 0.1944$, $\lambda_6 = 0.1528$, $\lambda_7 = 0.0833$, $\lambda_8 = 0.1111$ can be obtained using simple ranking method [1, 3] for corresponding criteria consequence of the experts' assessments: $Q_1 \succ Q_5 \succ (Q_2 = Q_6) \succ Q_8 \succ Q_7 \succ Q_3 \succ Q_4$.

Using (3) for calculation of the global criterion, we can get the usefulness of alternatives (Table 1).

The results of global criteria $Q(E_i)$ calculation using the LCM and simple ranking method for weight coefficients are represented in Table 2.

Table 2. Global criteria for all alternative decisions

	E_1	E_2	E_3	E_4	E_5	E_6	E_7
$Q(E_i)$	7.70833	7.51389	7.66667	6.98611	7.19444	7.27778	7.13889

Ranking row of alternatives $E_1, E_2, \ldots, E_7$ (Table 2) can be presented as:

$$E_1 \succ E_3 \succ E_2 \succ E_6 \succ E_5 \succ E_7 \succ E_4. \tag{4}$$

Thus, the first alternative decision E_1 (Kaa IoT platform) is the most rational (4) for specialized IoT systems and networks.

4 Soft Computing Approach for IoT Platform Selection

Appropriate MCDM methods have some limitations: the need to take into account weight coefficients of the criteria, the definition of the Pareto-optimal set of alternative decisions, etc. [1, 9, 11]. The implementation of the soft computing approach based on Mamdani-type fuzzy logic inference engine looks more promising, from the authors' point of view, for selection of the specialized IoT platform [17–19]. In this case, previously proposed criteria $Q_1, Q_2, \ldots, Q_8$ can act as input signals (coordinates) or factors $X = \{x_1, x_2, \ldots, x_8\}$ of the fuzzy logic inference system.

The structure $y = f(x_1, x_2, \ldots, x_8)$ of the fuzzy logic inference system for selection of the specialized IoT platform includes 3 fuzzy subsystems and has 8 input coordinates $X = \{x_j\}, j = 1, \ldots, 8$, and one output y, which are interconnected (by common properties) by the fuzzy dependencies $y_1 = f_1(x_1, x_2, x_3, x_4)$, $y_2 = f_2(x_5, x_6, x_7, x_8)$ and $y = f_3(y_1, y_2)$ of appropriate rule bases (RBs) of 3 subsystems [20–22].

To estimate input coordinates $X = \{x_j\}, j = 1, \ldots, 8$, three linguistic terms (LTs) with a triangular form of membership function (MF), in particular, *low* (L), *medium* (M), and *high* (H), are chosen. Five LTs $y_1, y_2 \in \{L, LM, M, HM, H\}$ are chosen for evaluation of intermediate coordinates. The output coordinate y has 7 LTs $y \in \{VL, L, LM, M, HM, H, VH\}$. Partial sets of rules of RBs for the first $y_1 = f_1(x_1, x_2, x_3, x_4)$ and for the third $y = f_3(y_1, y_2)$ subsystems are given in Table 3.

In the example discussed earlier, using MCDM methods for selecting of IoT platform before the system's work, it was necessary to clearly indicate the number of alternatives and criteria, and also to form a decisions matrix with expert estimates

Table 3. Partial sets of rules of RBs for the first and for the third fuzzy subsystems

№ of rule	Subsystem $y_1 = f_1(x_1, x_2, x_3, x_4)$					№ of rule	Subsystem $y = f_3(y_1, y_2)$		
	x_1	x_2	x_3	x_4	y_1		y_1	y_2	y
1	L	L	L	L	L	1	L	L	VL
...		...		...		2	L	LM	L
37	M	M	L	L	LM	...		...	...
38	M	M	L	M	M	13	M	M	M
...		...		...		...		...	...
81	H	H	H	H	H	25	H	H	VH

(Table 1). Using the Mamdani-type algorithm to develop the fuzzy logic inference system for selection of IoT platform, we eliminate the need to form weight coefficients for the criteria. In this case, this soft computing approach allows us to get rid of the limitations on the number of alternatives.

The author's software application (Fig. 1) allows evaluating IoT platforms based on certain input coordinates (factors) $X = \{x_1, x_2, \ldots, x_8\}$ with the use of the fuzzy logic inference engine. The number of IoT platforms is not limited to 7 $(E_1, E_2, \ldots, E_7)$ as in the case of MCDM methods application (Fig. 1). The corresponding software application provides a preview of the RBs in the system's real-time mode. For example, estimates have been made of 10 $(E_1, E_2, \ldots, E_{10})$ IoT platforms (all data stored in the system) that are ranked without the use of the threshold value (for example, *Thr* = 7.2) and with its use (Table 4, Fig. 1). Table 4 contains 7 $(E_1, E_2, \ldots, E_7)$ IoT platforms mentioned above and 3 additional IoT platforms: E_8 is the ThingWorks IoT platform, E_9 is the Cisco Cloud IoT platform, E_{10} is the Salesforce IoT platform.

Also, the software application (Fig. 1) provides the use of the author's proposed method of two-stage correction of fuzzy RBs [22] in case of change the dimension of the input vector (for example, if DM cannot estimate (NE) values of input coordinates). It allows to perform automatic correction of fuzzy rules in interactive mode without changing the structure of the fuzzy system, which provides an increase in the efficiency and speed of decision-making in various situations [20, 22].

Table 4. Evaluations and ranking of the IoT platforms

	E_1	E_2	E_3	E_4	E_5	E_6	E_7	E_8	E_9	E_{10}
Evaluations of the IoT platforms	7.8	7.4	7.5	6.6	7.15	7.25	7.0	5.9	6.5	5.3
Ranking without the threshold	$E_1 \succ E_3 \succ E_2 \succ E_6 \succ E_5 \succ E_7 \succ E_4 \succ E_9 \succ E_8 \succ E_{10}$									
Ranking with the threshold	$E_1 \succ E_3 \succ E_2 \succ E_6$									

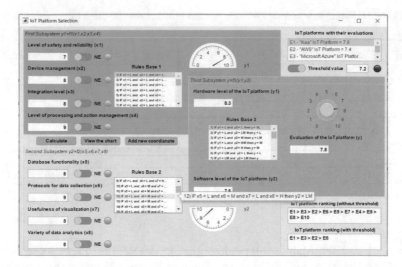

Fig. 1. Fuzzy logic inference system for the selection of specialized IoT platform

The threshold value is determined by the expert on the basis of her experience and requirements to the necessary operating conditions of the IoT platform. Chart of the IoT platforms ranking (Fig. 1, Table 4) using the threshold value ($Thr = 7.2$) is presented in Fig. 2.

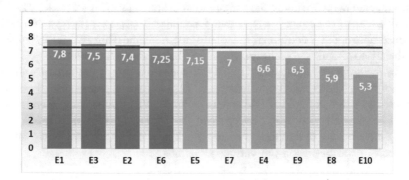

Fig. 2. Chart of the IoT platforms ranking using the threshold value

5   Conclusions

The necessity of using the MCDM methods for selection of specialized IoT platform is concerned with the complexity of taking into account all features, opportunities, and services provided by IoT platform developers. Besides, an incorrectly selected platform may lead to reduction of the reliability and safety of the IoT systems. The authors illustrate the solving MCDM problem using the linear convolution method with simple ranking approach to forming weight coefficients for criteria. This enables DM to analyze the results of the IoT platforms ranking, depending on the conditions and requirements to IoT systems.

Since MCDM methods have some limitations, the authors try to show the perspectives of the soft computing implementation (using, in particular, Mamdani-type fuzzy logic inference engine) for selection of the most rational specialized IoT platform.

The use of fuzzy inference engine of the Mamdani type allows users to choose the best specialized platform with any increase (decrease) in the number of alternative platforms without changing the model and structure of the system. This significantly simplifies the decision-making process (compared with the method of linear convolution) when changing the dimension of the set of alternatives.

Considered approaches to solving described problem can be easily integrated into various tasks, in particular, for solving vehicle routing problems [20], choosing a transport company [23], assessing the level of cooperation within the consortium "University—IT Company" [24], and others. In future research, it is necessary to compare obtained results with another fuzzy logic based MCDM approaches, for instance based on the fuzzy TOPSIS [25, 26], which allows to determine the distance from each alternative to the ideal and worst-case solutions.

References

1. Katrenko, A.V., Pasichnyk, V.V., Pas'ko, V.P.: Decision Making Theory. BHV Publishing Group, Kyiv (2009). (in Ukrainian)
2. Kondratenko, Y.P., Klymenko, L.P., Sidenko, I.V.: Comparative analysis of evaluation algorithms for decision-making in transport logistics. In: Jamshidi, M., Kreinovich, V., Kacprzyk, J. (eds.) Advance Trends in Soft Computing. Studies in Fuzziness and Soft Computing, vol. 312, pp. 203–217. Springer, Cham (2014). https://doi.org/10.1007/978-3-319-03674-8_20
3. Zaychenko, Y.P.: Decision making theory. NTUU "KPI", Kyiv (2014). (in Ukrainian)
4. Hussain, F.: Internet of Things: Building Blocks and Business Models. Springer, Cham (2017). https://doi.org/10.1007/978-3-319-55405-1
5. Javed, A.: IoT Platforms. Apress, Berkeley (2016). https://doi.org/10.1007/978-1-4842-1940-9_12
6. Keramidas, G., Voros, N., Hubner, M.: Components and Services for IoT Platforms. Springer, Cham (2017). https://doi.org/10.1007/978-3-319-42304-3
7. Kondratenko, Y., Kozlov, O., Korobko, O., Topalov, A.: Complex industrial systems automation based on the Internet of Things implementation. In: Bassiliades, N., et al. (eds.) International Conference on ICTERI 2017, Kyiv, Ukraine, pp. 164–187 (2018)
8. Guth, J., Breitenbucher, U., Falkenthal, M., Leymann, F., Reinfurt, L.: Comparison of IoT platforms architectures: a field study based on a reference architecture. In: IEEE Conference on Cloudification of the Internet of Things (IoT) (2016). https://doi.org/10.1109/ciot.2016.7872918
9. Rotshtein, A.P.: Intelligent Technologies of Identification: Fuzzy Logic, Genetic Algorithms, Neural Networks. Universum Press, Vinnitsya (1999). (in Russian)

10. Kondratenko, Y.P., Kondratenko, N.Y.: Reduced library of the soft computing analytic models for arithmetic operations with asymmetrical fuzzy numbers. Int. J. Comput. Res. **23** (4), 349–370 (2016)
11. Piegat, A.: Fuzzy Modeling and Control. Springer, Heidelberg (2001)
12. Lorkowski, J., Kreinovich, V.: Decision making under uncertainty and restrictions on computation resources: from heuristic to optimal techniques. In: Bounded Rationality in Decision Making Under Uncertainty: Towards Optimal Granularity. Studies in Systems, Decision and Control, vol. 99. Springer, Cham (2018)
13. Trunov, A.N.: An adequacy criterion in evaluating the effectiveness of a model design process. Eastern-Eur. J. Enterp. Technol. **1**(4(73)), 36–41 (2015)
14. Mineraud, J., Mazhelis, O., Su, X., Tarkoma, S.: A gap analysis of Internet-of-Things platforms. Comput. Commun. **90**, 5–16 (2016). https://doi.org/10.1016/j.comcom.2016.03.015
15. Razzaque, M., Milojevic-Jevric, M., Palade, A., Clarke, S.: Middleware for Internet of Things: a survey. IEEE Internet Things J. **3**(1), 70–95 (2016)
16. Uckelmann, D., Harrison, M., Michahelles, F.: Architecting the Internet of Things. Springer, Berlin (2011). https://doi.org/10.1007/978-3-642-19157-2
17. Kondratenko, Y.P., Simon, D.: Structural and parametric optimization of fuzzy control and decision making systems. In: Zadeh, L., et al. (eds.) Recent Developments and the New Direction in Soft-Computing Foundations and Applications, vol. 361. Springer, Cham (2018). https://doi.org/10.1007/978-3-319-75408-6_22
18. Kacprzyk, J., Zadrozny, S., Tre, G.: Fuzziness in database management systems: half a century of developments and future prospects. Fuzzy Sets Syst. **281**, 300–307 (2015). https://doi.org/10.1016/j.fss.2015.06.011
19. Zadeh, L.: Fuzzy Sets. Inf. Control **8**(3), 338–353 (1965)
20. Kondratenko, Y.P., Sidenko, I.V.: Decision-making based on fuzzy estimation of quality level for cargo delivery. In: Zadeh, L., et al. (eds.) Recent Developments and New Directions in Soft Computing. Studies in Fuzziness and Soft Computing, vol. 317, pp. 331–344. Springer, Cham (2014). https://doi.org/10.1007/978-3-319-06323-2_21
21. Maslovskyi, S., Sachenko, A.: Adaptive test system of student knowledge based on neural networks. In: IEEE International Conference on IDAACS, Warsaw, Poland, pp. 940–945 (2015). https://doi.org/10.1109/idaacs.2015.7341442
22. Kondratenko, Y.P., Sidenko, I.V.: Method of actual correction of the knowledge database of fuzzy decision support system with flexible hierarchical structure. J. Comput. Optim. Econ. Finance **4**(2/3), 57–76 (2012)
23. Solesvik, M., Kondratenko, Y., Kondratenko, G., Sidenko, I., Kharchenko, V., Boyarchuk, A.: Fuzzy decision support systems in marine practice. In: Proceedings of the IEEE International Conference on Fuzzy Systems, Naples, Italy (2017). https://doi.org/10.1109/fuzz-ieee.2017.8015471
24. Kondratenko, G., Kondratenko, Y., Sidenko, I.: Fuzzy decision making system for model-oriented academia/industry cooperation: university preferences. In: Berger-Vachon, C. (ed.) Complex Systems: Solutions and Challenges in Economics, Management and Engineering, vol. 125, pp. 109–124. Springer, Cham (2018). https://doi.org/10.1007/978-3-319-69989-9_7
25. Buyukozkan, G., Cifci, G.: A novel hybrid MCDM approach based on fuzzy DEMATEL, fuzzy ANP and fuzzy TOPSIS to evaluate green suppliers. Expert Syst. Appl. **39**(3), 3000–3011 (2012). https://doi.org/10.1016/j.eswa.2011.08.162
26. Onut, S., Soner, S.: Transshipment site selection using the AHP and TOPSIS approaches under fuzzy environment. Waste Manag. **28**(9), 1552–1559 (2008)

How Intelligence Community Interprets Imprecise Evaluative Linguistic Expressions, and How to Justify this Empirical-Based Interpretation

Olga Kosheleva[1] and Vladik Kreinovich[2(✉)]

[1] Department of Teacher Education, University of Texas at El Paso, El Paso, TX 79968, USA
olgak@utep.edu

[2] Department of Computer Science, University of Texas at El Paso, El Paso, TX 79968, USA
vladik@utep.edu

Abstract. To provide a more precise meaning to imprecise evaluative linguistic expressions like "probable" or "almost certain", researchers analyzed how often intelligence predictions hedged by each corresponding evaluative expression turned out to be true. In this paper, we provide a theoretical explanation for the resulting empirical frequencies.

Keywords: Evaluative linguistic expressions
Numerical estimates of linguistic expressions · Intelligence community
Empirical frequencies · Theoretical explanation

1 How Intelligence Community Interprets Imprecise Evaluative Linguistics Expressions

Need to Interpret Imprecise Evaluative Linguistic Expressions. A large portion of expert's knowledge is formulated by experts who use imprecise evaluative expressions from natural language, such as "most probably", "small", etc.

We humans understand such evaluative expressions, but computers have big trouble understanding such a knowledge. Computers are designed to process numbers, not linguistic expressions. It is therefore necessary to develop techniques that would translate such evaluative expressions into the language of numbers. This need was one of the main motivations behind Lotfi Zadeh's idea of fuzzy logic; see, e.g., [1,7,9,12,14,21].

Zadeh's pioneering ideas inspired many techniques for assigning numerical values to different evaluative linguistic expressions; see, e.g., [13] and references therein.

© Springer Nature Switzerland AG 2019
O. Chertov et al. (Eds.): ICDSIAI 2018, AISC 836, pp. 81–89, 2019.
https://doi.org/10.1007/978-3-319-97885-7_9

Why Intelligence Community Needs to Interpret Imprecise Evaluative Linguistic Expressions in Numerical Terms. The ultimate objective of intelligence estimates is to make decisions. According to the decision theory analysis (see, e.g., [3, 4, 8, 11, 15]), a rational person should select an action that maximizes the expected value of a special *utility function* (a function that describes a person's preferences). To compute the expected value of the utility function, we need to know the probabilities of different possible events. Thus, to make a decision, we need to estimate the probabilities of different possible consequences of each action.

Based on different pieces of intelligence, intelligence analysts estimate the possibility of different scenarios. Their estimates usually come in terms of imprecise evaluative expressions from natural language such as "almost certain", "probable", etc. To use these estimates in decision making, it is therefore desirable to come up with a probability corresponding to each such evaluative expression.

How Intelligence Community Interprets Imprecise Evaluative Linguistic Expressions: Main Idea. A natural way to assign a probability value to each evaluative linguistic expression is as follows:

- we consider all situations in which the experts' prediction used the corresponding evaluative linguistic expression, and
- we consider the frequency with which these predictions turned out to be true.

For example, if among 40 predictions in which the experts used the expression "probable", the prediction turned out to be true in 30 cases, the corresponding frequency is $30/40 = 75\%$.

Main Idea: What We Expect. It is reasonable to expect that:

- the more confident the experts are,
- the higher should be the frequencies with which these predictions turn out to be right.

For example, we expect that for the cases when the experts were almost certain, the corresponding frequency would be higher than for situations in which the experts simply stated that the corresponding future event is probable.

Possibility to Go Beyond the Main Idea. It is worth mentioning that in situations where a sample is too small to provide a meaningful estimation of the frequency, we can use an alternative approach for providing numerical estimates for evaluative linguistic expression as described in [16]. In this alternative approach:

- We ask several experts to estimate the degree of confidence (subjective probability) corresponding to each of these expressions.
- For each expression, we then take the average of degrees provided by different experts as the (subjective) probability corresponding to this evaluative linguistic expression.

– The standard deviation of these degrees can then be used as gauging the accuracy of the average-based probability estimate.

How Intelligence Community Interprets Imprecise Evaluative Linguistic Expressions: The Resulting Interpretation. The analysis along the lines of the above-described main idea was indeed undertaken at the US Central Intelligence Agency (CIA), under the leadership of Sherman Kent; see, e.g., [5,6] (see also [2,19,20]). This analysis has shown that the imprecise evaluative linguistic expressions describing expert's degree of certainty can be divided into seven groups. Within each group, evaluative expressions have approximately the same frequency. The frequencies corresponding to a typical evaluative expression from each group are described in Table 1.

Table 1. Empirical frequencies corresponding to different evaluative linguistic expressions

certain	100%
almost certain	93%
probable	75%
chances about even	50%
probably not	30%
almost certainly not	7%
impossible	0%

Here are the groups of evaluative linguistic expressions as described in [5,6]:

– the group containing the expression "almost certain" also contained the following expressions:
  • virtually certain,
  • all but certain,
  • highly probable,
  • highly likely,
  • odds (or chances) overwhelming;
– the group containing the expression "possible" also contained the following expressions:
  • conceivable,
  • could,
  • may,
  • might,
  • perhaps;
– the group containing the expression "50-50" also contained the following expressions:
  • chances about even,
  • chances a little better (or less) than even;

- improbable,
- unlikely;
- the group containing the expression "probably not" also contained the following expressions:
    - we believe that not,
    - we estimate that not,
    - we doubt,
    - doubtful;
- the group containing the expression "almost certainly not" also contained the following expressions:
    - virtually impossible,
    - almost impossible,
    - some slight chance,
    - highly doubtful.

What Is Clear and What Is Not Clear About this Empirical Result.
The fact that we got exactly seven different categories is in perfect agreement with the well-known "seven plus minus two law" (see, e.g., [10,17]) according to which human usually divide everything into seven plus minus two categories – with the average being exactly seven.

What is not clear is why namely the above specific probabilities are associated with seven terms, and not, e.g., more naturally sounding equidistant frequencies

$$0, \frac{1}{6}, \frac{2}{6} \left(= \frac{1}{3} \right), \frac{3}{6} \left(= \frac{1}{2} \right), \frac{4}{6} \left(= \frac{2}{3} \right), \frac{5}{6}, 1.$$

What We Do in this Paper. In this paper, we provide a theoretical explanation for the above empirical frequencies.

2   Towards a Theoretical Explanation for Empirical Frequencies

We Make Decisions Based on Finite Number of Observations. Crudely speaking, expert's estimates are based on his/her past experience. At any given moment of time, an expert has observed a finite number of observations. Let us denote this number by n.

If the actual probability of an event is p, then, for large n, the observed frequency is approximately normally distributed, with mean $\mu = p$ and standard deviation

$$\sigma = \sqrt{\frac{p \cdot (1 - p)}{n}};$$

see, e.g., [18].

For two different processes, with probabilities p and p', the difference between the corresponding frequencies is also normally distributed, with mean $d \overset{\text{def}}{=} p - p'$ and standard deviation

$$\sigma_d = \sqrt{\sigma^2 + (\sigma')^2},$$

where σ is as above and

$$\sigma' = \sqrt{\frac{p' \cdot (1 - p')}{n}}.$$

In general, for a normal distribution, all the values are:

- within the 2-sigma interval $[\mu - 2\sigma, \mu + 2\sigma]$ with probability $\approx 90\%$;
- within the 3-sigma interval $[\mu - 3\sigma, \mu + 3\sigma]$ with probability $\approx 99.9\%$;
- within the 6-sigma interval $[\mu - 6\sigma, \mu + 6\sigma]$ with probability $\approx 1 - 10^{-8}$, etc.

Whatever level of confidence we need, for appropriate k_0, all the value are within the interval $[\mu - k_0 \cdot \sigma, \mu + k_0 \cdot \sigma]$ with the desired degree of confidence.

Thus:

- If $|p - p'| \leq k_0 \cdot \sigma_d$, then the zero difference between frequencies belongs to the k_0-sigma interval

$$[\mu - k_0 \cdot \sigma_d, \mu + k_0 \cdot \sigma_d]$$

and thus, it is possible that we will observe the same frequency in both cases.
- On the other hand, if $|p - p'| > k_0 \cdot \sigma_d$, this means that the zero difference between the frequencies is no longer within the corresponding k_0-sigma interval and thus, the observed frequencies are always different. So, by observing the corresponding frequencies, we can always distinguish the resulting probabilities.

Natural Idea. Since we cannot distinguish close probabilities, we have a finite number of distinguishable probabilities. It is natural to try to identify them with the above empirically observed probabilities.

From the Qualitative Idea to Precise Formulas. For each value p, the smallest value $p' > p$ which can be distinguished from p based on n observations is the value $p' = p + \Delta p$, where $\Delta p = k_0 \cdot \sigma_d$. When $p \approx p'$, we have $\sigma \approx \sigma'$ and thus,

$$\sigma_m \approx \sqrt{\frac{2p \cdot (1 - p)}{n}}.$$

So,

$$\Delta p = k_0 \cdot \sqrt{\frac{2p \cdot (1 - p)}{n}}.$$

By moving all the terms connected to p to the left-hand side of this equality, we get the following equality:

$$\frac{\Delta p}{\sqrt{p \cdot (1 - p)}} = k_0 \cdot \sqrt{\frac{2}{n}}. \tag{1}$$

By definition, the Δp is the difference between one level and the next one. Let us denote the overall number of levels by L. Then, we can associate:

- Level 0 with number 0,
- Level 1 with number $\dfrac{1}{L-1}$,
- Level 2 with number $\dfrac{2}{L-1}$,
- ...
- until we reach level $L - 1$ to which we associate the value 1.

Let $v(p)$ is the value corresponding to probability p. In these terms, for the two neighboring values, we get

$$\Delta v = \frac{1}{L-1},$$

thus $1 = (L - 1) \cdot \Delta v$, and the formula (1) takes the form

$$\frac{\Delta p}{\sqrt{p \cdot (1 - p)}} = k_0 \cdot \sqrt{\frac{2}{n}} \cdot (L - 1) \cdot \Delta v,$$

i.e., the form

$$\frac{\Delta p}{\sqrt{p \cdot (1 - p)}} = c \cdot \Delta v,$$

where we denoted

$$c \stackrel{\text{def}}{=} k_0 \cdot \sqrt{\frac{2}{n}} \cdot (L - 1).$$

The differences Δp and Δv are small, so we can approximate the above difference equation by the corresponding differential equation

$$\frac{dp}{\sqrt{p \cdot (1 - p)}} = c \cdot dv.$$

Integrating both sides, we conclude that

$$\int \frac{dp}{\sqrt{p \cdot (1 - p)}} = c \cdot v.$$

The integral in the left-hand side can be explicitly computed if we substitute $p = \sin^2(t)$ for some auxiliary quantity t. In this case, $dp = 2 \cdot \sin(t) \cdot \cos(t) \cdot dt$, and $1 - p = 1 - \sin^2 t = \cos^2(t)$, thus

$$\sqrt{p \cdot (1 - p)} = \sqrt{\sin^2(t) \cdot \cos^2(t)} = \sin(t) \cdot \cos(t).$$

Hence,

$$\frac{dp}{\sqrt{p \cdot (1 - p)}} = \frac{2 \sin(t) \cdot \cos(t) \cdot dt}{\sin(t) \cdot \cos(t)} = 2dt,$$

so

$$\int \frac{dp}{\sqrt{p \cdot (1 - p)}} = 2t,$$

and the above formula takes the form

$$t = \frac{c}{2} \cdot v.$$

Thus,

$$p = \sin^2(t) = \sin^2\left(\frac{c}{2} \cdot v\right).$$

We know that the highest level of certainty $v = 1$ corresponds to $p = 1$, so

$$\sin^2\left(\frac{c}{2}\right) = 1,$$

hence

$$\frac{c}{2} = \frac{\pi}{2}$$

and $c = \pi$.

Finally, we arrive at the following formula for the dependence on p on v:

$$p = \sin^2\left(\frac{\pi}{2} \cdot v\right).$$

In our case, we have 7 levels: Level 0, Level 1, ..., until we reach Level 6. Thus, the corresponding values of v are $\frac{i}{6}$. So:

– for Level 0, we have $v = 0$, hence

$$p = \sin^2\left(\frac{\pi}{2} \cdot 0\right) = 0;$$

– for Level 1, we have $v = \frac{1}{6}$, so we have

$$p = \sin^2\left(\frac{\pi}{2} \cdot \frac{1}{6}\right) = \sin^2\left(\frac{\pi}{12}\right) \approx 6.7\% \approx 7\%;$$

– for Level 2, we have $v = \frac{2}{6} = \frac{1}{3}$, so we have

$$p = \sin^2\left(\frac{\pi}{2} \cdot \frac{1}{3}\right) = \sin^2\left(\frac{\pi}{6}\right) = \sin^2(30°) = (0.5)^2 = 0.25;$$

– for Level 3, we have $v = \frac{3}{6} = \frac{1}{2}$, so we have

$$p = \sin^2\left(\frac{\pi}{2} \cdot \frac{1}{2}\right) = \sin^2\left(\frac{\pi}{4}\right) = \sin^2(45°) = \left(\frac{\sqrt{2}}{2}\right)^2 = 0.5;$$

– for Level 4, we have $v = \frac{4}{6} = \frac{2}{3}$, so we have

$$p = \sin^2\left(\frac{\pi}{2} \cdot \frac{2}{3}\right) = \sin^2\left(\frac{\pi}{3}\right) = \sin^2(60°) = \left(\frac{\sqrt{3}}{2}\right)^2 = \frac{3}{4} = 0.75;$$

– for Level 5, we have $v = \dfrac{5}{6}$, so we have

$$p = \sin^2\left(\frac{\pi}{2} \cdot \frac{5}{6}\right) = \sin^2\left(\frac{5\pi}{12}\right) \approx 0.93;$$

– finally, for Level 6, we have we have $v = 1$, hence

$$p = \sin^2\left(\frac{\pi}{2} \cdot 1\right) = 1^2 = 1;$$

Discussion. We have an *almost perfect* match.

The only difference is that, for Level 2, we get 25% instead of 30%. However, since the intelligence sample was not big, we can probably explain this difference as caused by the small size of the sample.

3   Conclusions

To gauge to what extend different future events are possible, experts often use evaluative linguistic expressions such as "probable", "almost certain", etc. Some predictions turn out to be true, some don't. A natural way to gauge the degree of confidence as described by a given evaluative expression is to analyze, out of all the prediction that used this expression, how many of them turned out to be true. Such an analysis was indeed performed by the intelligence community, and the corresponding empirical frequencies have been used to make expert's predictions more precise.

In this paper, we provide a theoretical explanation for the resulting empirical frequencies. This explanation is based on a natural probabilistic analysis of the corresponding situation.

Acknowledgments. This work was supported in part by the US National Science Foundation grant HRD-1242122.

The authors are greatly thankful to the anonymous referees for their valuable suggestions and to Dan Tavrov for his help.

References

1. Belohlavek, R., Dauben, J.W., Klir, G.J.: Fuzzy Logic and Mathematics: A Historical Perspective. Oxford University Press, New York (2017)
2. Central Intelligence Agency, Center for the Study of Intelligence: Sherman Kent and the profession of intelligent analysis, Washington, D.C. (2002)
3. Fishburn, P.C.: Utility Theory for Decision Making. Wiley, New York (1969)
4. Fishburn, P.C.: Nonlinear preference and utility theory. The John Hopkins Press, Baltimore (1988)
5. Kent, S.: Esimates and Influence, Studies in Intelligence. Central Intelligence Agency, Washington, D.C. (1968). https://www.cia.gov/library/center-for-the-study-of-intelligence/csi-publications/books-and-monographs/sherman-kent-and-the-board-of-national-estimates-collected-essays/4estimates.html

6. Kent, S.: Words of estimative probability, in [19]. https://www.cia.gov/library/center-for-the-study-of-intelligence/csi-publications/books-and-monographs/sherman-kent-and-the-board-of-national-estimates-collected-essays/6words.html
7. Klir, G., Yuan, B.: Fuzzy Sets and Fuzzy Logic. Prentice Hall, Upper Saddle River (1995)
8. Luce, R.D., Raiffa, R.: Games and Decisions: Introduction and Critical Survey. Dover, New York (1989)
9. Mendel, J.M.: Uncertain Rule-based Fuzzy Systems: Introduction and New Directions. Springer, Cham (2017)
10. Miller, G.A.: The magical number seven, plus or minus two: some limits on our capacity for processing information. Psychol. Rev. **63**(2), 81–97 (1956)
11. Nguyen, H.T., Kosheleva, O., Kreinovich, V.: Decision making beyond Arrow's "Impossibility Theorem", with the analysis of effects of collusion and mutual attraction. Int. J. Intell. Syst. **24**(1), 27–47 (2009)
12. Nguyen, H.T., Walker, E.A.: A First Course in Fuzzy Logic. Chapman and Hall/CRC, Boca Raton (2006)
13. Novák, V.: Evaluative linguistic expressions vs. fuzzy categories? Fuzzy Sets Syst. **281**, 81–87 (2015)
14. Novák, V., Perfilieva, I., Močkoř, J.: Mathematical Principles of Fuzzy Logic. Kluwer, Boston/Dordrecht (1999)
15. Raiffa, H.: Decision Analysis. Addison-Wesley, Reading (1970)
16. Rajati M.R., Mendel J.M.: Modeling linguistic probabilities and linguistic quantifiers using interval type-2 fuzzy sets. In: Proceedings of the 2013 Joint World Congress of the International Fuzzy Systems Association and the Annual Conference of the North American Fuzzy Information Processing Society IFSA/NAFIPS 2013, Edmonton, Canada, 24–28 June 2013, pp. 327–332 (2013)
17. Reed, S.K.: Cognition: Theories and Application. Wadsworth Cengage Learning, Belmont (2010)
18. Sheskin, D.J.: Handbook of Parametric and Nonparametric Statistical Procedures. Chapman and Hall/CRC, Boca Raton (2011)
19. Steury, D.P. (ed.). Sherman Kent and the Board of National Estimates. Central Intelligence Agency, Center for the Study of Intelligence, Washington, D.C. (1994)
20. Tetlock, P.E., Gardner, D.: Superforecasting: The Art and Science of Prediction. Broadway Books, New York (2015)
21. Zadeh, L.A.: Fuzzy sets. Inf. Control **8**, 338–353 (1965)

How to Explain Empirical Distribution of Software Defects by Severity

Francisco Zapata[1], Olga Kosheleva[2], and Vladik Kreinovich[3(✉)]

[1] Department of Industrial, Manufacturing, and Systems Engineering,
University of Texas at El Paso, El Paso, TX 79968, USA
fazg74@gmail.com
[2] Department of Teacher Education, University of Texas at El Paso,
El Paso, TX 79968, USA
olgak@utep.edu
[3] Department of Computer Science, University of Texas at El Paso,
El Paso, TX 79968, USA
vladik@utep.edu

Abstract. In the last decades, several tools have appeared that, given a software package, mark possible defects of different potential severity. Our empirical analysis has shown that in most situations, we observe the same distribution or software defects by severity. In this paper, we present this empirical distribution, and we use interval-related ideas to provide an explanation for this empirical distribution.

Keywords: Software defects · Severity distribution
Empirical distribution · Theoretical explanation · Interval techniques

1 Empirical Distribution of Software Defects by Severity

Automatic Detection and Classification of Defects. Software packages have defects of different possible severity. Some defects allow hackers to enter the system and can thus, have a potentially high severity. Other defects are minor and maybe not be worth the effort needed to correct them. For example, if we declare a variable which is never used (or we declare an array of too big size, so that most of its elements are never used), this makes the program not perfect, but does not have any serious negative consequences other than wasting some computer time on this declaration and wasting some computer memory; see, e.g., [1–4, 7, 11–14].

In the last decades, several tools have appeared that, given a software package, mark possible defects of different potential severity; see, e.g., [9].

Usually, software defects which are worth repairing are classified into three categories by their relative severity [9]:

– software defects of very high severity (they are also known as *critical*);
– software defects of high severity (they are also known as *major*); and
– software defects of medium severity.

© Springer Nature Switzerland AG 2019
O. Chertov et al. (Eds.): ICDSIAI 2018, AISC 836, pp. 90–99, 2019.
https://doi.org/10.1007/978-3-319-97885-7_10

This is equivalent to classifying *all* the defects into four categories, where the fourth category consists of *minor* defects, i.e., defects which are not worth repairing.

Cautious Approach. The main objective of this classification is not to miss any potentially serious defects. Thus, in case of any doubt, a defect is classified into the most severe category possible.

As a result, the only time when a defect is classified into medium severity category is when we are absolutely sure that this defect is not of high or of very high severity. If we have any doubt, we classify this defect as being of high or very high severity.

Similarly, the only time when a defect is classified as being of high severity is when we are absolutely sure that this defect is not of very high severity. If there is any doubt, we classify this defect as being of very high severity.

In particular, in situations in which we have no information about severity of different defects, we should classify all of them as of very high severity. As we gain more information about the consequences of different defects, we can start assigning some of the discovered defects to medium or high severity categories. However, since by default we classify a defect as having high severity:

- the number of defects classified as being of very high severity should still be the largest,
- followed by the number of defects classified as being of high severity,
- and finally, the number of defects classified as being of medium severity should be the smallest of the three.

Empirical Results. Since software defects can lead to catastrophic consequences, it is desirable to learn as much as possible about different defects. In particular, it is desirable to know how frequently one meets defects of different severity.

To find out the empirical distribution of software defects by severity, it is necessary to study a sufficiently large number of detects. This is a challenging task: while many large software packages turn out to have defects, even severe defects, such defects are rare, and uncovering each severe defect in a commercially used software is a major event.

At first glance, it may seem that, unless we consider nor-yet-released still-being tested software, there is no way to find a sufficient number of defects to make statistical analysis possible. However, there is a solution to this problem: namely, legacy software, software that was written many years ago, when the current defect-marking tools were not yet available. Legacy software works just fine – if it had obvious defects, they would have been noticed during its many years of use. However, it works just fine only in the original computational environment, and when the environment changes, many hidden severe defects are revealed.

In some cases, the software package used a limited size buffer that was sufficient for the original usage, but when the number of users increases, the buffer becomes insufficient to store the jobs waiting to be performed. Another typical situation is when a C program, which should work for all computers, had some initial bugs that were repaired by hacks that explicitly took into account that in those days, most machines used 32-bit words. As a result, when we run the supposedly correct program on a 64-but machine, we get many errors.

Such occurrences are frequent. As a result, when hardware is upgraded – e.g., from 32-bit to 64-bit machines – software companies routinely apply the defect-detecting software packages to their legacy code to find and repair all the defects of very high, high, and medium severity. Such defects are not that frequent, but even if in the original million-lines-of codes software package, 99.9% of the lines of code are absolutely flawless, this still means that there may be a thousand of severe defects. This is clearly more than enough to provide a valid statistical analysis of distribution of software defects by severity.

So, at first glance, we have a source of defects. However, the problem is that companies – naturally – do not like to brag about defects in their code – especially when a legacy software, software used by many customers, turns out to have thousands of severe defects. On the other hand, companies are interested in knowing the distribution of the defects – since this would help them deal with these defects – and not all software companies have research department that would undertake such an analysis. In view of this interest, we contacted software folks from several companies who allowed us to test their legacy code and record the results – on the condition that when we publish the results, we would not disclose the company names. (We are not even allowed to disclose which defect-detecting tool was used – because many companies use their own versions of such tools.)

Interestingly, for different legacy software packages, as long as they were sufficiently large (and thus containing a sufficiently large number of severe defects), the distribution of defects by severity was approximately the same. In this paper, we illustrate this distribution on three typical cases. These cases, sorted by the overall number of defects, are presented in Table 1.

Table 1. Defects of different severity: three typical cases

	Case 1	Case 2	Case 3
Total number of defects	996	1421	1847
Very high severity defects	543	738	1000
High severity defects	320	473	653
Medium severity defects	133	210	244

Analysis of the Empirical Results: General Case. In all the cases, the numbers of very high, high, and medium severity defects can be approximately described by the ratio 5:3:1. In other words:

– the proportion of software defects of very high severity is close to

$$\frac{5}{5+3+1} = \frac{5}{9} \approx 56\%;$$

– the proportion of software defects of high severity is close to

$$\frac{3}{5+3+1} = \frac{3}{9} \approx 33\%;$$

and
– the proportion of software defects of medium severity is close to

$$\frac{1}{5+3+1} = \frac{1}{9} \approx 11\%.$$

Let us show it on the example of the above three cases.

Case 1. In this case, $\frac{1}{9} \cdot 996 = 110\frac{2}{3}$, so we should:

– observe $\frac{1}{9} \cdot 996 = 110\frac{2}{3} \approx 111$ medium severity defects;

– observe $\frac{3}{9} \cdot 996 = 3 \cdot \left(\frac{1}{9} \cdot 996\right) = 3 \cdot 110\frac{2}{3} = 332$ high severity defects, and

– observe $\frac{5}{9} \cdot 996 = 5 \cdot \left(\frac{1}{9} \cdot 996\right) = 5 \cdot 110\frac{2}{3} = 553\frac{1}{3} \approx 553$ very high severity defects.

As we can see from Table 2, the actual numbers of defects of different severity are very close to these numbers. The match is up to 20% accuracy, which for the problem of predicting number of software defects is not so bad.

Table 2. Case 1: comparing actual and predicted numbers of defects of different severity

	Actual number	Predicted number
Very high severity defects	543	553
High severity defects	320	332
Medium severity defects	133	111

Case 2. In this case, $\frac{1}{9} \cdot 1421 = 157\frac{8}{9}$, so we should:

– observe $\frac{1}{9} \cdot 1421 = 157\frac{8}{9} \approx 158$ medium severity defects;

– observe $\frac{3}{9} \cdot 1421 = 3 \cdot \left(\frac{1}{9} \cdot 1421\right) = 3 \cdot 157\frac{8}{9} = 471\frac{8}{3} \approx 474$ high severity defects, and

– observe $\frac{5}{9} \cdot 1421 = 5 \cdot \left(\frac{1}{9} \cdot 1421 \right) = 5 \cdot 157\frac{8}{9} = 785\frac{40}{9} \approx 789$ very high severity defects.

As we can see from Table 3, the actual numbers of defects of different severity are very close to these numbers. The match is also up to $\approx 20\%$ accuracy.

Table 3. Case 2: comparing actual and predicted numbers of defects of different severity

	Actual number	Predicted number
Very high severity defects	738	789
High severity defects	473	474
Medium severity defects	210	158

Case 3. In this case, $\frac{1}{9} \cdot 1847 = 205\frac{2}{9}$, so we should:

– observe $\frac{1}{9} \cdot 1847 = 205\frac{2}{9} \approx 205$ medium severity defects;
– observe $\frac{3}{9} \cdot 1847 = 3 \cdot \left(\frac{1}{9} \cdot 1847 \right) = 3 \cdot 205\frac{2}{9} = 615\frac{6}{9} \approx 616$ high severity defects, and
– observe $\frac{5}{9} \cdot 1847 = 5 \cdot \left(\frac{1}{9} \cdot 1847 \right) = 5 \cdot 205\frac{2}{9} = 1025\frac{10}{9} \approx 1026$ very high severity defects.

As we can see from Table 4, the actual numbers of defects of different severity are very close to these numbers. The match is also up to 20% accuracy.

Table 4. Case 3: comparing actual and predicted numbers of defects of different severity

	Actual number	Predicted number
Very high severity defects	1000	1026
High severity defects	653	616
Medium severity defects	244	206

2   How to Explain the Empirical Distribution

What We Want: A Brief Reminder. We want to find the three frequencies:

– the frequency p_1 of defects of medium severity;
– the frequency p_2 of defects of high severity, and

– the frequency p_3 of defects of very high severity.

All we know is that $p_1 < p_2 < p_3$.

What We Do. We will use ideas related to interval uncertainty and interval computations (see, e.g., [6,8,10]) to explain the above empirical dependence.

First Idea: Let Us Use Intervals Instead of Exact Numbers. In principle, these frequencies can somewhat change from one example to another – as we have seen in the above examples. So, instead of selecting single values p_1, p_2, and p_3, we should select three *regions* of possible values, i.e., we should select:

– an interval $\left[\underline{F}_1, \overline{F}_1\right]$ of possible values of p_1;
– an interval $\left[\underline{F}_2, \overline{F}_2\right]$ of possible values of p_2; and
– an interval $\left[\underline{F}_3, \overline{F}_3\right]$ of possible values of p_3.

To guarantee that $p_1 < p_2$, we want to make sure that every value from the first interval $\left[\underline{F}_1, \overline{F}_1\right]$ is smaller than or equal to any value from the second interval $\left[\underline{F}_2, \overline{F}_2\right]$. To guarantee this, it is sufficient to require that the largest value $\overline{F}_1$ from the first interval is smaller than or equal to the smallest value of the second interval: $\overline{F}_1 \leq \underline{F}_2$.

Similarly, to guarantee that $p_2 < p_3$, we want to make sure that every value from the second interval $\left[\underline{F}_2, \overline{F}_2\right]$ is smaller than or equal to any value from the third interval $\left[\underline{F}_3, \overline{F}_3\right]$. To guarantee this, it is sufficient to require that the largest value $\overline{F}_2$ from the first interval is smaller than or equal to the smallest value of the third interval: $\overline{F}_2 \leq \underline{F}_3$.

First Idea Expanded: Let Us Make These Intervals as Wide as Possible. We decided to have intervals of possible values of p_i instead of exact values of the frequencies. To fully follow this idea, let us make these intervals as wide as possible, i.e., let us make sure that it is not possible to increase one of the intervals without violating the above inequalities.

This means that we should have no space left between $\overline{F}_1$ and $\underline{F}_2$ – otherwise, we can expand either the first or the second interval. We should therefore have

$$\overline{F}_1 = \underline{F}_2.$$

Similarly, we should have no space left between $\overline{F}_2$ and $\underline{F}_3$ – otherwise, we can expand either the second or the third interval. We should therefore have

$$\overline{F}_2 = \underline{F}_3.$$

Also, we should have $\underline{F}_1 = 0$ – otherwise, we can expand the first interval.

As a result, we get the division of the interval $[0, F]$ of possible frequencies into three sub-intervals:

– the interval $\left[0, \overline{F}_1\right]$ of possible values of the frequency p_1;
– the interval $\left[\overline{F}_1, \overline{F}_2\right]$ of possible values of the frequency p_2; and
– the interval $\left[\overline{F}_2, F\right]$ of possible values of the frequency p_3.

Second Idea: Since We Have to Reason to Take Intervals of Different Widths, Let Us Take them Equal. We have no a priori reason to assume that the three intervals have different widths. Thus, it is reasonable to assume that these three intervals have the exact same width, i.e., that

$$\overline{F}_1 = \overline{F}_2 - \overline{F}_1 = F - \overline{F}_2.$$

From the equality $\overline{F}_2 - \overline{F}_1 = \overline{F}_1$, we conclude that $\overline{F}_2 = 2\overline{F}_1$. Now, from the condition that $F - \overline{F}_2 = \overline{F}_1$, we conclude that

$$F = \overline{F}_2 + \overline{F}_1 = 2\overline{F}_1 + \overline{F}_1 = 3\overline{F}_1.$$

So, we have the following three intervals:

- the interval $\left[0, \overline{F}_1\right]$ of possible values of the frequency p_1;
- the interval $\left[\overline{F}_1, 2\overline{F}_1\right]$ of possible values of the frequency p_2; and
- the interval $\left[2\overline{F}_1, 3\overline{F}_1\right]$ of possible values of the frequency p_3.

Third Idea: Which Value from the Interval Should We Choose. We would like to select a single "typical" value from each of the three intervals.

If we know the probability of different values from each interval, we could select the average value. We do not know these probabilities, so to use this approach, we need to select one reasonable probability distribution on each interval.

A priori, we have no reason to believe that some values from a given interval are more probable than others. Thus, it is reasonable to conclude that all the values within each interval are equally probable – i.e., that on each of the three intervals, we have a uniform distribution.

Comment. This conclusion can be viewed as a particular case of *Laplace Indeterminacy Principle* – and of its natural generalization, the Maximum Entropy approach; see, e.g., [5].

Now, We Are Ready to Produce the Desired Probabilities. For the uniform distribution on an interval, the mean value, as one can clearly check, is the midpoint of the interval. So:

- as the estimate for p_1, we select the midpoint of the first interval $\left[0, \overline{F}_1\right]$, i.e., the value

$$p_1 = \frac{0 + \overline{F}_1}{2} = \frac{\overline{F}_1}{2};$$

- as the estimate for p_2, we select the midpoint of the second interval $\left[\overline{F}_1, 2\overline{F}_1\right]$, i.e., the value

$$p_2 = \frac{\overline{F}_1 + 2\overline{F}_1}{2} = 3 \cdot \frac{\overline{F}_1}{2};$$

- finally, as the estimate for p_1, we select the midpoint of the third interval $\left[2\overline{F}_1, 3\overline{F}_1\right]$, i.e., the value

$$p_3 = \frac{2\overline{F}_1 + 3\overline{F}_1}{2} = 5 \cdot \frac{\overline{F}_1}{2}.$$

Conclusion. We see that $p_2 = 3p_1$ and $p_3 = 5p_1$. So, we indeed have an explanation for the empirical ratios 1:3:5 between the frequencies of software flaws of different severity.

3   What if We Had a More Detailed Classification – Into More than Three Severity Levels?

Formulation of the Problem. In the above analysis, we used the usual subdivision of all non-minor software defects into three levels: of medium severity, of high severity, and of very high severity. However, while such a division is most commonly used, some researchers and practitioners have proposed a more detailed classification, when, e.g., each of the above three levels is further subdivided into sub-categories.

To the best of our knowledge, such detailed classification have not yet been massively used in software industry. Thus, we do not have enough data to find the empirical distribution of software defects by sub-categories. However, while we do not have the corresponding empirical data, we can apply our theoretical analysis to come up with reasonable values of expected frequencies. Let us do it.

Analysis of the Problem. We consider the situation when each of the three severity levels is divided into several sub-categories. Let us denote the number of sub-categories in a level by k. Then, overall, we have $3k$ sub-categories of different level of severity.

Similar to the previous section, it is reasonable to conclude that each of these categories correspond to intervals of possible probability values:

$$\left[0, \overline{F}_1\right], \left[\overline{F}_1, \overline{F}_2\right], \left[\overline{F}_2, \overline{F}_3\right], \ldots, \left[\overline{F}_{3k-1}, \overline{F}_{3k}\right].$$

Also, similarly to the previous section, it is reasonable to require that all these intervals are of the same length. Thus, the intervals have the form

$$\left[0, \overline{F}_1\right], \left[\overline{F}_1, 2\overline{F}_1\right], \left[2\overline{F}_1, 3\overline{F}_1\right], \ldots, \left[(3k-1)\cdot\overline{F}_1, (3k)\cdot\overline{F}_1\right].$$

As estimates for the corresponding frequencies, similar to the previous section, we take midpoints of the corresponding intervals:

$$\frac{\overline{F}_1}{2}, 3\cdot\frac{\overline{F}_1}{2}, 5\cdot\frac{\overline{F}_1}{2}, \ldots, \frac{6k-1}{2}\cdot\frac{\overline{F}_1}{2}.$$

Results of the Analysis. This analysis shows that the frequencies of detects of different levels of severity follow the ratio 1:3:5:...:(6k − 1).

This Result Consistent with Our Previous Findings. If for each of three main severity levels, we combine all the defects from different sub-categories of this level, will we still get the same ratio 1:3:5 as before?

The answer is "yes". Indeed, by adding up the k sub-categories with lowest severity, we can calculate the total frequency of medium severity defects as

$$(1 + 3 + 5 + \ldots + (2k - 5) + (2k - 3) + (2k - 1)) \cdot \frac{\overline{F}_1}{2}.$$

In the above sum of k terms, adding the first and the last terms leads to $2k$; similarly, adding the second and the last but one terms lead to $2k$, etc. For each pair, we get $2k$. Out of k terms, we have $\dfrac{k}{2}$ pairs, so the overall sum is equal to

$$1 + 3 + 5 + \ldots + (2k - 5) + (2k - 3) + (2k - 1) = \frac{k}{2} \cdot (2k) = k^2.$$

Thus, the total frequency of medium severity defects is $k^2 \cdot \dfrac{\overline{F}_1}{2}$.

Similar, the total frequency of high severity defects is equal to

$$((2k + 1) + (2k + 3) + \ldots + (4k - 3) + (4k - 1)) \cdot \frac{\overline{F}_1}{2},$$

where similarly, the sum is equal to

$$(2k + 1) + (2k + 3) + \ldots + (4k - 3) + (4k - 1) = \frac{k}{2} \cdot (6k) = 3k^2.$$

Thus, the total frequency of high severity defects is $3k^2 \cdot \dfrac{\overline{F}_1}{2}$.

The total frequency of very high severity defects is similarly equal to

$$((4k + 1) + (4k + 3) + \ldots + (6k - 3) + (6k - 1)) \cdot \frac{\overline{F}_1}{2},$$

where similarly, the sum is equal to

$$(4k + 1) + (4k + 3) + \ldots + (6k - 3) + (6k - 1) = \frac{k}{2} \cdot (10k) = 5k^2.$$

Thus, the total frequency of high severity defects is $5k^2 \cdot \dfrac{\overline{F}_1}{2}$.

These frequencies indeed follow the empirical ratio 1:3:5, which means that our analysis is indeed consistent with our previous findings.

What if We Have 4 or More Original Severity Levels? Another possible alternative to the usual 3-level scheme is to have not 3, but 4 or more severity levels. In this case, if we have L levels, then a similar analysis leads to the conclusion that the corresponding frequencies should follow the ratio

$$1:3:5:\ldots:(2L - 1).$$

Acknowledgments. This work was supported in part by the US National Science Foundation grant HRD-1242122.

The authors are greatly thankful to the anonymous referees for their valuable suggestions and to Dan Tavrov for his help.

References

1. Ackerman, A.F., Buchwald, L.S., Lewski, F.H.: Software inspections: an effective verification process. IEEE Softw. **6**(3), 31–36 (1989)
2. D'Ambros M., Bacchelli A., Lanza, M.: On the impact of design flaws on software defects. In: Proceedings of the 10th International Conference on Quality Software, QSIC 2010, Zhangjiajie, China, 14–15 July 2010, pp. 23–31 (2010)
3. Harter, D.E., Kemerer, C.F., Slaughter, S.A.: Does software process improvement reduce the severity of defects? A longitudinal field study. IEEE Trans. Softw. Eng. **38**(4), 810–827 (2012)
4. Javed, T., Durrani, Q.S.: A study to investigate the impact of requirements instability on software defects. ACM SIGSOFT Softw. Eng. Notes **29**(3), 1–7 (2004)
5. Jaynes, E.T., Bretthorst, G.L.: Probability Theory: The Logic of Science. Cambridge University Press, Cambridge (2003)
6. Jaulin, L., Kiefer, M., Didrit, O., Walter, E.: Applied Interval Analysis, with Examples in Parameter and State Estimation, Robust Control, and Robotics. Springer, London (2001)
7. Kapur, P.K., Kumar, A., Yadav, K., Khatri, S.K.: Software reliability growth modelling for errors of different severity using change point. Int. J. Reliab. Qual. Saf. Eng. **14**(4), 311–326 (2007)
8. Mayer, G.: Interval Analysis and Automatic Result Verification. de Gruyter, Berlin (2017)
9. The MITRE Corporation: Common weakness enumeration: a community-developed list of software weakness types. https://cwe.mitre.org/cwss/cwss_v1.0.1.html. Accessed 2 Mar 1018
10. Moore, R.E., Kearfott, R.B., Cloud, M.J.: Introduction to Interval Analysis. SIAM, Philadelphia (2009)
11. Ostrand, T.J., Weyuker, E.J.: The distribution of faults in a large industrial software system. ACM SIGSOFT Softw. Eng. Notes **27**(4), 55–64 (2002)
12. Sullivan, M., Chillarege, R.: Software defects and their impact on system availability – a study of field failures in operating systems. In: Proceedings of the 21st International Symposium on Fault-Tolerant Computing, Montreal, Canada, 25–27 June 1991, pp. 2–9 (1991)
13. Sullivan, M., Chillarege, R.: A comparison of software defects in database management systems and operating systems. In: Proceedings of the 22nd International Symposium on Fault-Tolerant Computing, Boston, Massachusetts, 8–10 July 1992, pp. 475–484 (1992)
14. Zheng, J., Williams, L., Nagappan, N., Snipes, W., Hudepohl, J.P., Vouk, M.A.: On the value of static analysis for fault detection in software. IEEE Trans. Softw. Eng. **32**(4), 240–253 (2006)

Applications of Data Science to Economics. Applied Data Science Systems

Tobacco Spending in Georgia: Machine Learning Approach

Maksym Obrizan[1](✉)(iD), Karine Torosyan[2,3](iD),
and Norberto Pignatti[2,4](iD)

[1] Kyiv School of Economics, Kyiv, Ukraine
mobrizan@kse.org.ua
[2] International School of Economics at TSU, Tbilisi, Georgia
[3] Global Labor Organization, Geneva, Switzerland
[4] IZA Institute of Labor Economics, Bonn, Germany

Abstract. The purpose of this study is to analyze tobacco spending in Georgia using various machine learning methods applied to a sample of 10,757 households from the Integrated Household Survey collected by GeoStat in 2016. Previous research has shown that smoking is the leading cause of death for 35–69 year olds. In addition, tobacco expenditures may constitute as much as 17% of the household budget. Five different algorithms (ordinary least squares, random forest, two gradient boosting methods and deep learning) were applied to 8,173 households (or 76.0%) in the train set. Out-of-sample predictions were then obtained for 2,584 remaining households in the test set. Under the default settings, a random forest algorithm showed the best performance with more than 10% improvement in terms of root-mean-square error (RMSE). Improved accuracy and availability of machine learning tools in R calls for active use of these methods by policy makers and scientists in health economics, public health and related fields.

Keywords: Tobacco spending · Household survey · Georgia
Machine learning

1 Introduction

Earlier estimates of the very high premature mortality from smoking-related conditions indicated high smoking rates prevailing in the former Soviet Union. Peto et al. in [14] projected that during the 1990s tobacco would cause about 30% of all deaths for those of 35–69 years of age, making it the largest cause of premature death, with about 5 million deaths occurring in the former USSR.

After the collapse of the Soviet Union, smoking prevalence reduced in some of the former USSR republics. For example, a study of eight countries of the Former Soviet Union reports a significantly lower smoking prevalence among men in 2010 compared with 2001 for four countries, but not Georgia [15].

Even nowadays, smoking remains a serious health problem in Georgia. A national household survey of 1,163 adults in Georgia in 2014 reports smoking prevalence rates

© Springer Nature Switzerland AG 2019
O. Chertov et al. (Eds.): ICDSIAI 2018, AISC 836, pp. 103–114, 2019.
https://doi.org/10.1007/978-3-319-97885-7_11

of 54.2% in men and of 6.5% in women [3]. Secondhand smoke exposure (SHEs) is also quite high, with 54.2% reporting SHEs at home.

Smoking is not only bad for health but also for household budgets. In [17], a range of country estimates of total expenditures on tobacco is reported. For example, the US is on the lower bound in terms of percent with nearly 4% of average expenditures being spent on tobacco products in smoking households [6]. In monetary terms, however, this amount is more than $1,000 annually. China is on the upper bound of the range in terms of percent with current smokers spending 60% of personal income and 17% of household income [11].

The additional burden of smoking on individuals and their families lies in the fact that expenditure on tobacco has "crowding out" effect on household expenditures, especially pronounced in poorer developing country households facing income constraints [6, 17].

In [2], the association is explored between tobacco expenditures and food expenditures at a household level in rural Java, Indonesia. Their findings suggest that 70% of the expenditures on tobacco products are financed by a reduction in food expenditure. By displacing food expenditures in favor of tobacco expenditures smoking has significantly negative impacts on the nutritional and health status of smokers' family members, and especially children [2]. This specific impact on child nutrition is not unique to developing countries. It has been shown that children of low-income smokers in Canada had a poorer diet than low-income children of non-smokers [13].

Displacement of non-food expenditures is also alarmingly sizeable. It has been found that in some developing countries, male cigarette smokers spend more than twice as much on cigarettes as per capita expenditure on clothing, housing, health and education combined [9]. It is even more alarming that the types of non-food expenditures that are often sacrificed to tobacco related purchases include medical care and education expenses for family members, thus undermining human capital accumulation and long-term wellbeing of the household and its members [18].

Diversion of household's financial resource to support tobacco consumption should be taken into account when assessing household wellbeing. For example, poverty levels are underestimated when spending on tobacco products is included in measures of household wealth. It is important to separate tobacco expenditures from overall household budget to properly gauge the prevalence of poverty. In [10], it is reported that 1.5% of household in Vietnam who were above the food poverty line would drop below that line if their tobacco expenditures were taken into account. Moreover, if the amount spent on tobacco products was used to buy food instead, 11.2% of current food-poor in Vietnam would emerge out of poverty. In addition, as the poor spend disproportionally more on tobacco products, tobacco expenditures contribute to inequality. In the same review of Vietnam's case [10], it is reported that after separating tobacco spending from overall household budget, the Gini coefficient increases from 0.34 to 0.43 in urban and from 0.27 to 0.32 in rural areas.

Discussion above makes it clear that tobacco use measured in terms of household expenditure on tobacco products and its share in overall household budget, represent important outcome variables that, in addition to prevalence rate of smoking, deserves careful examination by researchers and policy makers. Given that even a small diversion of resources of poor families who live at or below the edge of poverty can have a significant impact on their wellbeing [4], it becomes very important to estimate the economic burden of tobacco use on household budget. This is especially true for developing countries, which suffer heavily from both wide spread of tobacco consumption and low level of living standards among large groups of population.

Previous studies of household expenditures on tobacco products in Georgia (and five other New Independent States) indicate high opportunity costs of smoking for household finances, especially for the poor [8]. This study reveals that rich households spend more in absolute terms, while poor households devote much higher shares of their monthly household consumption on tobacco products, which is especially problematic given high rates of poverty and smoking in transition countries.

In this paper we aim to provide the most recent evidence on household tobacco spending in Georgia using the Integrated Household Survey collected by GeoStat in 2016. In addition to standard linear regression, we employ a number of recent machine learning methods. From a policy and scientific perspective, it is important to select the estimation methods that give the best performance in terms of out-of-sample predictions of prevalence of tobacco use and household spending on tobacco products, both in absolute terms and as the share of overall household budget. This added accuracy can help tracking better changes in smoking habit and expenditures as Georgia is implementing its active tobacco control policies[1].

The rest of the paper is organized as follows. First, we describe the GeoStat dataset used in this study. We provide descriptive statistics on key indicators of tobacco spending by Georgian households. Next, we develop a linear regression model which serves as a natural benchmark for future comparisons. Finally, we apply machine learning methods and compare their performance with OLS in terms of root-mean-square error (RMSE). The last section concludes.

2 Data and Methods

2.1 GeoStat Data

To analyze the tobacco spending in Georgia we use the Integrated Household Survey (IHS) for 2016 collected by GeoStat.[2] The Integrated Household Survey is a rotating panel with each household being interviewed for four consecutive quarters and then being replaced by a new household. Sampling strategy employed by GeoStat aims to maintain a random sample representative of the population by ten regions and settlement type (rural and urban). Altogether there are nine regions, which cover various geographical areas of the country and the capital city of Tbilisi. IHS surveys offer a rich selection of household and individual level variables, including housing, demographic,

---

[1] Georgian government implemented/increased excise taxes on tobacco products in September 2013, January 2015 and July 2015. As of May of 2018, Georgia has introduced ban on public smoking of tobacco, including electronic cigarettes and Hookah.

[2] Data are publicly available at http://www.geostat.ge/index.php?action=meurneoba&mpid=1&lang=eng.

income and expenditure, health and education related variables. Detailed discussion of GeoStat IHS data is beyond the scope of this paper and can be found, for example, in [16].

Data on tobacco expenditures come from GeoStat file "tblShinda03_1", which includes daily data on consumption of food and non-food goods and services by household over one week. The consumption data includes the amounts in local currency (Georgian Lari, GEL) paid for different products defined by GeoStat's version of COICOP. Total tobacco spending is defined by combining household expenditures (in current GEL) on filter cigarettes (COICOP code of 22111), unfiltered cigarettes (22112), cigars (22121), tutuni[3] (22131) and other tobacco products (22132). Share of tobacco expenditures in total household expenditures is computed by dividing expenditure on tobacco by total household expenditure in the same period.

The final dataset for 2016 includes information on 10,757 households, which are divided for the purposes of this study into 8,173 households (or 76.0%) in the train set and 2,584 households in the test set. The train set is used for model building, while the test set is employed for out-of-sample predictions for model comparison. 2,665 out of 5,508 (or 32.6%) households in the train set report non-zero expenditures on tobacco products. 876 out of 2,584 (or 33.9%) households in the test set report non-zero expenditures on tobacco products. The equality of means test cannot reject the null of equal means share of households reporting non-zero amounts on tobacco products. It is also reassuring that the share of households with non-zero tobacco expenditures is very close to the estimated smoking prevalence rate reported in other studies (such as 36%, provided in [1]).

Figure 1 shows tobacco spending patterns in Georgia for households with smokers based on IHS 2016. While the majority of such households spend less than 20 GEL (approximately 7.60 Euros) per week on tobacco products (top right graph), this relatively small amount may constitute as much as 40 or more percent of household income for sizable chunks of population (top left graph). Households with higher expenditures have a lower share of tobacco spending (bottom left graph). Finally, higher tobacco expenditures in GEL imply (weakly) higher share spent on tobacco products by a household.

2.2 Benchmark OLS Regressions

We begin with benchmark models estimated via ordinary least squares (OLS). We assess spending on tobacco products along three different dimensions. First, we estimate a linear probability model for a discrete variable "Any Tobacco Spending" taking value of 100 if household reports any tobacco expenditures and zero otherwise. Second, we estimate via OLS a model for "% of Tobacco in Spending" capturing household expenditures on tobacco products (including households with zero expenditures on tobacco). Finally, we estimate a linear regression for "Tobacco Spending, GEL" capturing expenditure on tobacco products in the local Georgian currency (also including households with zero expenditures on tobacco).

---

[3] Row tobacco used for pipe and hand rolling.

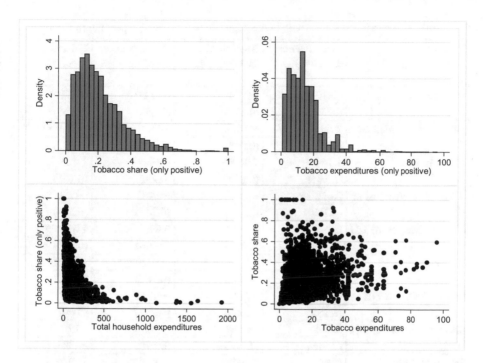

Fig. 1. Tobacco spending in Georgia based on IHS 2016. *Source: Authors' calculations based on GeoStat data.*

Since expenditure data are only available at the household but not the individual level, we have to be creative in defining potential correlates of tobacco expenditures. Specifically, for most variables we use average shares of household characteristics such as age, marital status, attained education level and profession. For example, the variable "Secondary special school" is the average proportion of individuals in the household who have completed secondary special school. This variable will be equal to 2/3 for a household with three members, two of whom completed secondary special school. In addition, we include number of males in the household, household size, measures of household income, characteristics of the local job market, region of Georgia and dummies for quarters 2, 3 and 4.

Results in Table 1 indicate that tobacco spending increases in average household age (but at a diminishing rate) and with a higher share of married members; for better educated households (somewhat surprisingly); for certain occupations; in Kvemo Kartli and Samtskhe-Javakheti regions. Spending on tobacco is lower for Azerbaijani households and households with fewer members, households with internally displaced people (IDP) and recent movers (relative to those who always live in the same place); for rural households and in Adjara a.r., Guria and Imereti and Racha-Lechkh regions. To provide one illustration, one additional male in a household is associated with 7.4 percentage points higher probability of "Any Tobacco Spending," 2.0 percentage points higher share of tobacco spending in total expenditures and 1.5 additional GEL spent on tobacco. Other coefficients can be interpreted similarly.

Table 1. Descriptive statistics and OLS regressions for tobacco spending in Georgia

Variable	Mean (standard deviation)	Any tobacco spending (0/1)	% of tobacco in spending	Tobacco spending, GEL
HH mean age	46.103	0.963***	0.312***	0.084**
	(17.203)	(0.175)	(0.050)	(0.035)
HH mean age squared	2421.387	−0.012***	−0.004***	−0.001***
	(1741.145)	(0.002)	(0.000)	(0.000)
Males in HH	1.709	7.360***	1.966***	1.429***
	(1.165)	(0.810)	(0.232)	(0.161)
Share of Azerbaijani in HH	0.072	−4.476*	−0.221	−1.485***
	(0.258)	(2.597)	(0.744)	(0.518)
Share of Married in HH	0.485	6.285***	0.690	1.391***
	(0.337)	(2.409)	(0.690)	(0.480)
HH size	3.516	−1.144**	−0.729***	−0.172
	(1.894)	(0.560)	(0.160)	(0.112)
Share of IDPs in HH	0.029	−4.583	−1.740*	−1.475**
	(0.155)	(3.464)	(0.992)	(0.691)
Share of Movers in HH	0.564	−6.852***	−1.604***	−1.011***
	(0.316)	(1.921)	(0.550)	(0.383)
Secondary special school	0.362	7.294***	2.690***	1.019***
	(0.350)	(1.907)	(0.546)	(0.380)
Hand craft school	0.075	7.620**	3.008***	0.568
	(0.182)	(2.991)	(0.857)	(0.596)
Higher education	0.032	17.932***	5.098***	1.861**
	(0.123)	(4.214)	(1.207)	(0.840)
Senior officials and managers	0.014	17.228**	3.324	8.566***
	(0.073)	(7.458)	(2.137)	(1.487)
Technicians and associate professionals	0.034	9.588**	1.510	3.405***
	(0.117)	(4.661)	(1.335)	(0.929)
Plant and machine operators	0.021	23.545***	4.876***	6.046***
	(0.089)	(5.613)	(1.608)	(1.119)
Elementary occupations	0.028	13.381***	3.497**	2.244**
	(0.109)	(4.908)	(1.406)	(0.978)
Share of public sector employees	0.071	−8.467**	−2.605**	−2.048***
	(0.172)	(3.721)	(1.066)	(0.742)
Labor Inc., other HH adults	0.023	60.510***	−3.192	23.953***
	(0.043)	(14.336)	(4.107)	(2.858)
Non-lab income, HH	3.711	0.191**	0.013	0.111***
	(5.878)	(0.080)	(0.023)	(0.016)
Local LFP	0.700	80.201***	12.886***	12.160***
	(0.090)	(14.144)	(4.052)	(2.820)

(*continued*)

Table 1. (*continued*)

Variable	Mean (standard deviation)	Any tobacco spending (0/1)	% of tobacco in spending	Tobacco spending, GEL
Rural	0.614	−21.869***	−3.068***	−3.223***
	(0.487)	(3.102)	(0.889)	(0.618)
Shida Kartli	0.077	1.791	3.600***	−0.681
	(0.267)	(2.311)	(0.662)	(0.461)
Kvemo Kartli	0.115	7.960***	2.631***	1.623***
	(0.319)	(2.457)	(0.704)	(0.490)
Samtskhe-Javakheti	0.062	4.718	1.286	2.108***
	(0.242)	(3.097)	(0.887)	(0.617)
Adjara a.r.	0.074	−11.248***	−4.370***	−3.263***
	(0.262)	(2.538)	(0.727)	(0.506)
Guria	0.062	−7.040***	−2.550***	−2.424***
	(0.242)	(2.692)	(0.771)	(0.537)
Samegrelo-Zemo Svaneti	0.097	7.910***	0.444	−0.328
	(0.296)	(2.299)	(0.658)	(0.458)
Imereti, Racha-Lechkh	0.164	−6.612***	−1.961***	−1.780***
	(0.370)	(2.118)	(0.607)	(0.422)
Intercept		−33.569***	−6.357**	−5.923***
		(10.822)	(3.100)	(2.157)
Observations	10757	8173	8173	8173
Adjuster R-squared		0.113	0.083	0.112

*Notes: *p < 0.1, **p < 0.05, ***p < 0.01. Models are estimated for 8,173 households in the train set. Independent variables represent share of individuals with certain marital status and education in the household. Only coefficients significant at 1% in at least one regression are shown to save space. All models also include a share of Armenian, divorced, widowed, disabled individuals; a share of individuals with no schooling or minimal education, four broad definitions of profession; local unemployment rate and households living in Tbilisi and Mtskheta-Mtianeti.*

Source: Authors' calculations based on GeoStat data.

While the OLS models are in general adequate, their predictive power is quite limited as they tend to explain only 8.3 to 11.3% of the variation in dependent variables. Hence, in the next section we turn to machine learning methods in order to improve the predictive power of the models measured by RMSE in a test set of 2,584 households.

3 Machine Learning Approach to Tobacco Spending

Machine learning algorithms are widely used in business, scientific and policy applications nowadays. For the purposes of this study we will employ free R packages "h2o" [12] and "xgboost" [7]. The great advantage of "h2o" package is that it allows running of multiple machine learning algorithms from one R package. This makes machine learning tools accessible even to researchers with limited programming experience.

For example, just nine lines of code will run a random forest algorithm in R (provided train_data and test_data are already loaded):

```
library(h2o)
h2o.init()
trainHex <- as.h2o(train_data)
testHex <- as.h2o(test_data)
features<-
colnames(train_data)[!(colnames(train_data)%in%c("Tobacco
_","train_set"))]
RF_Hex    <-    h2o.randomForest(x=features,    y="Tobacco_",
training_frame=trainHex)
pred.RF_ <- as.data.frame(h2o.predict(RF_Hex,testHex))
resid.RF_ = test_data$Tobacco_ - pred.RF_
RMSE.RF_ <- sqrt(mean(resid.RF_^2))
```

It is enough to replace line six with

```
GBM_Hex    <-    h2o.gbm(x=features,    y="Tobacco_",    train-
ing_frame=trainHex)
```

to run a gradient boosting algorithm, and so on.

Next we illustrate the application of machine learning algorithms to tobacco spending by Georgian households. Five different algorithms (ordinary least squares, random forest, two gradient boosting methods and deep learning) were applied to 8,173 households (or 76.0%) in the train set. In addition, we report the results of H2O's AutoML which is a simple wrapper function used for automatic training and parameter tuning within a pre-specified time.[4] Out-of-sample predictions were then obtained for 2,584 remaining households in the test set.

Table 2 reports RMSE, improvement in RMSE over OLS in % as well as time spent on executing codes in R for three variables of interest—whether household has any tobacco spending, the share of tobacco spending in total expenditure and total expenditure on tobacco products in GEL.

The results in Table 2 clearly show the advantage of machine learning algorithms over OLS even in such a small sample and with default settings. Almost all algorithms perform better than OLS in all three models. Random forest is the best performing algorithm under default settings: out-of-sample RMSE is lower by 10.6% compared to OLS in a model for whether household has any tobacco spending; RMSE is lower by 7.7% compared to OLS in a model for share of tobacco spending; RMSE is lower by

---

[4] For details refer to http://h2o-release.s3.amazonaws.com/h2o/master/3888/docs-website/h2o-docs/automl.html.

Table 2. Relative performance of different machine learning algorithms in terms of RMSE for 2,584 households in the test set

Model	RMSE	Δ RMSE vs OLS, %	Time, seconds
Any tobacco spending (0/1)			
OLS	0.450	0.000	1.196
Random forest	0.402	−10.605	8.810
GBM	0.435	−3.162	4.704
Deep Learning	0.446	−0.763	12.298
XGBoost	0.436	−2.926	1.260
Automatic ML	0.421	−6.267	
% of tobacco in spending			
OLS	12.898	0.000	1.173
Random forest	11.901	−7.729	12.111
GBM	12.593	−2.366	4.560
Deep Learning	12.575	−2.502	13.760
XGBoost	12.896	−0.017	1.471
Automatic ML	12.219	−5.261	
Tobacco spending, GEL			
OLS	8.791	0.000	1.095
Random forest	8.125	−7.577	10.895
GBM	8.547	−2.774	4.671
Deep Learning	8.680	−1.262	13.328
XGBoost	8.840	0.553	1.061
Automatic ML	8.265	−5.981	

Notes: Models are first estimated for 8,173 households in the train set (not shown to save space). RMSE is computed for 2,584 households in the test set to evaluate the quality of out-of-sample predictions. GBM stands for Gradient Boosting Machine. XGBoost stands for eXtreme Gradient Boosting. Automatic ML stands for automatic machine learning command in "h2o" package. All models except for XGBoost are estimated using "h2o" package in R [12]. XGBoost is estimated using "xgboost" package in R [7]. All models are estimated using the default parameters except for XGBoost, which does not have default for "nrounds" (set to 2000).

7.6% compared to OLS in a model for total tobacco spending in GEL. Automatic machine learning (limited to 20 min of execution time in each of three models) is the second-best algorithm with default settings.

Given that random forest outperforms other algorithms under default settings we are also interested in variable importance as discussed in [5]. Specifically, on the first step the out-of-bag error for each observation is recorded and averaged over the forest. Next, values of each feature are permuted and out-of-bag error is re-computed. Importance score is then the average difference in out-of-bag error before and after the permutation over all trees normalized by the standard deviation of the differences. Table 3 shows ten the most important features for all three models ranked by scaled

Table 3. Variable importance in random forest

Any tobacco spending (0/1)			% of tobacco in spending			Tobacco spending, GEL		
Variable	S.I.	%	Variable	S.I.	%	Variable	S.I.	%
HH mean age squared	1.00	9.08%	HH mean age	1.00	9.25%	HH mean age	1.00	7.92%
Males in HH	1.00	9.03%	HH mean age squared	0.98	9.11%	Non-lab income, HH	0.96	7.64%
HH mean age	0.91	8.25%	Non-lab income, HH	0.69	6.43%	Labor Inc., other HH adults	0.85	6.76%
Non-lab income, HH	0.67	6.04%	Males in HH	0.64	5.88%	Males in HH	0.78	6.15%
Local unemployment	0.59	5.31%	Local unemployment	0.60	5.56%	HH mean age squared	0.77	6.09%
Local LFP	0.56	5.08%	Secondary special school	0.55	5.10%	HH size	0.60	4.77%
Labor Inc., other HH adults	0.50	4.53%	Local LFP	0.54	4.98%	Local unemployment	0.59	4.71%
Share of Movers in HH	0.49	4.47%	Share of Movers in HH	0.51	4.72%	Share of Movers in HH	0.56	4.42%
Secondary special school	0.47	4.30%	Share of Married in HH	0.44	4.07%	Secondary special school	0.55	4.37%
HH size	0.43	3.94%	HH size	0.36	3.37%	Local LFP	0.55	4.32%

Notes: Models are estimated for 8,173 households in the train set. S.I. stands for scaled importance.

importance. While almost the same features are the most important for error reduction in all three cases, their relative ranking is somewhat different.

We would like to stress that these substantial improvements in predictive accuracy compared to OLS were achieved with default parameters in all algorithms. It is needless to say that fine-tuning of the parameters together with cross-validation may substantially improve the predictive accuracy of machine learning algorithms. In addition, the time spent on executing the algorithms is also very reasonable (but may increase in bigger samples). Hence, these results call for active use of modern machine learning tools in practical applications and scientific research, even by experts with limited programming experience in R.

4   Conclusions

In this paper we provide the most recent estimates of tobacco spending by households in Georgia in 2016. In the first stage, we describe the extent of tobacco spending by households and identify its important predictors using ordinary least square regressions. Given the limited predictive ability of the OLS model, we next turn to machine learning methods.

Random forest and automatic machine learning lead to a substantial reduction of RMSE of up to 10.6% even with default settings without any fine-tuning of hyper-parameters. This improved accuracy together with the ease of use of machine learning tools, which are readily available in R calls for active application of these methods by policy makers and researchers in health economics, public health and related fields.

References

1. Bakhturidze, G., Peikrishvili, N., Mittelmark, M.: The influence of public opinion on tobacco control policy-making in Georgia: perspectives of governmental and non-governmental stakeholders. Public participation in tobacco control policy-making in Georgia. Tob. Prev. Cessat. 2(1), 1 (2016)
2. Block, S., Webb, P.: Up in smoke: tobacco use, expenditure on food, and child malnutrition in developing countries. Econ. Dev. Cult. Chang. 58(1), 1–23 (2009)
3. Berg, C.J., Topuridze, M., Maglakelidze, N., Starua, L., Shishniashvili, M., Kegler, M.C.: Reactions to smoke-free public policies and smoke-free home policies in the Republic of Georgia: results from a 2014 national survey. Int. J. Public Health 61(4), 409–416 (2016)
4. de Beyer, J., Lovelace, C., Yürekli, A.: Cover essay: poverty and tobacco. Tob. Control 10 (3), 210–211 (2001)
5. Breiman, L.: Random forests. Mach. Learn. 45(1), 5–32 (2001)
6. Busch, S., Jofre-Bonet, M., Falba, T., Sindelar, J.: Burning a hole in the budget: tobacco spending and its crowd-out of other goods. Appl. Health Econ. Health Policy 3(4), 263–272 (2004)
7. Chen T., He T., Benesty M., Khotilovich V., Tang Y.: xgboost: EXTREME Gradient Boosting. R package version 0.6.4.1. https://CRAN.R-project.org/package=xgboost (2018)
8. Djibuti, M., Gotsadze, G., Mataradze, G., Zoidze, A.: Influence of household demographic and socio-economic factors on household expenditure on tobacco in six New Independent States. BMC Public Health 7(1), 222 (2007)
9. Efroymson, D., Ahmed, S., Townsend, J., Alam, S.M., Dey, A.R., Saha, R., Dhar, B., Sujon, A.I., Ahmed, K.U., Rahman, O.: Hungry for tobacco: an analysis of the economic impact of tobacco consumption on the poor in Bangladesh. Tob. Control 10, 212–217 (2001)
10. Efroymson, D., Pham, H.A., Jones, L., FitzGerald, S., Le Thuand, T., Le Hien, T.T.: Tobacco and poverty: evidence from Vietnam. Tob. Control 20(4), 296–301 (2011)
11. Gong, Y.L., Koplan, J.P., Feng, W., Chen, C.H., Zheng, P., Harris, J.R.: Cigarette smoking in China: prevalence, characteristics, and attitudes in Minhang District. JAMA 274(15), 1232–1234 (1995)
12. The H2O.ai Team: h2o: R Interface for H2O. R package version 3.16.0.2. (2017). https://CRAN.R-project.org/package=h2o
13. Johnson, R., Wang, M.Q., Smith, M., Connolly, G.: The association between parental smoking and the diet quality of low-income children. Pediatrics 97(3), 312–317 (1996)
14. Peto, R., Boreham, J., Lopez, A.D., Thun, M., Heath, C.: Mortality from tobacco in developed countries: indirect estimation from national vital statistics. Lancet 339(8804), 1268–1278 (1992)
15. Roberts, B., Gilmore, A., Stickley, A., Rotman, D., Prohoda, V., Haerpfer, C., McKee, M.: Changes in smoking prevalence in 8 countries of the former Soviet Union between 2001 and 2010. Am. J. Public Health 102(7), 1320–1328 (2012)
16. Torosyan K., Pignatti N., Obrizan M.: Job Market Outcomes of IDPs: The Case of Georgia. IZA DP No. 11301 (2018)

17. Wang, H., Sindelar, J.L., Busch, S.H.: The impact of tobacco expenditure on household consumption patterns in rural China. Soc. Sci. Med. **62**(6), 1414–1426 (2006)
18. Xin, Y., Qian, J., Xu, L., Tang, S., Gao, J., Critchley, J.A.: The impact of smoking and quitting on household expenditure patterns and medical care costs in China. Tob. Control **18** (2), 150–155 (2009)

Explaining Wages in Ukraine: Experience or Education?

Valentyna Sinichenko[1], Anton Shmihel[1], and Ivan Zhuk[2($\boxtimes$)]

[1] Kyiv School of Economics, Kyiv, Ukraine
{vkalonova, ashmihel}@kse.org.ua
[2] Igor Sikorsky Kyiv Polytechnic Institute, Kyiv, Ukraine
zis96@ukr.net

Abstract. In this article, we analyze a large database of job vacancies in Ukraine, webscrapped from Work.ua website in January–February 2017. The obtained dataset was processed with bag-of-words approach. Exploratory data analysis revealed that experience and city influence wages. For example, wages in the capital are much higher than in other cities. To explain variation in wages, we used three models to predict wages: multiple linear regression, decision tree and random forest; the latter has demonstrated the best explanatory power. Our work has confirmed the old finding by Mincer that experience is an important variable that explains wages. In fact, this factor was the most informative. Education, however, was an unimportant factor to determine wages. English, teamwork, sales skills, car driving and programming languages are the skills for which modern employers are willing to pay.

Keywords: Wage · Predicting · Ukraine · Exploratory data analysis
Multiple linear regression · Random forest

1 Introduction

Why do different people get different income? What drives the difference? Could someone increase his or her future income by observing such information?

In our work we want to predict wages in Ukraine using vacancies description. Our research could be useful to workers who want to know how much to bid at interview. Students might also be interested to know what skills obtain to get higher income.

On the other hand, companies could be interested as well. This information could be used for planning purposes and decision making.

2 Literature Overview

Paper [1] is a classical work on income differences. The model developed by the author is presented below:

© Springer Nature Switzerland AG 2019
O. Chertov et al. (Eds.): ICDSIAI 2018, AISC 836, pp. 115–124, 2019.
https://doi.org/10.1007/978-3-319-97885-7_12

$$\ln(w) = \beta_0 + \beta_1 s + \beta_2 e + \beta_3 e^2 + \varepsilon,$$

where w is wage, s is years of schooling, e is years of experience, and ε is a residual with zero expected value.

Mincer [1] developed a framework describing two economic concepts:

1. A pricing equation (wage function) showing how the labor market rewards productive characteristics like schooling and work experience.
2. Positive returns to schooling and work experience.
3. Diminishing returns to experience.

Lemieux [2] notes that model developed by Mincer did not perfectly suit the data. He notes that Mincer's model tends to underestimate experience the effect on income for young people. He proposed to add higher order polynomial to eliminate this effect.

Del Caprio, Kupets, Muller and Olefir [3], using data web scrapped from HeadHunter website, have shown what skills are valuable to employers in modern Ukraine. We have made a little step further by developing a model that shows how Ukrainian labor market remunerates each skill.

3 Data Description

Our analysis was performed on Ukrainian vacancies wage data from the work.ua website. This dataset was web scrapped and provided by Viktoria Malchenko, a Kyiv School of Economics alumni (class of 2017). The dataset covers 1-month period from January 16, 2017 to February 15, 2017. The data have been collected for five sectors: agriculture, trade, IT, administration, and top management. We translated the original dataset using onlinedoctranslator.com. The dataset contains 14 thousand observations and contains the following variables:

- name — vacancy name;
- company — company/company name;
- city — city where potential employee would work;
- experience — years of working experience required;
- education — level of education required;
- sector — industry, where potential employee would work;
- wage — wage amount proposed;
- full_part — full time or part time employment.

Before applying any models to the data, several rearrangements and adjustments were performed such as:

- observations that missed wage data were dropped;
- URLs, punctuation signs (such as comma, dash, etc.), and white spaces were eliminated;
- all the text was turned to lower case (to obtain consistency in comparison among words);
- typical English stopwords were removed (no much sense for analysis);
- stemming was performed (based on roots of words).

4 Exploratory Data Analysis

Word cloud in Fig. 1 visually represents frequency of words used in vacancies from our data. The most frequent word is "Sales," which potentially indicates high demand for such skills. Moreover, "experience," "manager," "knowledge," "working," "kiev," "skills," and "development" are also used in vacancies descriptions relatively frequently.

A boxplot in Fig. 2 indicates that there is no significant difference in median wages among education levels.

Fig. 1. Word cloud built among words used in vacancies description

As expected, top management sector has the highest median wage level, but the wage distribution among sector is higher as well. It is worth noting that there is no significant difference among median wage in administrative, agriculture, IT, and trade sectors (Fig. 3).

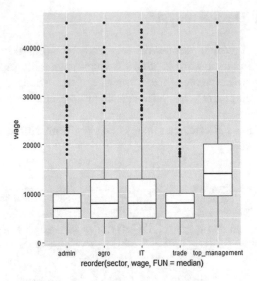

Fig. 2. Boxplot showing differences in wage among sectors

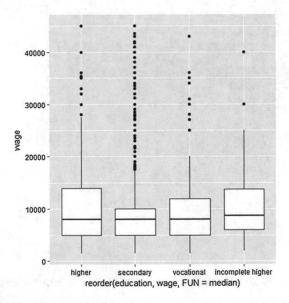

Fig. 3. Boxplot showing differences in wage among educational levels

As expected (Fig. 4), median wage is relatively higher in Kyiv due to higher business activity in the capital of Ukraine. Also, median wages in cities with population more than one million people are higher as well.

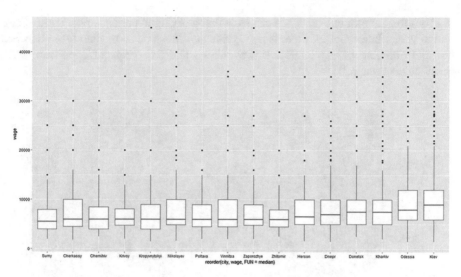

Fig. 4. Boxplot showing differences in wage among big Ukrainian cities

AM Holding is the company that posted the highest number of vacancies. Colors in the word cloud in Fig. 5 represent relative frequencies of each word in vacancies description.

Fig. 5. Word cloud built among employees' names

From the boxplot in Fig. 6, we can state that working experience positively affects median wage. Interestingly, there is no significant difference in median wage between zero and one year of experience. However, two and five years of experience have higher median wage.

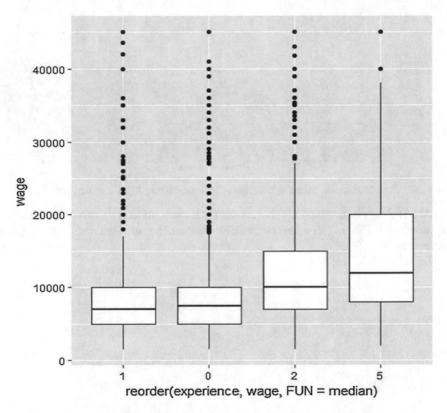

Fig. 6. Boxplot showing differences in wage among different years of experience

5   Methodology and Estimation Results

In our research, we have used the so-called "bag-of-words" approach, where each word is treated independently. First, we have transformed vacancy descriptions into a dataframe, where columns show frequencies of each word in each vacancy. We have estimated three models to predict wages: multiple linear regression, regression trees, and random forest.

Multiple linear regression model is appealing, because it estimates the sign of the direction of effect of different factors on wage. We have tried to fit a high-order polynomial in experience suggested by [2], however, the coefficients before the second- and third-dergee polynomials were statistically insignificant.

Figure 7 shows the list of variables that proved to be significant in the multiple linear regression. Thus, jobs abroad, in Kyiv and in Odesa pay significantly more. Young men and women are paid less. Knowledge of English, teamwork, sales skills, and car driving are the skills, for which modern employers are willing to pay.

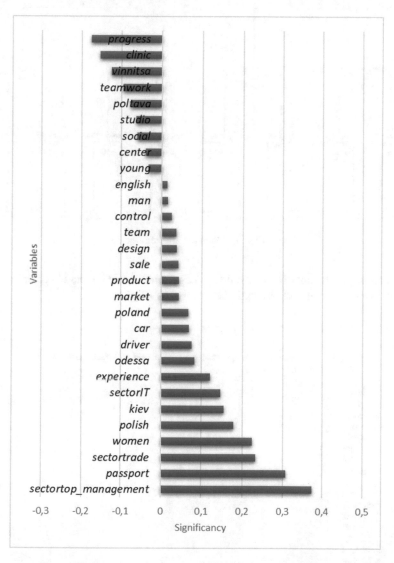

Fig. 7. Variables significance in multiple linear regression model

Figure 8 represents a decision tree model with experience being the root node. Conditions contain fractional values. For example, for experience, it takes place because in some advertisements experience is not required (i.e., experience factor for these advertisements is equal to zero).

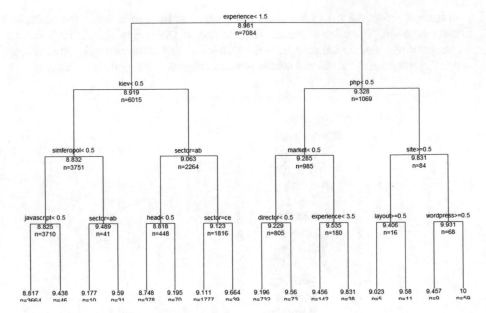

Fig. 8. Decision tree model

According to the decision tree model, "experience," "Kyiv," and "sector" variables explain the greatest part of the variation of wages (Fig. 9).

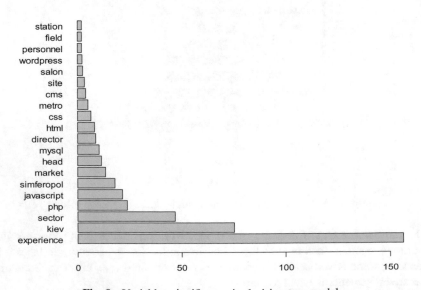

Fig. 9. Variables significance in decision tree model

Lastly, we have tried a random forest approach. According to the model, again, experience and capital city are the most important variables (Fig. 10).

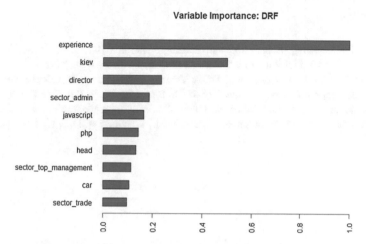

Fig. 10. Variables significance in decision random forest model

Model fits are reported in Table 1, with random forest demonstrating the best fit.

Table 1. Root mean squared error (RMSE) for each model

Model	RMSE/1000
MLR	40.6
Decision tree	8.6
Random forest	8.0

6 Conclusion

Our work has confirmed the old finding by Mincer [1] that experience is an important variable that explains wages. Skills and sector are also important determinants, which confirms the findings of Del Caprio, Kupets, Muller and Olefir [3]. According to our research, modern employers are willing to pay for knowledge of English, ability to work in teams, and to drive a car.

Our research has revealed a surprising finding. The level of education is an insignificant variable to define wage level.

In our research, we have used the so-called "bag-of-words" approach, where each word is treated independently of its context. This is too mechanic to treat human speech as a collection of words. As a potential to expand the predictive power of our models, one may transform the text data into n-grams (combinations of n words that follow one another, such as "cold call"). Each n-gram has a potential to have a better meaning that each of its component. However, the size of the dataset will grow exponentially. As a result, additional computational power is needed to expand the scope of the project.

References

1. Mincer, J.: Schooling, Experience, and Earnings. National Bureau of Economic Research, New York (1974). Distributed by Columbia University Press, New York
2. Lemieux, T.: The Mincer Equation, Thirty Years after Schooling Experience, and Earnings. Center for Labor Economics, University of California-Berkeley, Berkeley (2003)
3. Del Carpio, X., Kupets, O., Muller, N., Olefir, A.: Skills for a Modern Ukraine. Directions in Development - Human Development. World Bank, Washington, D.C. (2017)

On the Nash Equilibrium in Stochastic Games of Capital Accumulation on a Graph

Ruslan Chornei[✉]

National University of Kyiv-Mohyla Academy, Kyiv, Ukraine
r.chornei@ukma.edu.ua

Abstract. This paper discusses some applications of the theory of controlled Markov fields defined on some finite undirected graph. This graph describes a system of "neighborhood dependence" of the evolution of a random process described by local and synchronously change of state of vertices, depending on the decisions made in them. The main attention is paid to solving the problem of finding a Nash equilibrium for stochastic games of capital accumulation with many players.

Keywords: Markov decision process · Stochastic game
Nash equilibrium · Capital accumulation · Random fields
Local interaction

1 Introduction

The theory of managing random processes and fields is fairly extensive and bulk. The application of its theoretical positions gives the ability to describe processes of different nature. In [1] deals with some of the problems that arise during the solution many applied problems of economics, recognition, sociology, biology, disaster modeling. The main purpose of this work is The study of the application of controlled random fields given on a finite non-oriented graph described in [2], to the problem of finding the equilibrium of the Nash for stochastic capital accumulation games with many players. Economic games of this type, also known as game mining resources, introduced in the work of Levhari and Mirman [3], have been extensively studied in recent years. The results include the theorems for the existence of a stationary equilibrium for a deterministic version of this game in Sundaram [4] and two variants of stochastic games with symmetric players proposed by Majumdar and Sundaram [5] and Dutta and Sundaram [6].

The direct impetus for writing this article was the work of Balbus, Nowak [7] and Więcek [8] based on the results of Amir [9]. In the [7] paper we consider a symmetric capital accumulation game with a nonzero sum, while the work [8] extends the results of [5,6] to a non-symmetric case. This generalization was achieved at the expense of some additional structural assumptions (continuity and convexity of the law of motion between states, limited activity space of the

O. Chertov et al. (Eds.): ICDSIAI 2018, AISC 836, pp. 125–133, 2019.
https://doi.org/10.1007/978-3-319-97885-7_13

players). However, this allowed the author to show some important features of stationary equilibrium strategies, such as continuity, monotony, and property of Lipschitz.

In this paper we summarize the results of Balbus, Nowak [7] and Wiȩcek [8] on a multidimensional model: players are concentrated at the vertices of a finite graph that determines their local interaction. Such a generalization makes it possible to use in real economic models with "neighborhood dependence" (for example, see [10, 11]) for describe the process of capital accumulation.

The recent book of Voit's puts (among others) these models under the title The Statistical Mechanics of Financial Markets [12]. Most parts of the book are on random walks and scaling limits, so merelyon the macroscopic behavior of the market models. But there is also a chapter on Microscopic Market Models that fit in the class of models we are working on. The author considers (discrete) populations of agents who buy and sell units of one stock and by observing the prices over time, they adjust their trading decisions and behavior. Depending on the assumptions, these agents behave Markovian in time, or act with a longer memory. With Markovian behavior for some examples, a behaviour occurs that resembles Ising models or spin glass models. We believe that this is a promising class of models in this area of applications, especially when the restriction on being close to the original models of statistical physics is relaxed. Then we are able to obtain more versatile models with more general state spaces in the flavor of Kelly's spatial processes or in the realm of general stochastic networks. If this is successful, then the question of optimization obviously enters the stage.

2   General Statement of the Problem

The material in this section summarizes and modifies the results of work [2].

For systems with locally interacting coordinates the interaction structure is defined via an undirected finite neighborhood graph $\Gamma = (V, B)$ with set of vertices (nodes) V and set of edges $B \subset V^2 \setminus \text{diag}(V)$. Denote by $\{k, j\}$ the edge of the graph connecting vertices k and j. The neighborhood of vertex k is the node set $N(k) = \{j \mid \{k, j\} \in B\}$. The complete neighborhood of vertex k is $\widetilde{N}(k) = N(k) \cup \{k\}$, i. e. the neighborhood of the vertex k, and including k. For any $K \subset V$ we define the neighborhood $N(K) = \bigcup_{k \in K} N(k) \setminus K$, and the complete neighborhood $\widetilde{N}(K) = N(K) \cup K$.

Let X be an interval containing zero called the state space or the set of all possible capital stocks, $X = [0; \infty)$.

A random variable ξ defined on $\langle \Omega, \mathcal{F}, \mathsf{P} \rangle$ with values in X is called a (discrete) random process over X.

Throughout the paper we fix the underlying probability space $\langle \Omega, \mathcal{F}, \mathsf{P} \rangle$ where all random variables are defined on.

We now describe the $|V|$-person stochastic games which we shall investigate in this paper. The players are located at the vertices of the interaction graph Γ, the edges represent the connections between the players. The set of actions (control values) usable at control instants is $A(x) = \underset{i \in V}{\times} A_i(x)$ over Γ, where

$A_i(x) = [0; a(x)]$, $x \in X$, is a set of possible actions (decisions) for vertex i. The quantity $a(x)$ is the consumption capacity of every player in state x.

We also define a Borel set

$$\Delta = \big\{(x,a) \colon x \in X, a \in A\big\}.$$

Definition 1. *(1) Let α_i^t denote the action chosen by the player (decision maker) at node i at time t, $\alpha^t := (\alpha_i^t, i \in V)$ the joint decision vector of all players at time $t \in \mathbb{N}$.*

(2) A strategy (or policy) π for game with interacting components is defined as vector of coordinate policies $\pi = (\pi_i, \ i \in V)$, where for node i $\pi_i = \{\pi_i^0, \ldots, \pi_i^t, \ldots\}$ is a sequence of transition probabilities

$$\pi_i^t = \pi_i^t(\cdot \mid x^0, a^0, \ldots, x^{t-1}, a^{t-1}, x^t).$$

So for any history $h^t = (x^0, a^0, \ldots, x^{t-1}, a^{t-1}, x^t)$ of the system up to time t π_i^t is a probability measure on A_i which are measurably dependent on the history $h^t = (x^0, a^0, \ldots, x^{t-1}, a^{t-1}, x^t)$ up to time t.

We assume that the decisions of the players are synchronized according to the following scheme, which is a synchronous control kernel. Note however, that this form of sychronization is a standard assumption for decision making in discrete time sequential games. Even games with perfect information may be considered as allowing at each decision instant for one of the customers only one "decision", see [13, Sect. 5], which formally results in the general synchronization.

$$P\big(\alpha^t \in \underset{i \in V}{\times} B_i \mid \xi^0 = x^0, \alpha^0 = a^0, \ldots, \alpha^{t-1} = a^{t-1}, \xi^t = x^t\big)$$

$$= \prod_{i \in V} P\big(\alpha_i^t \in B_i \mid \xi^0 = x^0, \alpha^0 = a^0, \ldots, \alpha^{t-1} = a^{t-1}, \xi^t = x^t\big)$$

$$= \prod_{i \in V} \pi_i^t(B_i \mid x^0, a^0, \ldots, a^{t-1}, x^t),$$

$B_k \in \mathfrak{A}_k$, $a^s \in A$, $x^s \in X$, where $\mathfrak{A}_k$ is the sigma-field of Borel subsets of A_k for all $k \in V$.

We now introduce various forms of structured strategies.

Definition 2. *(1) If at any time $t = 0, 1, \ldots =: \mathbb{N}^+$ the decision α_i^t for node i is made according to the probability π_i^t on basis of the history of the complete neighborhood $\widetilde{N}(i)$ of i only, i. e. on the basis of $h_i^t = \big(x^0, a_{\widetilde{N}(i)}^0, \ldots, x^{t-1}, a_{\widetilde{N}(i)}^{t-1}, x^t\big)$, and if $\pi_i^t\big(A_i^t(x) \mid h_i^t\big) = 1$, $x^s \in X$, $a_i^s \in A_i^s(x^s)$, then the sequence of transition probabilities (decisions) $\pi_i = \{\pi_i^t, t \in \mathbb{N}^+\}$ is called admissible local strategy for vertex i, and $\pi = (\pi_i, i \in V)$ is said to be locally admissible.*

(2) An admissible local strategy $\pi = (\pi_i, \ i \in V)$ is called admissible local Markov strategy if

$$\pi_i^t\big(\cdot \mid x^0, a_{\widetilde{N}(i)}^0, \ldots, x^{t-1}, a_{\widetilde{N}(i)}^{t-1}, x^t\big) = \pi_i^t(\cdot \mid x^t), \qquad i \in V.$$

(3) An admissible local Markov strategy $\pi = (\pi_i, i \in V)$ is called admissible stationary local strategy if $\pi_i^{t'}(\cdot \mid x) = \pi_i^{t''}(\cdot \mid x)$, $i \in V$, for all t' and t'' and all x.

(4) An admissible deterministic (nonrandomized) stationary local strategy $\pi = (\pi_i, i \in V)$ is called admissible stationary local strategy if $\pi_i^t(\cdot \mid x)$, $i \in V$, are one-point measures on $A_i^t(x)$, $i \in V$, for each $x \in X$.

The class of all admissible local strategies is denoted by LS; the subclass of admissible local Markov strategies by LS_M. By $LS_S \subseteq LS_M$ we denote the class of admissible stationary local strategies with time invariant restriction sets: $\pi^t = \pi^{t'}$ for all $t, t' \in \mathbb{N}$ and $A^t(x) = A(x) = \underset{i \in V}{\times} A_i(x)$ for all t, and by $LS_D \subseteq LS_S$ the set of admissible deterministic stationary local strategies.

If it is clear from the context we shall henceforth omit the "admissible" in the description of the strategies under consideration.

The law of motion of the system is characterized by a set of time invariant transition probabilities. Whenever the state of the system is $\xi^t = x^t$ and decision $\alpha^t = a^t$ is taken, then the transition probability is

$$P\{\xi^{t+1} \in C \mid \xi^t = x^t, \alpha^t = a^t\} =: Q(C \mid x^t, a^t),$$

and therefore independent of the past given the present (generalized) state. We further assume, that this transition probability is independent of t, the motion is homogeneous in time.

Applying a control policy to a time dependent Markov random process we shall call the duplex (ξ, π) a controlled version of ξ using strategy π. The controlled process $\xi = (\xi^t)$ in general is not Markovian because at first the sequence (α^t) of decisions depends not only on the actual local states $x_{\widetilde{N}(i)}^t$, $i \in V$, but on the previous local controls $a_{\widetilde{N}(i)}^0, \ldots, a_{\widetilde{N}(i)}^{t-1}$ as well. An immediate consequence of the Definition 2 will be, that if Markov strategy π are applied to control the time dependent Markov random field then we obtain a Markov process ξ from (ξ, π).

Definition 3. *A duplex (ξ, π) is called a controlled stochastic process with locally interacting synchronous components with respect to the finite interaction graph $\Gamma = (V, B)$ if $\xi = (\xi^t : t \in \mathbb{N}^+)$ is a stochastic process with state space X, $\pi = (\pi_i, i \in V)$ are admissible local strategies, and the transitions of ξ are determined as given below in (1). If the strategies π are stationary local Markov then the duplex (ξ, π) is called a controlled Markov process with locally interacting synchronous components and has a time invariant transition law. The rationale behind the construction of this transition law is along the following lines.*

$$P\{\xi^{t+1} \in C \mid \xi^0 = x^0, \alpha^0 = a^0, \ldots, \xi^{t-1} = x^{t-1}, \alpha^{t-1} = a^{t-1}, \xi^t = y, \alpha^t = a\}$$

$$\overset{(1)}{=} P\{\xi^{t+1} \in C \mid \xi^0 = x^0, \ldots, \xi^{t-1} = x^{t-1}, \xi^t = y, \alpha^t = a\}$$

$$\overset{(2)}{=} P\{\xi^{t+1} \in C \mid \xi^t = y, \alpha^t = a\}$$

$$\overset{(3)}{=} Q(C \mid y, a), \qquad y \in X, \quad a_j \in A_j(y). \tag{1}$$

ξ will then shortly be called a controlled time dependent random field.

At every time instant $t \in \mathbb{N}$ the players make decisions on the basis of the complete information about the state history and of all preceding decisions of their full neighborhoods. After the joint decision of all players the i-th player (at the i-th vertex) receives a (positive or negative) payoff $u_i\big(x, a_{\widetilde{N}(i)}\big)$, $x \in X$, $a_j \in A_j(x)$, $j \in \widetilde{N}(i)$, where we assume that $u_i \colon X \times X^{|\widetilde{N}(i)|} \to X$ is a bounded continuous utility or payoff function for every player. Thus, for each strategy π and any discount factor $\beta_i \in (0; 1)$ the n-stage discounted expected utility of player i is

$$\varphi_i^n(x, \pi) = \mathsf{E}_x^\pi \sum_{t=1}^n \beta_i^{t-1} u_i\big(x^t, a_{\widetilde{N}(i)}^t\big),$$

where E_x^π means the expectation operator with respect to the probability measure P_x^π. The discounted utility of player i over the infinite future is

$$\varphi_i(x, \pi) = \lim_{n \to \infty} \varphi_i^n(x, \pi).$$

The problem is to find the Nash equilibrium in the following sense.

Definition 4. Let $\pi_{\pi_i \leftrightarrow \pi_i'}$ denotes the strategy of the system where all the players $j \in V \setminus \{i\}$ follow the strategy π_j and i-th player follows the strategy π_i'.
A strategy π is called a Nash equilibrium in the discounted stochastic game iff no unilateral deviations from it are profitable, that is, for each $x \in X$,

$$\varphi_i(x, \pi) \geqslant \varphi_i\big(x, \pi_{\pi_i \leftrightarrow \pi_i'}\big)$$

for every player i and any π'. The functions $V_i(\pi) := \varphi_i(x, \pi)$, $i \in V$, are called the players' value functions for optimally. Nash equilibria for the finite horizon games are defined similarly.

3 Main Results

If $\widetilde{N}(i) = \{i\}$ for all $i \in V$, and if we restrict ourselves to considering deterministic stationary local strategies (LS_D), then all the results of the existence and uniqueness of symmetric strategies as in [7] are easily taken into account in our case.

We assume that utility functions u_i, transition probability Q and consumption-capacity functions A_i satisfy the following:

A1: Utility functions u_i for any $i \in V$ utility functions depend on their arguments such that

$$u_i\big(x, a_{\widetilde{N}(i)}\big) = u_i\bigg(x, a_i, \sum_{j \in N(i)} a_j \bigg).$$

In addition, $u_i\big(x, \cdot, \sum_{j \in N(i)} a_j\big)$ is nondecreasing concave function and $u_i(x, a_i, \cdot)$ is nonincreasing convex function such that $u(0; 0; 0) = 0$.

A2: Q is a transition probability from X to X with cumulative distribution function $F(\cdot \mid x - \sum_{i \in V} a_i)$, i. e.

$$Q([0;y] \mid x, a) := Q\left([0;y] \mid x - \sum_{i \in V} a_i\right) = F\left(y \mid x - \sum_{i \in V} a_i\right).$$

We assume that for each $y \geqslant 0$ $F(y \mid \cdot)$ is a nonincreasing continuous convex function on X and $F(0 \mid 0) = 1$.

A3: For $i \in V$ $A_i(x) = [0; K_i(x)]$ and K_i is nondecreasing, uniformly bounded above by some constant $C_i \in [0; +\infty)$, and satisfies $K_i(0) = 0$ for all i, $\sum_{i \in V} K_i(x) \leqslant x$ for all $x \in X$.

Theorem 1. *Every controlled stochastic process with locally interacting synchronous components satisfying **A1**–**A3** has a nonrandomized Markov Nash equilibrium $\pi \in LS_D$.*

Proof. Let $B(X)$ be the space of all bounded nonnegative Borel measurable functions $v \colon X \to \mathbb{R}$ such that $v(0) = 0$. Let $\pi \in LS_D$. Put

$$(T_\pi v_i)(x) = u_i\left(x, \pi_i(x), \sum_{j \in N(i)} \pi_j(x)\right) + \beta_i \int_X v_i(x') dF(x' \mid y), \quad i \in V \quad (2)$$

where $y = x - \sum_{i \in V} \pi_i(x)$. Since $v_i(0) = 0$, from **A3**, it follows that $T_\pi v_i \in B(X)$. With any $v_i \in B(X)$, $i \in V$, and $x \in X$, we associate the one-stage discounted stochastic game in which the payoff function for every player i is

$$u_i^1(v_i, x, a) = u_i\left(x, a_i, \sum_{j \in N(i)} a_j(x)\right) + \beta_i \int_X v_i(x') dF\left(x' \mid x - \sum_{i \in V} a_i\right). \quad (3)$$

According to standard dynamic programming arguments we start with studying finite horizon stochastic games. To find Nash equilibria we use the backward induction procedure [14].

Let n be a horizon of the game. Let $\pi_i^1(x) := K_i(x)$ for each $x \in X$ and $i \in V$. Then

$$v_i^1(x) = \max_{a_i \in A_i(x)} u_i\left(x, a_i, \sum_{j \in N(i)} a_j\right) = u_i\left(x, K_i(x), \sum_{j \in N(i)} K_j(x)\right).$$

If $v_i^0(x) := 0$ for each $x \in X$ and $i \in V$, then $\pi^1(x)$ is a Nash equilibrium in the one-stage game, v^1 is the equilibrium payoff for each player and $v_i^1 = T_{\pi^1} v_i^0$.

Similarly we can define $\pi^2, \ldots, \pi^n \in LS_D$ and $v^2, \ldots, v^n \in B(X)$ as follows $v_i^k(x) := (T_{\pi^k} v_i^{k-1})(x)$, $k = 2, \ldots, n$. Then optimal n-step Markov strategies for the player i $\pi_i^{(n)} = (\pi_i^n, \pi_i^{n-1}, \ldots, \pi_i^1)$.

For infinite horizon discounted stochastic game we have

$$v_i^\star(x) = \lim_{n\to\infty} v_i^n(s) = \lim_{n\to\infty} (T_{\pi^n} v_i^{n-1})(x)$$

$$= u_i\left(x, \pi_i^\star(x), \sum_{j\in N(i)} \pi_j^\star(x)\right) + \beta_i \int_X v_i^\star(x')dF\left(x' \mid x - \sum_{i\in V} \pi_i^\star(x)\right)$$

$$= \max_{a_i\in[0;A_i(x)]} \left\{ u_i\left(x, a_i, \sum_{j\in N(i)} \pi_j^\star(x)\right)\right.$$

$$\left. + \beta_i \int_X v_i^\star(x')dF\left(x' \mid x - a_i - \sum_{j\in V\setminus\{i\}} \pi_j^\star\right)\right\}, \quad i \in V.$$

Example 1. Let it be given graph $\Gamma = (V, B)$ with $V = \{v_1, v_2, v_3\}$, $B = \{(v_1, v_2), (v_2, v_3)\}$. Then $\widetilde{N}(1) = \{v_1, v_2\}$, $\widetilde{N}(2) = \{v_1, v_2, v_3\}$, $\widetilde{N}(3) = \{v_2, v_3\}$. Put $X = [0; 1]$,

$$u_i\left(x, a_i, \sum_{j\in N(i)} a_j\right)$$

$$= \frac{x}{3}\left[1 - \left(\frac{x}{3} - a_i\right) \sum_{j\in N(i)} a_j - \frac{1}{2}\left(\frac{x}{3} - a_i\right)^2 \left(1 - \sum_{j\in N(i)} a_j\right)^2\right], \quad i \in V,$$

$K_i(x) = \frac{x}{3}$ for all i and

$$F(x \mid y) = x\left(1 + y(x - 1) + \frac{1}{2}y^2(x - 1)^2\right)$$

for $x \in X$, $y = x - \sum_{j\in V} a_j > 0$. Clearly, that $v_i^1(x) = \frac{x}{3}$ for all i and $\pi^1 = \left(\frac{x}{3}, \frac{x}{3}, \frac{x}{3}\right)$.

In order to find v^2 and π^2, we need to solve the system of equations

$$v_i^2(x) = \max_{a_i\in[0;\frac{x}{3}]} \left\{\frac{x}{3}\left[1 - \left(\frac{x}{3} - a_i\right) \sum_{j\in N(i)} a_j - \frac{1}{2}\left(\frac{x}{3} - a_i\right)^2 \left(1 - \sum_{j\in N(i)} a_j\right)^2\right]\right.$$

$$\left. + \beta_i \int_0^1 \frac{x'}{3} d\left(x'\left(1 + y(x' - 1) + \frac{1}{2}y^2(x' - 1)^2\right)\right)\right\}$$

$$= \max_{a_i\in[0;\frac{x}{3}]} \left\{\frac{x}{3}\left[1 - \left(\frac{x}{3} - a_i\right) \sum_{j\in N(i)} a_j - \frac{1}{2}\left(\frac{x}{3} - a_i\right)^2 \left(1 - \sum_{j\in N(i)} a_j\right)^2\right]\right.$$

$$\left. + \frac{\beta_i}{3}\left(\frac{1}{2} + \frac{1}{6}y - \frac{1}{24}y^2\right)\right\}, \quad i = 1; 2; 3.$$

4   Conclusion

The Nash equilibrium-situation gives each player at least his guaranteed gain, although Nash equilibrium-strategies may not be cautious and the more Pareto-optimal (i.e., "The Prisoner's Dilemma"). Moreover, if each Nash equilibrium situation is Pareto-optimal, then the coexistence of several different Pareto situations creates a struggle for leadership, which makes it impossible to find "optimal" strategies. Therefore, the problem class considered in the work is only a small particle of a large problem. After the introduction of the concept of equilibrium by Nash, many specialists tried to formulate additional conditions for choosing a single equilibrium. Some of these concepts were proposed by the Nobel Prize in Economics in 1994 by Harsanyi and Selten [15] and by the Nobel Prize in Economics in 2005 Aumann [16] and Schelling [17].

References

1. Knopov, P.S., Samosonok, A.S.: On Markov stochastic processes with local interaction for solving some applied problems. Cybern. Syst. Anal. **47**(3), 346–363 (2015)
2. Chornei, R.K., Daduna, H., Knopov, P.S.: Control of Spatially Structured Random Processes and Random Fields with Applications. Springer, Boston (2006)
3. Levhari, D., Mirman, L.: The great fish war: an example using a dynamic Cournot-Nash solution. ACA Trans. **11**, 322–334 (1980)
4. Sundaram, R.K.: Perfect equilibrium in non-randomized strategies in a class of symmetric dynamic games. J. Econ. Theory. **47**, 153–177 (1989)
5. Majumdar, M., Sundaram, R.K.: Symmetric stochastic games of resource extraction: existence of nonrandomized stationary equilibrium. In: Raghavan, T.E.S., Ferguson, T.S., Parthasarathy, T., Vrieze, O.J. (eds.) Stochastic Games And Related Topics. Theory and Decision Library (Game Theory, Mathematical Programming and Operations Research), vol. 7, pp. 175–190. Springer, Dordrecht (1991)
6. Dutta, P.K., Sundaram, R.K.: Markovian equilibrium in a class of stochastic games: existence theorems for discounted and undiscounted models. Econ. Theory **2**, 197–214 (1992)
7. Balbus, L., Nowak, A.S.: Construction of nash equilibria in symmetric stochastic games of capital accumulation. Math. Meth. Oper. Res. **60**, 267–277 (2004)
8. Więcek, P.: Continuous convex stochastic games of capital accumulation. In: Nowak, A.S., Szajowski, K. (eds.) Advances in Dynamic Games. Annals of the International Society of Dynamic Games, vol. 7, pp. 111–125. Birkhäuser (2005)
9. Amir, R.: Continuous stochastic games of capital accumulation with convex transitions. Games Econ. Behav. **15**, 111–131 (1996)
10. David, P.A., Foray, D.: Percolation structures, Markov random fields. the economics of EDI standards diffusions. In: Pogorel, G. (ed.) Global Telecommunications Strategies and Technological Changes. North-Holland, Amsterdam (1993). First version: Technical report, Center for Economical Policy Research, Stanford University (1992)
11. David, P.A., Foray, D., Dalle, J.M.: Marshallian externalities and the emergence and spatial stability of technological enclaves. Econ. Innov. New Technol. **6**(2–3), 147–182 (1998). First version: Preprint, Center for Economical Policy Research, Stanford University (1996)
12. Voit, J.: The Statistical Mechanics of Financial Markets. Theoretical and Mathematical Physics, 3rd edn. Springer, Heidelberg (2005)
13. Federgruen, A.: On n-person stochastic games with denumerable state space. Adv. Appl. Probab. **10**, 452–471 (1978)
14. Rieder, U.: Equilibrium plans for nonzero-sum Markov games. In: Moeschlin, O., Pallaschke, D. (eds.) Game Theory and Related Topics, pp. 91–102. North-Holland (1979)

15. Harsanyi, J.C., Selten, R.: A General Theory of Equilibrium Selection in Games. MIT Press, Cambridge (1988)
16. Aumann, R.: Lectures on Game Theory. Underground Classics in Economics. Westview Press, Boulder (1989)
17. Schelling, T.: The Strategy of Conflict. Harvard University Press, Cambridge (1980)

Indoor and Outdoor Air Quality Monitoring on the Base of Intelligent Sensors for Smart City

Andrii Shelestov[1,2,3], Leonid Sumilo[2,3]([✉]), Mykola Lavreniuk[1,2,3],
Vladimir Vasiliev[3], Tatyana Bulanaya[4,5], Igor Gomilko[4,5],
Andrii Kolotii[1,2,3], Kyrylo Medianovskyi[1,2], and Sergii Skakun[6]

[1] Space Research Institute NASU-SSAU, Kyiv, Ukraine
andrii.shelestov@gmail.com
[2] Igor Sikorsky Kyiv Polytechnic Institute, Kyiv, Ukraine
shumilo.leonid@gmail.com
[3] EOS Data Analytics, Kyiv, Ukraine
[4] Noosphere Engineering School, Dnipro, Ukraine
[5] Oles Honchar Dnipro National University, Dnipro, Ukraine
[6] University of Maryland, College Park, MD, USA

Abstract. People experience the problems of air quality every day, either inside or outdoors. The best solution to mitigate the problem inside the buildings is to open opening the windows. It is not just the most efficient, but also the cheapest solution. However, opening windows might only worsen the situation in the room in the case of excess air pollutants in the big cities. Consequently, one should use another method of improving air quality inside. Often people cannot recognize whether air quality is good enough inside, therefore, there is a need for a system which could monitor the air conditions inside and outside the buildings, analyze it and give recommendations for improving the air quality. Air quality monitoring is one of the important topics of the SMURBS/ERA-PLANET project within the European Commission's Horizon-2020 program. This study addresses the problem of using remote sensing data and Copernicus ecological biophysical models for air quality assessment in the city, and proposes the intelligent solution based on indoor and outdoor sensors for air quality monitoring controlled by a fuzzy logic decision block. We are planning to implement the distributed system in the framework of the Smart City concept in Kyiv (Ukraine) within the SMURBS project.

Keywords: Air quality index · Sensors · Amazon Web Service
Fuzzy logic · Adaptive neuro-fuzzy inference system · Smart City
SMURBS

© Springer Nature Switzerland AG 2019
O. Chertov et al. (Eds.): ICDSIAI 2018, AISC 836, pp. 134–145, 2019.
https://doi.org/10.1007/978-3-319-97885-7_14

1 Introduction

The Group on Earth Observation (GEO) has started the biggest initiative related to geospatial data usage for different social benefit areas since 2005[1]. Inter-governmental and Ministerial bodies from different countries and international organizations supported these activities. The year of 2015 marked the second decade of GEOSS (Global Earth Observation System of Systems) development, and a large number of projects and programs in different domains of environmental monitoring and Earth observation have generated an enormous amount of data and products on different aspects related to environmental quality and sustainability.

Now Big Data generated by in-situ or satellite platforms are being collected and archived by many systems within the European Copernicus Program. It increasingly makes difficult the sharing data to stakeholders and policy makers for supporting key economic and societal sectors. A number of the European projects are dedicated to define social benefit areas and, in particular within "Biodiversity and ecosystem sustainability" and "Disaster resilience"[2].

The main goal of the ERA-PLANET project is to deploy the European Research Area in the domain of Earth Observations (see footnote 2). This HORIZON 2020 project should be a Europe's impact into GEO activities and GEOSS system development and support. All these activities are interdisciplinary, depend heavily on Copernicus data, and should provide more accurate and comprehensive information and the new level of the decision-making support for key societal benefit areas.

One of the most important goals is a strong support of the GEOSS-related technological approaches. There are different strands within the ERA-PLANET project. One of them is the SMURBS project ("SMart URBan Solutions for air quality, disasters and city growth" Strand[3]), which is dedicated to smart cities solutions development, resilient societies, resource efficiency, environmental state estimations/management and human well-being provision at the city level.

Ukraine is one of the active participants within the ERA-PLANET and particularly SMURBS. Kyiv was selected as one of the pilot cities for technologies development and validation within the SMURBS project. This experience will be further expanded to other cities in Europe.

Pure natural air without impurities like dust, harmful aerosols, gases and vapors is essential for normal human life. The quality of both outdoor and indoor air is important, especially within large cities. The data on air quality provided by Copernicus have a low frequency and low accuracy. That is why, it is very important to have sensors, which can be used for real chemical composition estimation of the air. In our study, we show this by comparing Copernicus data with measurements of the Central Geophysical

---

[1] Group on Earth Observation, https://www.earthobservations.org/.

[2] The European Network of ovserving our Changing Planet, http://www.era-planet.eu.

[3] Smart Urban Solutions for Air Quality, Disasters and City Growth, http://www.smurbs.eu.

Observatory (CGO), which is responsible for air quality in Kyiv City. Another option is to deploy our own sensor solution and to validate different in-situ data sources.

It is necessary to monitor the concentration of CO_2, in addition temperature and humidity controlling [1, 2]. At the end of 1980's the average level of CO_2 in the air was around 350 ppm (ppm—part per million, 0.03%), while nowadays it is about 400 ppm on average. A person feels absolutely fine, when CO_2 level is about 700 ppm or below. At the level of 1000 ppm, one feels that air is getting stale and stuffy. When its level reaches 2500 ppm, it gets stuffier and stuffier (despite the fact that oxygen level changes slightly, decreases from 20% down to 19.75%), and further it gets worse and leads to the appearance of above-mentioned symptoms of oxygen deprivation. In other words, when one thinks that it's stuffy around and has a lack of oxygen, in fact, there's a huge level of carbon dioxide in the air. Thus, carbon dioxide is one of the factors, which influences working capacity.

In this paper, we discuss the smart solutions for air quality monitoring on the base of smart sensors. The Noosphere Engineering School has an experience in smart sensors development and the proposed solution is built on this background. In particular, the prototype of indoor air quality sensor has been developed and tested.

The proposed system consists of indoor and outdoor air quality sensors connected to a fuzzy logic based decision making system. A microcontroller system collects data from these sensors, processes it and conveys with PC-information to the cloud-based storage. They are coordinated by a cloud-based managing system, where data are collected, stored and analyzed. An intelligent controller will analyze the data and suggest a smart solution for indoor aeration. We plan to use semiconductor, electro-chemical and optical sensors of different companies. The Amazon Web Services (AWS) is used as a cloud-based solution for Big data processing and analysis.

2   Indoor Air Quality Monitoring

To assess CO_2 concentration, we have developed a device with the block structure shown in Fig. 1. Except for measuring carbon dioxide concentration, this device provides values for pressure, temperature and humidity. Besides that, collected data can be sent to a personal computer or mobile phone via Bluetooth interface.

The change of CO_2 concentration within the room during an hour and a half is presented in Fig. 2. Concentration of CO_2 increases rapidly in the room without airing as depicted in Fig. 2.

Analysis of known sources of information [1–4] allows us to conclude the following. On average, one person increases CO_2 level for 300 ppm in 100 m^3 during 1 h. Airing the room helps to normalize CO_2 concentration very quickly (usually within 5 min).

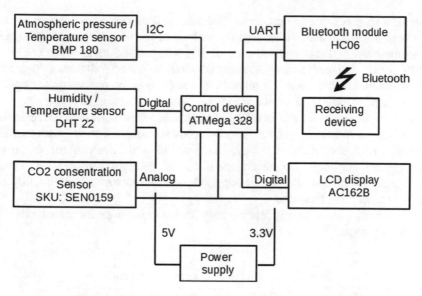

Fig. 1. Structural scheme for proposed air pollution measurement device

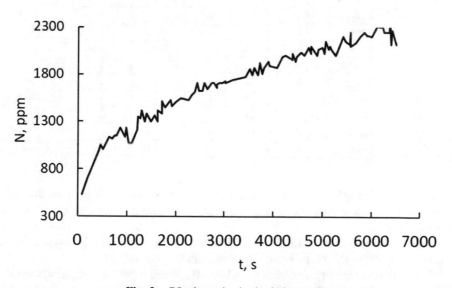

Fig. 2. CO_2 dynamics in the indoor area

3 Outdoor Air Quality Monitoring

The quality of the outdoor air is estimated by using an air quality index (AQI). AQI is a value used by governmental agencies in order to inform people about current air pollution level. If the index increases, a significant part of population will face with serious health consequences. Every country has its own AQIs according to different

138   A. Shelestov et al.

national standards. Table 1 summarizes air quality indicators used in different countries [3], where air contaminants measured for analysis of the air quality are represented.

As indicated in Table 1, to create a device, which could monitor the outdoor air quality, we need to analyze the following compounds: nitrogen dioxide (NO_2), ozone (O_3), particulate matter up to 2.5 microns in size ($PM_{2.5}$), particulate matter up to 10 microns in size (PM_{10}), sulfur dioxide (SO_2) and carbon monoxide (CO).

Nitrogen dioxide is characterized by adverse health effects. At high concentrations, it can affect the functioning of the lungs, exacerbates asthma and increases mortality.

Studies have shown that short-term exposure to peak levels of ozone can temporarily affect the lungs, the respiratory tract, and the eyes, and can also increase susceptibility to inhaled allergens. Long term exposure to ozone has primarily been found to reduce lung function.

PM2.5 can penetrate deeply into the lung, irritate and corrode the alveolar wall, and consequently impair lung function.

Table 1. Air quality indicators in different countries.

Country	NO_2	O_3	$PM_{2.5}$	PM_{10}	SO_2	CO	NH_3	Pb
Canada	+	+	+					
Hong-Kong	+	+	+	+	+			
China	+	+	+	+	+	+		
India	+	+	+	+	+	+	+	+
Mexico	+	+	+	+	+	+		
Singapore	+	+	+	+	+	+		
South Korea		+	+					
Great Britain	+	+	+	+	+			
Europe	+	+		+				
USA	+	+		+	+	+		

Sulfur dioxide affects human health when it is breathed in. It irritates the nose, throat, and airways to cause coughing, wheezing, shortness of breath, or a tight feeling around the chest.

Carbon monoxide is harmful when breathed because it displaces oxygen in the blood and deprives the heart, brain, and other vital organs of oxygen.

For outdoor air quality monitoring we have developed a device measuring the most of indicators mentioned above. The block structural diagram of this device is shown in Fig. 3.

The main functionality of each sensor is based on definite electrical dependencies. Thus, the MICS-4514 (sensor, which we are using in our system) sensor's typical temperature dependencies of the electrical conductivity at various CO concentrations in the air are shown in Fig. 4. When the temperature of the sample increases, its sensitivity is initially increased; however, at a temperature of about 550 K, the sensitivity decreases.

Also, the kinetic dependencies of electrical conductivity were measured as the data of the gas-sensing sensor. Typical kinetic dependencies based on different CO concentration values in the air are shown in Fig. 5.

Similar characteristics are taken for all other sensors. Sensor data allows us to get air quality.

The developed system functions using the following logic: inside and outside sensors measure indicators of the air quality. A decision making block is also located inside. All the sensors are connected to this block and it has an access to the data storage and the user interface (Fig. 8). Local multi-sensor systems are coordinated by a cloud-based managing system where all the data are sent, stored and analyzed.

Outside sensor measures indicators of nitrogen dioxide (NO_2), ozone (O_3), particulate matter up to 2.5 microns in size ($PM_{2.5}$), particulate matter up to 10 microns in size (PM_{10}), sulfur dioxide (SO_2) and carbon monoxide (CO). The general indoor sensors measure CO_2 concentration in addition to temperature and humidity.

Obtaining data on the air quality in large cities as Kyiv faces two major challenges. The first issue is representativeness and reliability of the collected data. In fact, the only analogue of the data from the installed air quality measurement sensors in the city are data from the model solutions of the Copernicus service. But because these data are updated once a day only, in reality it's impossible to accurately reflect the state of the air in the city in real time, that is, data are losing their relevance.

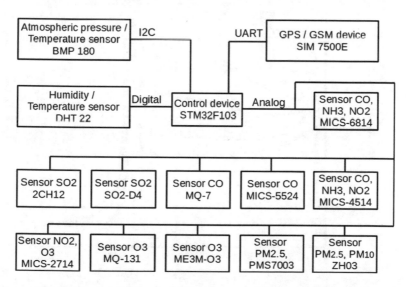

Fig. 3. Structural diagram for the proposed AQI measurement device

Using measurements of the CGO from January 1st until January 22nd, we analyzed the data provided by the Copernicus service for the Kyiv city. We will consider nitrogen dioxide and sulfur dioxide concentration indicators mg/m^3 (Figs. 6 and 7). As a result, we received a very low correlation between these data, and since the data of the CGO can be considered as the ground truth measurement data, we came to the conclusion that low reliability of Copernicus data for Kyiv.

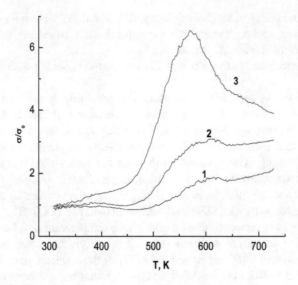

Fig. 4. The MICS-4514 sensor's temperature dependencies of the normalized electrical conductivity at CO concentrations in the air, ‰: 1–10; 2–50; 3–80

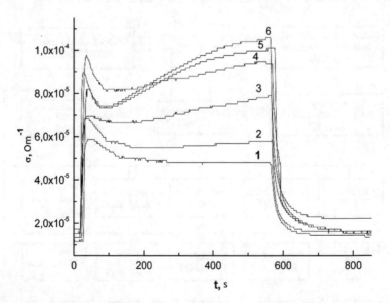

Fig. 5. The MICS-4514 sensor's kinetics of electrical conductivity at a temperature of 580 K at CO concentrations in the air, ‰: 1–10; 2–20; 3–30; 4–50; 5–70; 6–80

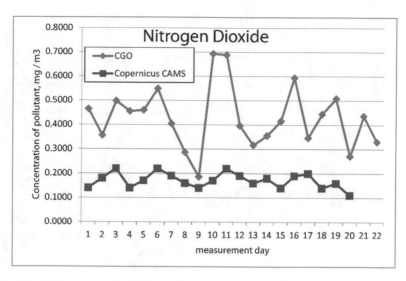

Fig. 6. Kyiv NO2 concentration with trend lines, bottom line—Copernicus data, top line—CGO measurements

4 Decision Making System

All sensors transfer collected measurements to a control and decision making block which sends information to a cloud-based storage AWS, where it is processed and analyzed [3]. The basis of the automatic decision making system will be the Sugeno-Takagi controller type with 12 inputs with information on current status of indoor and outdoor air quality indicators as input. Using adaptive neuro-fuzzy inference system output from the Sugeno-Takagi controller, the decision making block can inform the user [1] about the conditions of the air inside the room, in which the person is located, and outdoor, and give him/her an advice on the possibility of improving it (Fig. 9). This system will be useful for premises of any destination and will be able to improve the state of the microclimate of any workplace or living space, will help improve the well-being of people and inform them about dangerous air pollutants [4] in the building and on the street.

Outdoor scores from all established sensors will be used for implementation of the Kiev Smart City Air Monitoring System for decision making with help of fuzzy-ellipsoid method for robust estimation of the state of a grid system node [8]. Also, this methodology is useful in more global tasks of decision making and risk calculation in city and out of the city [6, 7].

The Decision Making System (Fig. 9) can utilize different intellectual methods for data processing, for instance deep learning [8, 9], neural network classification [10–12], risk assessment [7, 13]. These information technologies have been used successfully for satellite environmental monitoring [14, 15].

To provide the efficiency of distributed data processing we propose to use AWS as a cloud-based solution for managing system and enabling implementation for

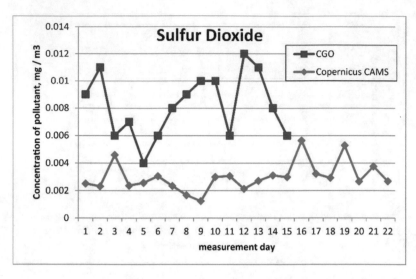

Fig. 7. Kyiv SO2 concentration with trend lines, bottom line—Copernicus data, top line—CGO measurements

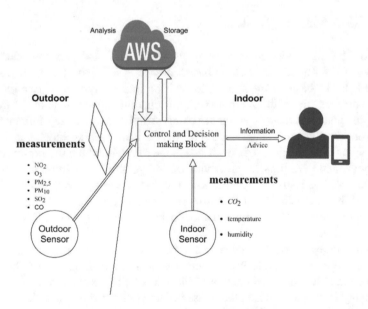

Fig. 8. A high level architecture of the air quality monitoring system with the user interface

geospatial information systems [16] and models [17, 18]. Since AWS provides fast access to satellite data, in the future the system will be extended by the use of satellite air quality products.

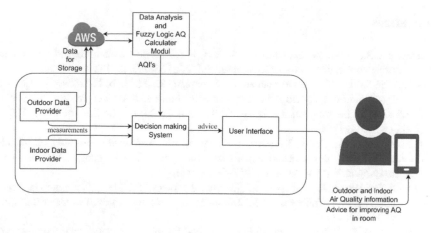

Fig. 9. Scheme of control and decision making block

5   Conclusions and Discussion

After Copernicus data validation we can conclude, that these data have spatial resolution which is not applicable to the measurement of air quality at the city level (at least at the moment). That is why, we believe that the installation of such sensors is the best solution to the problem of monitoring the quality of air in the city.

In this study, we proposed the intelligent solution based on indoor and outdoor sensors for air quality monitoring controlled by fuzzy logic decision block and have shown the need to use ground sensors to monitor air quality. The system is developed for air quality monitoring inside and outside the house, data analysis and providing recommendations for improving the air quality inside the room. The system consists of smart sensors as well as decision making system based on intelligent computations and implemented in Amazon cloud environment with Amazon Web Service technology. Air quality monitoring is one of the important topics of SMURBS/ERA-PLANET project within Horizon-2020 program. We are going to implement the distributed system in the framework of Smart City concept in Kyiv within SMURBS project.

On the other hand, each city is not autonomous system, which does not depend on any external conditions. So, satellite-derive products, for example on the base of Sentinel-5P data with 7×3.5 km spatial resolution, will be very useful in future for getting complex estimations of the air quality.

Acknowledgment. The authors acknowledge the funding received by ERA-PLANET (www. era-planet.eu), trans-national project SMURBS (www.smurbs.eu) (Grant Agreement No. 689443), funded under the EU Horizon 2020 Framework Program.

References

1. Abraham, S., Li, X.: A cost-effective wireless sensor network system for indoor air quality monitoring applications. Procedia Comput. Sci. **34**, 165–171 (2014)
2. Muhamad Salleh, N., Kamaruzzaman, S., Sulaiman, R., Mahbob, N.: Indoor air quality at school: ventilation rates and it impacts towards children. A review. In: 2nd International Conference on Environmental Science and Technology 2011, IPCBEE, vol. 6, pp. 418–422 (2011)
3. Pan, Y., Zhou, H., Huang, Z., Zeng, Y., Long, W.: Measurement and simulation of indoor air quality and energy consumption in two Shanghai office buildings with variable air volume systems. Energy Build. **35**, 877–891 (2003)
4. Orwell, R., Wood, R., Burchett, M., Tarran, J., Torpy, F.: The potted-plant microcosm substantially reduces indoor air VOC pollution: II. Laboratory study. Water Air Soil Pollut. **177**, 59–80 (2006)
5. Shelestov, A.Y.: Using the fuzzy-ellipsoid method for robust estimation of the state of a grid system node. Cybern. Syst. Anal. **44**(6), 847–854 (2008)
6. Kussul, N., Shelestov, A., Skakun, S., Kravchenko, O.: Data assimilation technique for flood monitoring and prediction. Institute of Information Theories and Applications FOI ITHEA (2008)
7. Skakun, S., Kussul, N., Shelestov, A., Kussul, O.: The use of satellite data for agriculture drought risk quantification in Ukraine. Geomat. Nat. Hazards Risk **7**(3), 901–917 (2016)
8. Mandl, D., Frye, S., Cappelaere, P., Handy, M., Policelli, F., Katjizeu, M.: Use of the earth observing one (EO-1) satellite for the namibia sensorweb flood early warning pilot. IEEE J. Sel. Top. Appl. Earth Obs. Remote Sens. **6**(2), 298–308 (2013)
9. Kussul, N., Lavreniuk, M., Skakun, S., Shelestov, A.: Deep learning classification of land cover and crop types using remote sensing data. IEEE Geosci. Remote Sens. Lett. **14**(5), 778–782 (2017)
10. Lavreniuk, M., Skakun, S., Shelestov, A., Yalimov, B., Yanchevskii, S., Yaschuk, D., Kosteckiy, A.: Large-scale classification of land cover using retrospective satellite data. Cybern. Syst. Anal. **52**(1), 127–138 (2016)
11. Skakun, S., Kussul, N., Shelestov, A., Lavreniuk, M., Kussul, O.: Efficiency assessment of multitemporal C-band Radarsat-2 intensity and Landsat-8 surface reflectance satellite imagery for crop classification in Ukraine. IEEE J. Sel. Top. Appl. Earth Obs. Remote Sens. **9**(8), 3712–3719 (2016)
12. Kussul, N., Lemoine, G., Gallego, F., Skakun, S., Lavreniuk, M., Shelestov, A.: Parcel-based crop classification in Ukraine using Landsat-8 data and Sentinel-1A data. IEEE J. Sel. Top. Appl. Earth Obs. Remote Sens. **9**(6), 2500–2508 (2016)
13. Lavreniuk, M., Kussul, N., Meretsky, M., Lukin, V., Abramov, S., Rubel, O.: Impact of SAR data filtering on crop classification accuracy. In: 2017 First Ukraine Conference on Electrical and Computer Engineering, pp. 912–916 (2017)
14. Kussul, N., Sokolov, B., Zyelyk, Y., Zelentsov, V., Skakun, S., Shelestov, A.: Disaster risk assessment based on heterogeneous geospatial information. J. Autom. Inf. Sci. **42**(12), 32–45 (2010)
15. Gallego, J., Kravchenko, A., Kussul, N., Skakun, S., Shelestov, A., Grypych, Y.: Efficiency assessment of different approaches to crop classification based on satellite and ground observations. J. Autom. Inf. Sci. **44**(5), 67–80 (2012)
16. Kussul, N., Skakun, S., Shelestov, A., Kussul, O.: The use of satellite SAR imagery to crop classification in Ukraine within JECAM project. In: IGARSS 2014, pp. 1497–1500 (2014)

17. Shelestov, A., Kravchenko, A., Skakun, S., Voloshin, S., Kussul, N.: Geospatial information system for agricultural monitoring. Cybern. Syst. Anal. **49**(1), 124–132 (2013)
18. Kogan, F., Kussul, N., Adamenko, T., Skakun, S., Kravchenko, O., Kryvobok, O., Shelestov, A., Kolotii, A., Kussul, O., Lavrenyuk, A.: Winter wheat yield forecasting in Ukraine based on Earth observation, meteorological data and biophysical models. Int. J. Appl. Earth Obs. Geoinf. **23**, 192–203 (2013)

Assessment of Sustainable Development Goals Achieving with Use of NEXUS Approach in the Framework of GEOEssential ERA-PLANET Project

Nataliia Kussul[1,2], Mykola Lavreniuk[1,2,3(✉)], Leonid Sumilo[1,2,3],
Andrii Kolotii[1,2,3], Olena Rakoid[4], Bohdan Yailymov[1,2,3],
Andrii Shelestov[1,2,3], and Vladimir Vasiliev[3]

[1] Space Research Institute NASU-SSAU, Kyiv, Ukraine
`nick_93@ukr.net`
[2] Igor Sikorsky Kyiv Polytechnic Institute, Kyiv, Ukraine
[3] EOS Data Analytics, Kyiv, Ukraine
[4] National University of Life and Environmental Sciences of Ukraine,
Kyiv, Ukraine

Abstract. In this paper, we propose methodology for calculating indicators of sustainable development goals within the GEOEssential project, that is a part of ERA-PLANET Horizon 2020 project. We consider indicators 15.1.1 Forest area as proportion of total land area, 15.3.1 Proportion of land that is degraded over total land area, and 2.4.1. Proportion of agricultural area under productive and sustainable agriculture. For this, we used remote sensing data, weather and climatic models' data and in-situ data. Accurate land cover maps are important for precisely land cover changes assessment. To improve the resolution and quality of existing global land cover maps, we proposed our own deep learning methodology for country level land cover providing. For calculating essential variables, that are vital for achieving indicators, NEXUS approach based on idea of fusion food, energy, and water was applied. Long-term land cover change maps connected with land productivity maps are essential for determining environment changes and estimation of consequences of anthropogenic activity.

Keywords: ERA-PLANET · Classification maps · Essential variables
GEOEssential

1 Introduction

Aimed at reaching Sustainable Development Goals (SDGs) accepted in Sendai Framework for 2015–2030, ERA-PLANET project implements use of earth observation data in tasks of Environmental Management[1,2]. In this paper, we present a methodology for calculating indicators of reaching the following goals: end hunger, achieve food security and improved nutrition, promote sustainable agriculture aimed on

---

[1] Group on Earth Observation, https://www.earthobservations.org/.
[2] The European Network of observing our Changing Planet, http://www.era-planet.eu.

© Springer Nature Switzerland AG 2019
O. Chertov et al. (Eds.): ICDSIAI 2018, AISC 836, pp. 146–155, 2019.
https://doi.org/10.1007/978-3-319-97885-7_15

ensuring sustainable food production systems and implementing resilient agricultural practices [1–5].

Our main targets are ensuring the conservation, restoration and sustainable use of inland freshwater ecosystems and their services; combating desertification by 2020; restoration of degraded lands by 2030. These goals can be indicated by proportion of agricultural area under productive and sustainable agriculture, the forest area as a proportion of total land area, and the proportion of land that is degraded over total land area.

For calculating indicators for these goals, we use land cover maps [6–9], NDVI trend productivity change map, essential variables by output of our NEXUS approach model, which include stochastic climate simulation model, and biophysical model for crop simulation WOFOST[3].

There are many global and open sources for land cover data, that require a time series of satellite and ground measurements that can be used in research purpose. But in this study, we used land cover maps, achieved by our machine learning methodology for Ukraine territory due to higher accuracy on a country scale and better spatial resolution in 10 m.

2 Data

In study of land cover change, we use classification maps built for territory of Ukraine, using methods of machine learning developed by us for classification of time series of Sentine-1 and Sentinel-2 images with 10 m spatial resolution. For the territory of Ukraine, 9 paths of Sentinel-1 were used. Each of these paths consist of several images and is constructed by merging for one date (Fig. 1). For learning and validation of model, we used in-situ data collected by us. More than 800 images were used for covering the territory of Ukraine with Sentinel-1A data during the vegetation season. The data amount used for land cover classification for 2016 and 2017 is over 29 Tb in total.

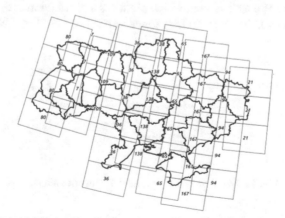

Fig. 1. Coverage of the territory of Ukraine by Sentinel-1A data with relative orbit number

---

[3] BioMA (Biophysical Models Applications) framework, http://bioma.jrc.ec.europa.eu/.

In case of use biophysical models and climatic models in NEXUS approach, we used time series of weather data provided by Copernicus and NASA-POWER that begin from year 1997, and that include daily minimal and maximal temperature, daily wind speed, daily precipitation, daily incoming global shortwave radiation, daily mean vapor pressure, and daily mean wind speed. All the data were collected from open data providers such as NASA Prediction of Worldwide Energy Resource and Copernicus for 100 km grid for Ukraine.

3 Methodology

3.1 Deep Learning Approach for Land Cover Classification

For providing more accurate land cover maps on the country level, we proposed methodology based on deep learning idea that is very popular nowadays in other tasks like voice recognition, face detection, buildings detection, and so on. The most popular deep learning approaches are convolutional neural networks (CNN), deep auto-encoders (DAE), recurrent neural network with Long Short-Term Memory (LSTM) model and deep belief networks (DBN). They outperform traditional classification approaches such as random forest (RF), support vector machines (SVM), and multi-layer perceptrons (MLP) [10–13].

Taking into account that land cover proving task is similar to visual pattern recognition tasks, in this paper, we propose a deep learning architecture in the form of ensembles of CNNs (4 neural networks) for optical and SAR time-series classification (Fig. 2) [14]. Each CNN from the committee is a 2-d CNN with 7 layers implemented within the Google's library TensorFlow. Different CNN has been trained on different subset of training data. In this experiment, two convolutional layers were used with kernel size 5×5 and 3×3, respectively. After each convolutional layer, a max-polling layer follows with kernel size 2×2. Finally, CNN ends with two fully connected layers and softmax function, which produces a distribution over the output class labels. The rectified linear unit (ReLU) function has been used in CNN architecture.

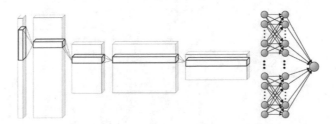

Fig. 2. Deep convolutional neural network architecture

All the images were preprocessed using standard technology. For optical images, we provided atmospheric correction and cloud masking using sen2cor tool, for SAR data we utilized radiometric correction, filtration, terrain correction, converting to backscatter coefficients with ESA SNAP toolbox. After preprocessing step, time series of optical and SAR data were formed based on in-situ data. Further, these data were divided into sets for training phase as inputs into classifier, for validation, and independent set for final test. During training phase, validation set was used for best hyper-parameters selection.

Taking into account large volume of satellite data that are necessary for land cover classification and in order to decrease the time of satellite imagery downloading, it makes sense to deploy a classification system in the cloud environment, for example, Amazon, where Sentinel-1 and Sentinel-2 data are already available for free. An architecture component diagram of the proposed classification system is shown in Fig. 3. With such cloud-based approach, geospatial research can be moved from data to analysis with way smaller delay from the beginning of the study.

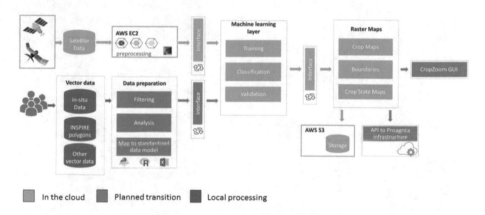

Fig. 3. Typical architecture component diagram for crop classification system

3.2   NEXUS Approach for Calculating Essential Variables

The core part of our NEXUS approach model that provides essential variables for food, water, and energy, is a WOFOST biophysical model, which is supplemented by stochastic climate simulation model and weather data from global models' products of Copernicus and NASA-POWER. We combine these results with models, NDVI productivities maps, and land cover maps in one data processing chain for essential variables calculating like in Fig. 4.

WOFOST Model with data used on a regular grid covering the territory of Ukraine is also named Crop Growth Modeling System (CGMS). During WOFOST model simulation of crop growth, we accept values of total biomass and leaf area index for every day, on levels of sub models like soil water content model, soil temperature model, etc. In this simulation, we also get values for water and energy essential variables like soil moisture, soil temperature evaporation and evapotranspiration [15, 16].

Stochastic climate simulation model is a statistical model based on Markov chains constructed using time series of weather data. In realization of this part, we face the problem of the lack of complete global data that cover the territory of Ukraine and contains all the necessary components such as daily minimal and maximal temperature, daily wind speed, daily precipitation, daily incoming global shortwave radiation, daily mean vapor pressure, and daily mean wind speed. Our weather data cover for Ukraine contain time series built from Copernicus and NASA-POWER data, starting from year 1997, and this is enough for use in stochastic climate model for long range statistical forecast of these weather parameters. These weather parameters not only are needed for WOFOST model simulation, they are separate essential variables for water and energy [17–23].

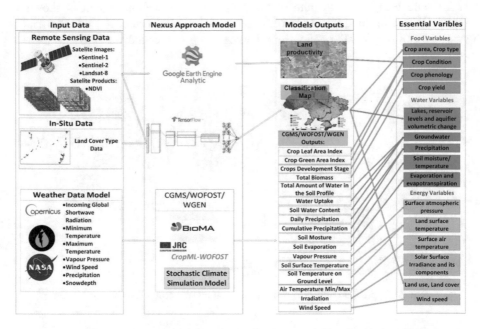

Fig. 4. Data processing chain for essential variables calculating

4   Results

In Table 1, results of cross-comparison of global land covers maps (ESA CCI-LC) and national land cover (SRI LC) maps is presented. In terms of User Accuracy (UA), Producers Accuracy (PA), and Overall Accuracy (OA), national datasets are more accurate than global products and possess higher correlation with national statistics data. In particular, difference between maps is obvious for grassland (Fig. 5). Also, our national land cover map has higher resolution compared to Global land cover map. Using our land cover classification maps for years 2000 and 2016, we detected all problematic areas for Ukrainian territory with deforestation. An example of deforestation for Kherson region is shown in Fig. 6.

Using total area of Ukraine, excluding inland water, we calculated indicator 15.1.1 Forest area as proportion of total land area by forest area calculated with our land cover maps and government statistics. We can see positive trend of forest area as a proportion of total land area indicator change by our classification maps and government statistics, and although the data for the 2000 s and 2010 s are very close to the statistics, our results differ for 2016.

Table 1. Comparison of user accuracy (UA), producer accuracy (PA) and overall accuracy for National land cover and Global land cover maps (ESA CCI) for 2000 and 2010 years

Year	Land cover national 2010		Land cover national 2000		Land cover dataset ESA CCI-LC-2010		Land cover dataset ESA CCI-LC-2000	
Class	PA, %	UA, %	PA, %	UA, %	PA, %	UA, %	PA, %	UA, %
Cropland	97.5	98.5	97.1	98.6	98.9	76.7	98.9	79
Tree-covered	97.2	97.4	98.8	98.4	84.9	95.6	86	97.1
Grassland	90.7	85.4	90.5	84.6	4.9	40.7	7.6	43.5
Other land	93.6	96.9	96.2	89.7	21.5	87.2	20.1	76.2
Water	99.5	99.8	99.5	99.9	96	99.6	96.4	99.8
Overall accuracy, %	97.5		97.7		85		87.3	

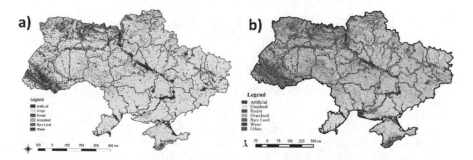

Fig. 5. Example of Global land cover maps (ESA CCI) with 300 m resolution for 2010 (a) and National land cover map for 2010 (b)

The JRC developed methodology for Land Productivity Dynamics (LPD) estimation based on NDVI profile derived from SPOT-VGT time series with course spatial resolution[4]. Taking into account that our national land cover maps have much higher spatial resolution, it is necessary to provide land productivity maps with the same resolution. Thus, we estimate productivity map for Ukraine territory for years

---

[4] JRC Science for Policy Report, http://publications.jrc.ec.europa.eu/repository/bitstream/JRC80540/lb-na-26500-en-n%20.pdf.

2010–2014 based on NDVI profile derived from Landsat images with simplified methodology. Further, we are going to extend of the JRC approach to Landsat and Sentinel-2 data.

Also, we get 15 essential variables for food, water, and energy that can be used for monitoring state of Ukrainian resources and indicators calculation. Food essential variables will be used for building new productivity map with help of classification map, replacing productivity map based on NDVI trend (Fig. 7). It will be used for calculating indicator 15.3.1 Proportion of land that is degraded over total land area and indicator 2.4.1. Proportion of agricultural area under productive and sustainable agriculture. At the same time, productivity map is sub indicator of achieving a zero level of land degradation.

Crop classification map were obtained on c4.4xlarge Amazon instance (500 Gb SSD drive, 16 threads CPU Intel(R) Xeon(R) CPU E5-2666 v3 @ 2.90 GHz and 30 GB RAM). This task takes approximately two weeks.

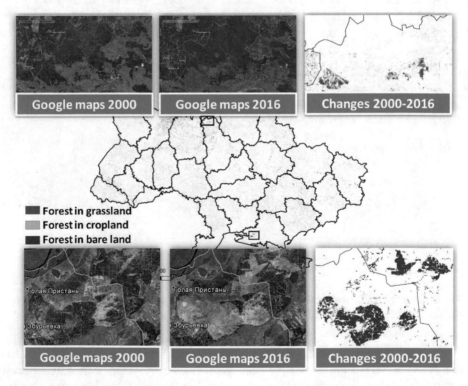

Fig. 6. Example of forest area changes from 2000 to 2016

Fig. 7. Productivity map for Ukraine territory for 2010–2014 years

5 Conclusions and Discussion

Accurate and precise land cover and land use maps with high resolution, built with use of remote sensing and in-situ data and deep learning approach, are essential for assessment sustainable development goals. Thus, to improve the resolution and quality of existing global land cover maps, we proposed our own deep learning methodology for country level land cover providing. NEXUS approach that combines three different spheres (food, water, and energy) was used for achieving workflow for fusion of different essential variables and obtaining indicators of approaching to selected goals. For this, we utilized WOFOST crop state model and different weather and climatic models' data, and as a result we obtained a possibility to calculate three indicators 15.1.1 Forest area as proportion of total land area, 15.3.1 Proportion of land that is degraded over total land area and indicator, and 2.4.1. Proportion of agricultural area under productive and sustainable agriculture. Based on this methodology, we could see the positive trend of indicator 15.1.1 for the Ukraine territory from 2000 to 2016.

References

1. Skakun, S., Kussul, N., Shelestov, A., Kussul, O.: The use of satellite data for agriculture drought risk quantification in Ukraine. Geomat. Nat. Hazards Risk **7**(3), 901–917 (2016)
2. Kussul, N., Sokolov, B., Zyelyk, Y., Zelentsov, V., Skakun, S., Shelestov, A.: Disaster risk assessment based on heterogeneous geospatial information. J. Autom. Inf. Sci. **42**(12), 32–45 (2010)

3. Shelestov, A.Y.: Using the fuzzy-ellipsoid method for robust estimation of the state of a grid system node. Cybern. Syst. Anal. **44**(6), 847–854 (2008)
4. Kussul, N., Shelestov, A., Skakun, S., Kravchenko, O.: Data assimilation technique for flood monitoring and prediction. Institute of Information Theories and Applications FOI ITHEA (2008)
5. Mandl, D., Frye, S., Cappelaere, P., Handy, M., Policelli, F., Katjizeu, M.: Use of the earth observing one (EO-1) satellite for the namibia sensorweb flood early warning pilot. IEEE J. Sel. Top. Appl. Earth Obs. Remote Sens. **6**(2), 298–308 (2013)
6. Lavreniuk, M., Skakun, S., Shelestov, A., Yalimov, B., Yanchevskii, S., Yaschuk, D., Kosteckiy, A.: Large-scale classification of land cover using retrospective satellite data. Cybern. Syst. Anal. **52**(1), 127–138 (2016)
7. Skakun, S., Kussul, N., Shelestov, A., Lavreniuk, M., Kussul, O.: Efficiency assessment of multitemporal C-band Radarsat-2 intensity and Landsat-8 surface reflectance satellite imagery for crop classification in Ukraine. IEEE J. Sel. Top. Appl. Earth Obs. Remote Sens. **9**(8), 3712–3719 (2016)
8. Kussul, N., Lemoine, G., Gallego, F., Skakun, S., Lavreniuk, M., Shelestov, A.: Parcel-based crop classification in Ukraine using Landsat-8 data and Sentinel-1A data. IEEE J. Sel. Top. Appl. Earth Obs. Remote Sens. **9**(6), 2500–2508 (2016)
9. Lavreniuk, M., Kussul, N., Meretsky, M., Lukin, V., Abramov, S., Rubel, O.: Impact of SAR data filtering on crop classification accuracy. In: 2017 First Ukraine Conference on Electrical and Computer Engineering, pp. 912–916 (2017)
10. Chen, Y., Lin, Z., Zhao, X., Wang, G., Gu, Y.: Deep learning-based classification of hyperspectral data. IEEE J. Sel. Top. Appl. Earth Obs. Remote Sens. **7**(6), 2094–2107 (2014)
11. Geng, J., Fan, J., Wang, H., Ma, X., Li, B., Chen, F.: High-resolution SAR image classification via deep convolutional autoencoders. IEEE Geosci. Remote Sens. Lett. **12**(11), 2351–2355 (2015)
12. Hu, F., Xia, G.S., Hu, J., Zhang, L.: Transferring deep convolutional neural networks for the scene classification of high-resolution remote sensing imagery. Remote Sens. **7**(11), 14680–14707 (2015)
13. Chen, Y., Zhao, X., Jia, X.: Spectral–spatial classification of hyperspectral data based on deep belief network. IEEE J. Sel. Top. Appl. Earth Obs. Remote Sens. **8**(6), 2381–2392 (2015)
14. Kussul, N., Lavreniuk, M., Skakun, S., Shelestov, A.: Deep learning classification of land cover and crop types using remote sensing data. IEEE Geosci. Remote Sens. Lett. **14**(5), 778–782 (2017)
15. Shelestov, A., Kravchenko, A., Skakun, S., Voloshin, S., Kussul, N.: Geospatial information system for agricultural monitoring. Cybern. Syst. Anal. **49**(1), 124–132 (2013)
16. Kogan, F., et al.: Winter wheat yield forecasting in Ukraine based on Earth observation, meteorological data and biophysical models. Int. J. Appl. Earth Observ. Geoinf. **23**, 192–203 (2013)
17. Kolotii, A., et al.: Comparison of biophysical and satellite predictors for wheat yield forecasting in Ukraine. Int. Arch. Photogram. Remote Sens. Spatial Inf. Sci. **40**(7), 39 (2015)
18. Kussul, N., Kolotii, A., Shelestov, A., Yailymov, B., Lavreniuk, M.: Land degradation estimation from global and national satellite based datasets within UN program. In: Intelligent Data Acquisition and Advanced Computing Systems: Technology and Applications (IDAACS), pp. 383–386 (2017)
19. Kussul, N., Kolotii, A., Adamenko, T., Yailymov, B., Shelestov, A., Lavreniuk, M.: Ukrainian cropland through decades: 1990–2016. In: Electrical and Computer Engineering (UKRCON), pp. 856–860 (2017)

20. Shelestov, A., Kolotii, A., Skakun, S., Baruth, B., Lozano, R.L., Yailymov, B.: Biophysical parameters mapping within the SPOT-5 Take 5 initiative. Eur. J. Remote Sens. **50**(1), 300–309 (2017)
21. Kussul, N., Shelestov, A., Skakun, S., Li, G., Kussul, O.: The wide area grid testbed for flood monitoring using earth observation data. IEEE J. Sel. Top. Appl. Earth Obs. Remote Sens. **5**(6), 1746–1751 (2012). Article Number 6236227
22. Kussul, N., Skakun, S., Shelestov, A., Kussul, O.: The use of satellite SAR imagery to crop classification in Ukraine within JECAM project. In: International Geoscience and Remote Sensing Symposium (IGARSS), pp. 1497–1500 (2014). Article Number 6946721
23. Kussul, N., Skakun, S., Shelestov, A., Kravchenko, O., Gallego, J.F., Kussul, O.: Crop area estimation in Ukraine using satellite data within the MARS project. In: International Geoscience and Remote Sensing Symposium (IGARSS), pp. 3756–3759 (2012). Article Number 6350500

An Efficient Remote Disaster Management Technique Using IoT for Expeditious Medical Supplies to Affected Area: An Architectural Study and Implementation

Vidyadhar Aski[✉], Sanjana Raghavendra, and Akhilesh K. Sharma

Manipal University Jaipur, Jaipur, India
{vidyadharjinnappa.aski,
akhileshkumar.sharma}@jaipur.manipal.edu,
sanjanar@muj.manipal.edu

Abstract. Creating technology enhanced optimized strategies to handle the subsequent healthcare issues emerged from natural calamities and disaster has now become more tranquil with the latest advancements in networking and low power electronics. Proposed system addresses an immediate action plan in accordance with the various adverse medical affronts during and post disaster events. The study focuses on developing strategies for disaster management in India, where flooding of rivers is one of the frequently occurring drastic event around the years in various regions resulting many people died due to unavailability of medical facilities. There are many studies conducted which proves that managing medical emergencies like cardiac seizure during such disasters where doctors unable to reach in these remotely affected areas due to damaged transportation systems by flood and they are struggling to provide prompt delivery of medical as well as relief to the remotely affected areas. Thus, it is gaining significant attention by disaster management organizations. Therefore, this study proposes a classic solution for handling various medical emergencies that occurs during and post disaster events. The system uses the drones to carry a weight of 1.5 kg to 2 kg medical kits to the affected areas. Miniaturized IoT based medical devices are designed with various Wireless Body Area Sensors (WBAS) and actuators. Along with a defibrillator unit and ECG analyzer. These medical kits are placed in a connected drone and this will drive to the affected areas aerially. All the devices are IoT enabled and are connected through the central cloud infrastructure of hospital by which medical experts can access the required medical parameters and convey the instructions through drone to a caretaker in real time to perform related events which may possibly postpone casualties.

Keywords: Disaster management · Drone · WBAS · Defibrillators
IoT · ECG analyzer

1 Introduction

The natural process of earth sometimes results in an unpropitious events known as natural hazards. Natural hazards like floods, tsunami, earthquake, drought, hurricanes leave behind catastrophic effects of deaths, economic fall, injuries, spread of disease

© Springer Nature Switzerland AG 2019
O. Chertov et al. (Eds.): ICDSIAI 2018, AISC 836, pp. 156–166, 2019.
https://doi.org/10.1007/978-3-319-97885-7_16

and property damage. This is when a natural hazard converts into a natural disaster. The statics claim that the evolution of technology has decreased the destructions and deaths caused by natural calamities over the last ten decades by almost 80 percent. The most vulnerable to natural hazards are the countries below the poverty line, about 81% people die in these disasters due to inadequacy of resources. The situation is becoming worse as about 325 million immensely poor people are expected to stay in the 49 most hazard-prone courtiers by 2030 [1].

The natural hazards are considered as out of human domain but we can obviously have control over the after effects of the disaster. Since the infrastructure supplies such as bridges, water, gas, communication cable, food, water, medical resources are disabled and the help through roads are highly impossible. In such situations communication through airways is the only possible way to help people stuck in the disastrous area. A self-determining flying is becoming very popular now a days known as Drones or unmanned aerial vehicle (UAV). Drones are acknowledged for its radio controlled system which is always leaned on first person review (FPV) with the help of a camera that streams a live feed directory to the controller, it is therefore not needed to be in optical axis as it can be controlled from certain kilometers away.

The reason why real time monitoring system is so important is that, if a person is cemented in a situation where he/she needs the immediate medical assistance and there is not enough time to take that person to hospital, at that time someone around that patient can give a call to the hospital (base station) and explain about the situation, then based on that information and information on the longitudinal and latitudinal values of the call, a doctor can send the aid in the drone by activating the real time monitoring system.

The drones can be controlled from the base station based on Global Positioning System (GPS) which can tell the current area of the drone, Geographic Information System (GIS) which can analyze, manage, and store the current spatial or geographic area and Airborne Synthetic Aperture Radar (SAR) which generates high resolution remote sensing imagery using the path, etc. [2]. After the drone reaches the desired destination the doctors can assist from the base station itself and anyone around/with the patient can follow the instructions and give the assistance to the patient [24].

The same can be done at the time of natural calamities. The system we are proposing is of a health based Drone or ambulance Drone which can be sent in the situation of a natural disaster from the base station to the destination station and real time health monitoring can be done using that Drone. The Drone will be designed in such a way that it can consist of all the health related sensors like temperature sensor, pulse rate sensor, ECG, defibrillator, spirometer. These sensors will be connected to a Raspberry pi computer. We will also put GPRS/GSM so that the Drone need not be dependent on the Wi-Fi. We further have a GPS also to track the location of the user who is in need of the medication (Drone). Then we are going to use a Surface Mounting Device (SMD) for flexible electronics, in order to decrease the area consumption by hardware tools. This was about the hardware part, now in software part we are going to store all the data in the cloud database so that we can fetch the data whenever it is required. Procedure of system will be such that – a person who is in need of immediate medical assistance can request for the medical Drone. After receiving the request, the Drone can be sent to the location of the caller with the help of GPS within very less time. The person who could have been in great loss if he/she would have been taken to hospital or could not reach hospital in time can be saved with the help of medical Drone. All this combination of combining the hardware and software tools that too in real time can be possible because of Internet of Things. IoT has made our lives simpler, quicker and easier to use.

2 Related Work

Many researchers have carried out various experiments in the field of deployment of Drones in healthcare sectors for providing IoT based instant medical support for remote areas. Most works focus on designing the Drone Ambulance for domestic applications within the limited geographic range. Initial moments after an accident during disaster are critical and required urgent attention. It is extremely necessary to provide right care to avert acceleration [25]. Thus its utmost essential for dispatching a medically equipped Drone to the affected areas based on positioning details obtained through received emergency call. The medical devices installed in the Drone are made of various biomedical sensors and advanced processor like Raspberry Pi. Using FPGA, Flexible electronics, surface mounting devices and flexible battery techniques we can miniaturize the device size and make it compact enough to place inside light weight drone. Centralized cloud infrastructure acts as a connection agent and it helps in establishing connection to the accidental area and base station's hospital. End to End data security is ensured through MQTT and JSON encryption to prevent third party interpretations. In this proposed work the technologies like WBAN, IoT, Cloud infrastructure, Drone are used for providing immediate medical attention to the accidents occurred in areas which are most affected by natural calamities like Flood, Earthquakes etc. biomedical data is being visualized, analyzed, interpreted by medical practitioner in real-time. Many methodologies are proposed to meet aforesaid requirements. Some works in this repute are discussed below.

Dhivya et al. presents Quadcopter based technology for an emergency Healthcare [25]. The work consists of several low cost wireless sensors have been deployed in system to create biomedical monitoring system and it is placed in light weight drone for carrying real-time data transferring experiments.

Sehrawat et al. proposed work for designing a drone based surveillance system in military sensitive areas for human detection under any rubber items [26]. The technology uses built in Wi-vi sensors, infrared camera, GPS modules for quick tracking human being stuck under walls during terror attacks or disasters. This system also could be used in case of detecting human motion under walls.

Vattapparamban et al. made a survey on various aspects of Drones in future smart cities and issues related cyber security, public safety and privacy [27]. Author tried to compare existing infrastructures for Drone flying and possible cyber-attacks on drones by focusing more on security parameter.

Naser Hossein Motlagh et al. made a comprehensive survey on deployment of unmanned aerial vehicles and consequent legal issues associated with using these type of equipment as far as public safety and privacy is concerned. Author also envisioned an architecture for deploying UAVs in services like goods shipments and related key issues are also discussed. In this architecture author explained how effective for using a distributive network infrastructure. Like a network of UAVs is are in place with a satellite connection establishment and collective data can be exchanged between all the UAVs and its network components. The exchange of data concerning their status and updates which in turn improves resource utilization ratio in real time.

Scott et al. made a review on numerous emerging trends in deployment of Drones in Healthcare sector, be it for delivery of Blood from Blood bank, and be it shipment of heart transplantation etc., Author has also compared drone delivering different healthcare components [28].

Amendola et al. proposed a survey study based on usage of RFID technologies for IoT based healthcare network [29]. Author has also discussed possible challenges and emerging trends in RFID applications in healthcare filed. Different types of RFID tags are discussed with their optimal usage in healthcare data acquisition system. Author has also mentioned in above study about data processing unit and human behavior analysis for installation of RFID tags.

Yeh presents an IoT based healthcare system with Body Sensor Networks (BSN) [30]. Author explained his modular approach to deploying body sensors with a robust sensor area network. Also explained using Raspberry pi as a data processing unit with its built-in capabilities of transmitting data wirelessly over the cloud infrastructure.

Plageras et al. proposed a comparative analysis on different technologies of IoT based surveillance system for ubiquitous healthcare [31]. Author explained wireless sensor network can together form a sensor node and data from each sensor can be stored and retrieved from cloud through defined architectural sensor node.

3   Proposed Architecture

Many researchers have been worked on IoT healthcare domain and showcased their results mainly in the domain of particular IoT area like network security, wearable IoT healthcare device, sensor based body area network etc., Our work emphasizes more on robust healthcare kit design and can be carried by a drone arially and we can provide medical assist to the people having health affronts in the arias affected by natural calamities.

The proposed architecture of IoT enabled healthcare Drone deployment for areas rendered inaccessible by natural disaster is shown in Fig. 1. In the proposed architecture Wireless Body Area Sensor Network (WBASN) comprising of 5 sensors and actuators are considered. These sensors are prototyped first with through-hole components on breadboard and experimentations shall be carried. Later during deployment phase these through-hole components are fabricated on surface mounting devices with flexible PCBs using FPGA technology. Miniaturization can be achieved using flexible batteries as well. During natural calamities like Flood, Tsunami etc., Roads are filled with overflowed flood water thus by blocking the possible connectivity roads to hospitals. Proposed system helps in aforesaid scenarios by avoiding the transportation dependency by road just by giving a call to disaster management health centers and explaining proper locational details. Call tracing is done at base station to gather locational information by considering co-location information which help us to infer positional details like longitudinal and latitudinal values [32]. Once longitudinal and latitudinal values are inferred, Drone is programed in such a way that it finds its way out to reach to its destination. Once it reaches to the patient's place any care taker can handle this drone and get the medical kit out. Medical kit has defibrillator and sensor node. These devices are installed on patient's body under doctor's assistance over the camera and display unit on drone. Once the sensor plate is it's connected to patient's body then doctor will be able to observe different biomedical parameters in real-time to analyze these data and easily take quick decisions on patient's medical condition.

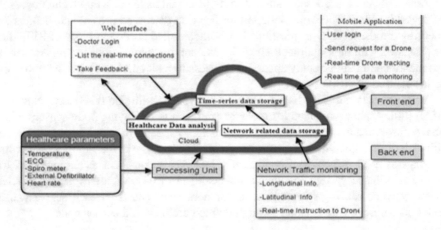

Fig. 1. IoT enabled Healthcare Drone deployment architectural diagram.

4   Flow Diagram

Following diagram shows the entire flow sequences of operations associated with proposed system. The sequences are explained as follows.

Request call from the calamity affected area is raised by users. Using embedded location tracing technique operator from base station will be able to collect location related information and same information is shared to Drone via Drone operational interface unit. Now Drone is equipped with necessary medical kits and sent to requested area. After Drone reaches to the patient's place, caretaker needs to handle drone and carry out necessary actions as suggested by doctor in real time. Now that the devices are installed and real time data flow starts to centralized cloud. Then doctor may start analyzing health data and he could prescribe necessary actions to be taken.

5   Modules of Proposed Architecture

Sensor Node. The sensor node is formed using various individual wireless sensors like temperature sensor, Pulse sensor, ECG sensor and Spirometer sensor.

Temperature Sensor. We will be using BM180 temperature sensor. It is high precision digital pressure sensor. The features include - long stability, high accuracy, low voltage electronics, ultra-low power and low altitude noise of 0.25 m at fast conversion time. The range of the pressure is from 300 hPa to 1100 hPa and supply voltage from 1.8 V to 3.6 V [17].

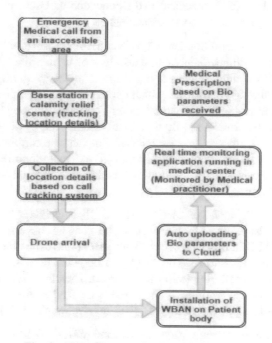

Fig. 2. Flow diagram of proposed technology.

Pulse Rate. We will be using SEN-11574 pulse rate sensor, which is user friendly. The feature includes- noise cancellation, power ranges is from 3 V to 5 V, it is amplification circuit with all embedded on a single chip [18].

Spirometer. To measure the functionality of the lungs spirometer is used. Spirometers helps in carrying PFTs tests that include in detecting the symptoms of lung problems and look after the course of chronic lung disease (such as asthma or chronic obstructive pulmonary disease (COPD)). These tests require a person to breathe in and out with high speed which may be dangerous for a person who had recently had a heart attack, chest surgery, abdominal surgery etc. We will be using hand-held wireless spirometer SP10 W, which displays the flow rate-volume and volume-time as charts along with the trend it tracks. The in-built memory allows you to delete, upload, and review data with ease. It intelligently alerts when volume and flow exceed their expected limitations. The power automatically shuts off when no process is detected within 60 s. It comes with a lithium battery than can be recharged [6]. The other features includes – the ratio of Forced Vital Capacity in one second (FVC1) and (FEV1%), wireless transmission, trend chart display, scaling and battery power display [13].

Electrocardiography (ECG). Sensor AD8232 is used to measure the electric activity of heart. The AD8232 is an integrated signal conditioning block for ECG and other bio potential measuring applications. It is designed to extract, amplify, and filter small bio potential signals in the presence of noisy conditions are some of its features which are created by motion or remote electrode placement. It has Leads off detection, LED indicator, 3.5 mm Jack for Biomedical Pad Connection or Use 3 pin head, shutdown pin, operating voltage −3.3 V and gives analogous output.

Defibrillator. A serious condition known as "ventricular fibrillation" occurs in human heart and if not treated immediately can cause loss of human life. Defibrillator is the solution for life threatening situations like cardiac arrest by giving electric current shock to the heart through chest wall which depolarizes considerable amount of heart muscles. Subsequently, the body's natural pacemaker in the sinoatrial node of the heart is able to re-establish normal sinus rhythm. Defibrillators can be three types according the usages- external, Trans venous, or implanted (implantable cardioverter-defibrillator. External defibrillators are known as automated external defibrillators (AEDs), it automates the diagnosis of treatable rhythms, which is a huge advantage that people with no training can also use it [8].

Automated External Defibrillator (AED). AED is light weighted and portable defibrillator which can also be used by the people who does not have medical knowledge. We will be using AED in this project. The pads that comes along the AED are used to give an electric shock if required. The shock can be given by placing the pads on the bare chest of the person suffering from cardiac arrest. Then the AED analyses the hearts rhythm, if the heart is beating properly then there is no need to give a shock, but in case the heart beat is not proper the AED will give the electric shock [3].

Embedded Technologies. Surface Mounting Devices (SMD). A SMD is a device whose components are directly mounted on the surface of printed circuit board (PCB), it is based on the technique known as surface mounting technique (SMT) which is considered better than the through-hole-technology (TMT) [15]. One more advantage of SMT is that it can be used both the sides when required. We will be using SMT for mounting all the electronic devices and sensors used in this project o flexible plastic such as polyimide, PEEK or transparent conductive polyester film [16]. This assembling technique of devices is known Flexible Electronics.

General Packet Radio Services (GPRS)/Global System for Mobile (GSM). We will be using SIM5215 which is a GSM/GPRS/EDGE and UMTS/HSDPA which runs on frequencies GSM 850 MHz, EGSM 900 MHz, DCS 1800 MHz and WCDMA 2100/900 MHz or it can also run on 1900/850 MHz. The module can be selected on the basis of the need of the different wireless communication. The best part about SIM5215 is that even with a small configuration of 36 * 26 * 4.5 mm and integrated functions it can fulfill any space requirement of user's application which includes PDA phone, industrial handhelds, smart phone, vehicle application, machine to machine etc. [14].

Global Positioning System (GPS). Through GPS we can find out the exact geolocation of the user. The GPS we are going to use is MT3339. It can achieve remarkable sensitivity, accuracy, time-to-first-fix (TTFF) and consumes very less power usage. MT3339 is high performance single chip GPS resolution that contains digital baseband, CMOS RF, ARM7 CPU and an embedded flash which is optional. MT3339 has up to 210 PRN channels. Out of which 66 are search channels and 22 are simultaneous tracking channels. Even in indoor MT3339 get track of satellite [9].

Flexible Battery. Having flexibility features in an electronic device is a great achievement in itself. These batteries are much smaller in size and also powerful as compared to normal batteries. The twisting, heating will not affect these batteries Infact they get charged and discharged easily when punched or twisted easily. We will be using Lithium Ceramic Battery (LCB) of ProLogium Technologies. It has high energy density and no leakage features along with that LCB is safe, flexible, incombustible, no swelling will occur, ultra-thin, and is ultra-safe [20].

Data Processing Unit and Cloud Infrastructure. Raspberry Pi-3. Bluetooth connectivity, wireless Local area network (WLAN), and powerful GPU/CPU are some additional features of pi3 which outshines pi2 with 50% more efficiency [5] and is 10 times faster than first generation Raspberry pi. Pi3 consists of - Broadcom BCM2387 chipset, 1.2 GHz Quad-Core ARM Cortex-A53, 4.1 Bluetooth, 802.11 b/g/n wireless LAN. Pi3 can run boots from micro SD card, Windows 10 IOT and can also run Linux operating system.

Cloud. The patient's information needs to be stored in database for data analysis purpose. This analysis can be very helpful in previously preparing for the types of problems people face during natural calamities. The database will have all the information regarding the type of natural disasters and the type of health problems occurred during that time. The suitable cloud platform for this purpose is IBM Bluemix. The IBM Bluemix helps in retrieving the real time data which has been collected by the sensors, also it easily connects to IoT cloud [20].

Data Processing Unit. Since raspberry pi does not contain analog to digital converter (ADC) unit. We need to interface analog sensors to raspberry pi through an explicit analog to digital converter circuit [33]. The output from ADC is given to GPIO pin of Raspberry Pi. Generic Delta Sigma ADC is explained herewith. One terminal of sensor is given as input for comparator. Comparator has got its another input from predefined sampled values from 1-bit Digital to analog converter (DAC) unit and both are compared and positive data sample is being now integrated and integrator output is then fed to an opamp at its inverting end. Thus at opamp we get our digital pulses and are fed directly to the GPIOs of micro controller unit (MCU).

6 Use Case Diagram (Web and Mobile Interface)

The use case diagram explains the action that user performs in collaboration with the doctor. The figure below explains that the user can login into the web application, user can register for a drone when required and on the other hand doctor can login into the web application and send the drone to the user. Then after receiving the drone user can use the drone for health monitoring, the doctor can also guide the user in monitoring by looking at the disguises sent by the user as our drone works in real time (Fig. 3).

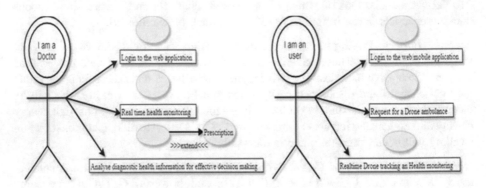

Fig. 3. Use case diagram.

7 Conclusion and Future Scope

In this article, we first have introduced a robust solution for providing healthcare solution in areas rendered inaccessible by natural disasters. The system comprises of hardware and computational unit. Hardware has various sensors for measuring bio parameters along with the equipment like AEDs and portable ECG analyses. All these components are designed compact enough to be carried by a drone. Drone is embedded with intelligent technologies to track the location autonomously with built-in demographic unit and microphone. Accordingly, our proposed system is suitable for monitoring healthcare data remotely and can provide immediate medical attentions.

In this project the drone is completely dependent on GPRS. As there is growing popularity for satellite communication, we can use satellite communication for transmitting the information. In satellite communication distance of each and every corner of earth can be covered easily, it is flexible, user can also have the control over the satellite network and cost is also not much, only the installation cost is more [21]. Drone trafficking can cause delay in the delivery of the drone, which can be a big disadvantage. To avoid this problem in the drones we can have distributed drone network in order to control the drone traffic. By doing this we can change the route of the drone in which there is congestion and send our drone to the desired location on time.

References

1. Child Fund International. https://www.childfund.org/Content/NewsDetail/2147489272/
2. Chi, T., Zhang, X., Chen, H., Tan, Y.: Research on information system for natural disaster monitoring and disaster. In: Geoscience and Remote Sensing Symposium (2003)
3. First Aid for free. http://www.firstaidforfree.com/how-does-an-aed-work/

4. Barometric Pressure sensor Arduino tutorial. http://www.theorycircuit.com/barometric-pressure-sensor-arduino-tutorial/
5. Raspberry Pi3 – Performance. http://www.trustedreviews.com/reviews/raspberry-pi-3-performance-and-verdict-page-3
6. Portable spirometer. http://www.medicalexpo.com/medical-manufacturer/portable-spirometer-30494.html
7. Automated Sensing System for Monitoring of Road Surface Quality by Mobile Devices. https://www.sciencedirect.com/science/article/pii/S1877042814000585
8. Introduction to defibrillator. https://en.wikipedia.org/wiki/Defibrillation
9. MT3339 all in one data sheet. https://s3-ap-southeast-1.amazonaws.com/mediatek-labs-imgs/downloads/
10. Davis Instruments. http://www.davis.com/blog/2014/03/24/advantages-and-disadvantages-of-infrared-thermometers/
11. MLX90614 family Melexis. https://www.sparkfun.com/datasheets/Sensors/Temperature/MLX90614_rev001.pdf
12. SP10W Spirometer Contec. http://www.contecmed.com/index.php?page=shop.product_details&flypage=flypage.tpl&product_id=181&category_id=21&option=com_virtuemart&Itemid=604_id=21&option=com_virtuemart&Itemid=604
13. Raspberry Pi Datasheet. http://docseurope.electrocomponents.com/webdocs/14ba/0900766b814ba5fd.pdf
14. SIM5215/SIM5216 hardware design. http://www.postar.com/public/uploads/20120326171149_704.pdf
15. Techopedia, Surface-Mount Device (SMD). https://www.techopedia.com/definition/12371/surface-mount-device-smd
16. Flexible Electronics. https://en.wikipedia.org/wiki/Flexible_electronics
17. BMP180 digital pressure sensor Datasheet. https://cdn-shop.adafruit.com/datasheets/BST-BMP180-DS000-09.pdf
18. Pulse Sensor SEN-11574. http://www.ekt2.com/pdf/412_ARDUINO_SENSOR_PULSE.pdf
19. ProLogium Lithium Ceramic battery profile. http://www.prologium.com.tw/upload/Download/20150302-16190635.pdf
20. Aski, V.J., Sonawane, S.S., Soni, U.: IoT enabled ubiquitous healthcare data acquisition and monitoring system for personal and medical usage powered by cloud application: an architectural overview. In: International Conference on Computing and Communication. Springer, Cham (2018)
21. Principles of Satellite communications. https://www.tutorialspoint.com/principles_of_communication/principles_of_satellite_communications.html
22. Introduction to Electrocardiographs. https://origin-www.maximintegrated.com/en/app-notes/index.mvp/id/4693
23. A closer look at Medical sensors. https://www.digikey.it/en/articles/techzone/2014/apr/a-closer-look-at-medical-sensor-solutions
24. Ambulance Drone. https://www.tudelft.nl/en/ide/research/research-labs/applied-labs/ambulance-drone/
25. Dhivya, A.J.A., Premkumar, J.: Quadcopter based technology for an emergency healthcare. In: The Proceedings of 2017 3rd International Conference on Biosignals, Images and Instrumentation (ICBSII), Chennai, 16–18 March 2017 (2017)
26. Sehrawat, A., Choudhury, T.A., Raj, G.: Surveillance drone for disaster management and military security. In: 2017 International Conference on Computing, Communication and Automation (ICCCA), pp. 470–475. IEEE, May 2017

27. Vattapparamban, E., Güvenç, İ., Yurekli, A. İ., Akkaya, K., Uluağaç, S.: Drones for smart cities: issues in cybersecurity, privacy, and public safety. In: 2016 International Wireless Communications and Mobile computing Conference (IWCMC), pp. 216–221. IEEE, September 2016
28. Scott, J., Scott, C.: Drone Delivery Models for Healthcare (2017)
29. Amendola, S., et al.: RFID technology for IoT-based personal healthcare in smart spaces. IEEE Internet Things J. **1**(2), 144–152 (2014)
30. Yeh, K.H.: A secure IoT-based healthcare system with body sensor networks. IEEE Access **4**, 10288–10299 (2016)
31. Plageras, A.P., Psannis, K.E., Ishibashi, Y., Kim, B.G.: IoT-based surveillance system for ubiquitous healthcare. In: IECON 2016-42nd Annual Conference of the IEEE Industrial Electronics Society, pp. 6226–6230, October 2016
32. Khazbak, Y., Cao, G.: Deanonymizing mobility traces with co-location information. In: 2017 IEEE Conference on Communications and Network Security (CNS). IEEE (2017)
33. Delta Sigma ADC. https://www.maximintegrated.com/en/app-notes/index.mvp/id/1870

Knowledge Engineering Methods. Ontology Engineering. Intelligent Educational Systems

Method of Activity of Ontology-Based Intelligent Agent for Evaluating Initial Stages of the Software Lifecycle

Tetiana Hovorushchenko[1]($\boxtimes$) (iD) and Olga Pavlova[2] (iD)

[1] Khmelnytskyi National University, Khmelnytskyi, Ukraine
tat_yana@ukr.net
[2] Khmelnytskyi Gymnasium No. 2, Khmelnytskyi, Ukraine
olya1607pavlova@gmail.com

Abstract. Importance of the task of automated evaluation of initial stages of the software lifecycle on the basis of software requirements specifications (SRS) analysis and the need for information technology of new generation for the software engineering domain necessitates the development of agent-oriented information technology for evaluating initial stages of the software lifecycle on the basis of ontological approach. The purpose of this study is the development of the method of activity of ontology-based intelligent agent for evaluating initial stages of the software lifecycle. The intelligent agent, which works on the basis of the developed method, evaluates the sufficiency of information in the SRS for assessing the non-functional software features—provides the conclusion about the sufficiency or insufficiency of information, the numerical evaluation of the level of sufficiency of information in the SRS for assessment of each non-functional feature in particular and all non-functional features in general, the list of attributes (measures) and/or indicators, which should be supplemented in the SRS for increasing the level of sufficiency of SRS information. During the experiments, the intelligent agent examined the SRS for the transport logistics decision support system and found that the information in this SRS is not sufficient for assessing the quality by ISO 25010 and for assessing quality by metric analysis.

Keywords: Software · Software requirements specification · Ontology
Intelligent agent · Ontology-based intelligent agent

1 Introduction

At the current stage of development and implementation of information technologies in various fields of human activity, the main problems that arise in the processing of large volumes of information, are the need of significant human resources for supporting the data analysis process, high computational complexity of available analysis algorithms, and rapid increase of volume of collected data. The use of known methods and tools for the processing such volumes of information doesn't justify expectations of developers, leads to over-utilization of resources, losses of significant information, and conflicts between customer's expectations and the results [1, 2]. The global trend in the

© Springer Nature Switzerland AG 2019
O. Chertov et al. (Eds.): ICDSIAI 2018, AISC 836, pp. 169–178, 2019.
https://doi.org/10.1007/978-3-319-97885-7_17

processing of large volumes of information is intellectualization of information processing (machine learning [3], cognitive computing [4], deep learning [5], semantic web [6]). However, specifics and features of subject domains, for which information technologies are being developed, significantly influence the content and methods of processing the information, therefore, the approach based on the study of characteristics and features of the subject domains remains justifiable, and the development of information technologies for specific domains is necessary [7].

Today, software engineering requires the focus of development and implementation of effective information technologies, since software bugs and failures can potentially lead to catastrophies that lead to human casualties, environmental disasters, significant time and financial losses.

Today, in the software engineering domain, a large number of software projects is implemented, statistics on successful and unsuccessful decisions and results are available, but there are no efficient information technologies that would facilitate obtaining all useful information from available statistics. Known information technologies presuppose human-machine interaction at all stages of information processing, i.e. all information is interpreted by a person. This often leads to losses of essential information and to emergence of bugs at early stages of the lifecycle. Therefore, the *pressing task* is development of information technologies of the new generation for software engineering domain, in which the person is eliminated from the activities of information processing and knowledge elicitation.

Nowadays, humanity is increasingly relying on software during solving complex problems, and the number of high-cost software projects is growing rapidly. But, as statistics show [8, 9], the share of challenged software projects is about half of all software projects, and the share of failed software projects is about 1/5 of all software projects. That is, successful software projects, which can be relied upon and which are worth spending money on, are only 1/3 of all software projects. A significant number of bugs is introduced into the software at initial stages of the software lifecycle. The vast majority of software-related crashes occurred due to bugs in the SRS [10]. Consequently, the success of the software project implementation (as delivered on time, on budget, and which have required features and functions) significantly depends on early stages of the software lifecycle, so the *pressing and important task* is an automated evaluation of the level of work at initial stages of the software lifecycle on the basis of the SRS analysis (in particular, evaluating sufficiency of information in the SRS as a key component of the success of software projects). This is why there is particular attention devoted to requirements, which characterize the non-functional software features (quality, reliability, dependability, complexity, performance, maintainability, safety, interoperability, security—according to ISO/IEC TR 19759: 2015 standard).

The urgency and importance of the task of automated evaluating initial stages of the software lifecycle on the basis of SRS analysis and the need of information technologies of the new generation for software engineering domain necessitates the development of agent-oriented information technology for evaluating the initial stages of the software lifecycle on the basis of ontological approach. *The task of this research is the development of the method of activity of ontology-based intelligent agent for evaluating initial stages of the software lifecycle.*

2 State of the Art of Ontology-Based Intelligent Agents for Software Engineering Domain

Advantages of using ontologies are the systematic approach to the study of the subject domain, the possibility of holistic presentation of known information about the subject domain, and identification of duplications and gaps in the knowledge. Ontologies are the key technology for development of the semantic web, since they provide access, understanding, and analysis of information by intelligent agents.

An intelligent agent (IA) is a system that monitors the environment, interacts with it, and whose behavior is rational in a sense that the agent understands the essence of its actions, and these actions are aimed at achieving a certain goal [11]. Such an agent can be a software system, a bot, and a service. During its activity, IA uses information obtained from the environment, analyzes it by comparing it with the facts already known to it, and, based on the results of the analysis, decides on further actions.

A number of studies is devoted to solving the task of developing ontology-based intelligent agents (OBIA) for software engineering domain. For instance, the authors of [12] research the application of ontologies for agent-oriented software engineering, present the tool that uses instantiated ontological designs to generate programming code and approves the advantages of the OBIA for software engineering domain. The paper [13] is devoted to eliminating the burden of ambiguity in gathered software requirements and to facilitating the communication for the stakeholders by OBIA implementation. The authors of [14] propose the development of OBIA for minimizing the semantic ambiguity during the development of SRS in Spanish language, for automatic obtaining of the basic elements of SRS, and for automatic building of KAOS goal diagrams. The paper [15] is devoted to applying the ontological agent-oriented models for formalizing initial application requirements with the purpose of reducing costs using the example of developing Ambient Assisted Living applications for patients with Parkinson disease. The authors of [16] propose the task-based support architecture according to the agent-oriented paradigm and ontology-driven design for decision support systems (using the example of clinical decision support systems in an emergency department), which allow for iterative translation of functional requirements into architectural components. The paper [17] is devoted to developing the framework for the formal representation, verification, and ensuring the functional correctness of resource-bounded context-aware systems into safety-critical domains in the form of OBIA.

The conducted analysis of known OBIA has shown that they don't solve the problem of automated quantitative evaluation of the level of work at initial stages of the software lifecycle on the basis of SRS analysis (in particular, evaluating sufficiency of information in SRS). Nowadays, there exist no standards or methods that contain formulas for computing this sufficiency.

The *only solution for evaluating the sufficiency of information in the SRS* is paper [18], which proposes theoretical and practical principles of the information technology of evaluating sufficiency of information on quality in SRS, and the ontology of "Software Engineering" subject domain (with the following parts: "Software Quality," "Software Quality. Metric Analysis," "Software Requirements Specification").

3   Method of Activity of Ontology-Based Intelligent Agent for Evaluating Initial Stages of the Software Lifecycle

Ontology-based intelligent agent for evaluating initial stages of the software lifecycle during its functioning will use ontologies as known facts, with which it will collate information from the SRS for real software. On the basis of obtained results, it will decide on the sufficiency of information in the SRS and on further actions.

Method of activity of ontology-based intelligent agent for evaluating initial stages of the software lifecycle consists of the following stages:

1. Collation and comparison of information in sections of the SRS for the real software (we believe that the structure of the SRS meets the requirements of the ISO 29148:2011 standard) with the base ontologies of the SRS sections that contain attributes (measures) and/or indicators, which are necessary for assessment of each non-functional software feature (these ontologies are contained in the knowledge base). Currently, this stage is implemented manually (the user of OBIA uses the base ontology of the subject domain "Software Engineering" (parts "Software Quality" and/or "Software Quality. Metric Analysis") and revises relevant sections of the specification for the presence of attributes (measures) and/or indicators specified in the base ontology). In the future, it is planned to automate this stage. A separate agent will be developed, which will conduct semantic analysis of the natural-language specifications for the search of attributes (measures) and/or indicators, i.e. extracts the input values from the natural-language specifications.

2. Identification of attributes (measures) and/or indicators, which are absent in the SRS for real software, i.e. the formation of the set of missing attributes (measures) and/or indicators (if the formed set is not empty, then the information in the SRS is insufficient for the non-functional features assessment—the more elements in this set, the lower the level of information sufficiency in the SRS).

3. Analysis (on the basis of components of the base ontology of the "Software Engineering" subject domain for every software non-functional features) of the influence of each element in the set of missing attributes (measures) and/or indicators on the non-functional software features and their subcharacteristics, and increase of counters of missing attributes (measures) and/or indicators for appropriate subcharacteristics and non-functional software characteristics.

4. Identification of non-functional software features and their subcharacteristics, which cannot be calculated on the basis of attributes (measures) and/or indicators from the SRS for real software. If counters of missing attributes (measures) and/or indicators for all subcharacteristics of a non-functional characteristic are equal to 0, then the SRS has sufficient information for assessment of the concrete non-functional feature. Otherwise, in the SRS, there are insufficient attributes (measures) and/or indicators for assessment of the concrete non-functional feature (with visualization of the gaps in knowledge), and there is a need to supplement this SRS with attributes (measures) and/or indicators (with specifying the attributes (measures) and/or indicators).

5. Numerical evaluation of the level of sufficiency of information in the SRS for assessing each non-functional software feature:

$$D_j = \left(k_j - \sum_{i=1}^{k_j} \frac{qm_i}{qn_i} \right) / k_j, \tag{1}$$

where k_j is the number of subcharacteristics of the jth non-functional characteristic ($j = 1, \ldots, 9$, since ISO/IEC TR 19759:2015 specifies 9 non-functional software features), qm_i is the number of missing (in the real SRS) attributes and/or indicators for the ith subcharacteristic of the jth non-functional feature, qn_i is the number of necessary attributes and/or indicators for the ith subcharacteristic of the jth non-functional feature (determined by base ontologies for every non-functional feature).

6. Numerical evaluation of the level of sufficiency of information in the SRS for assessing all non-functional software features:

$$D = \left(k - \sum_{i=1}^{k} \frac{qmc_i}{qnc_i} \right) / k, \tag{2}$$

where k is the number of non-functional software features ($k = 9$ according to ISO/IEC TR 19759:2015), qmc_i is the number of missing (in the real SRS) attributes and/or indicators for the ith non-functional feature, qnc_i is the number of necessary attributes and/or indicators for the ith non-functional feature (determined by base ontologies for every non-functional feature).

The developed method of activity of OBIA provides the conclusion about sufficiency or insufficiency of information in the real SRS and about the need for supplementing attributes and/or indicators in the SRS; evaluation of the level of sufficiency of information in the real SRS for assessing each non-functional feature; and assessing all non-functional features of the software.

Main results of the developed method are numerical evaluations of the level of sufficiency of the information in the SRS for assessing every non-functional software feature and for assessing all non-functional software features, equations for calculating of which are proposed for the first time.

4   Evaluating Sufficiency of Information in the SRS by Ontology-Based Intelligent Agent

For this experiment, OBIA uses, as known facts, base ontologies of the "Software Engineering" subject domain (with the following parts: "Software Requirements Specification (Quality Measures)", "Software Requirements Specification (Quality Indicators)", Figs. 1, 2, 3 and 4), which was developed in [18]. In these ontologies, quality measures (for assessing software quality according to ISO 25010:2011 and ISO 25023:2016) and quality indicators (for assessing software quality by metric analysis) are presented taking into account their division by sections of the SRS. Thereby, the developed ontologies are templates of the SRS in terms of presence of measures and/or indicators.

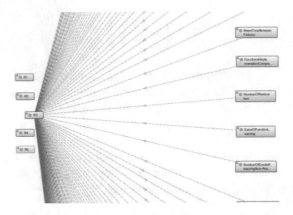

Fig. 1. Fragment of the ontology of the "Software Engineering" subject domain ("SRS (Quality Measures)" part)—template of the SRS in terms of presence of quality measures (according to ISO 29148:2011)

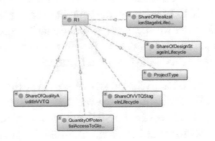

Fig. 2. Component of the ontology of the "Software Engineering" subject domain ("SRS (Quality Indicators)" part)—indicators present in Sect. 1 of the SRS according to ISO 29148:2011 (template of Sect. 1 of the SRS in terms of availability of quality indicators).

According to the developed method of activity of OBIA, the collation and comparison of the information of the 3rd section of the SRS of the transport logistics decision support system with the base ontology of the 3rd section of the SRS (Fig. 1) was made (3rd section of the SRS contains measures, which are necessary for assessing the software quality according to ISO 25010). As a result of this collation and comparison, it has been found that the following measures are missing in this SRS: Number of Functions, Product Size, Maximum Memory Utilization, IO Loading Limits, Mean Amount of Throughput, Number of Interface Elements, Number of Variables, Number of Data Items, Number of Controllability Requirements, Access Controllability, Number of Events Requiring Non-Repudiation Property, Number of Provided Authentification Methods, Number of Data Formats To Be Exchanged, Number of Data Structures, Ease of Installation, Number of Installation Steps.

Analysis of influence of each element in the set of missing measures on the software quality and its subcharacteristics, conducted on the basis of the base ontology of the "Software Engineering" subject domain ("Software Quality" part), developed in

[18], made it possible to calculate the numbers of missing measures for subcharacteristics of software quality. In addition, this analysis provided the conclusion that, on the basis of measures available in the SRS, it is impossible to assess such quality subcharacteristics as Functional Completeness, Functional Correctness, Functional Appropriateness, Maturity, Fault Tolerance, Time Behaviour, Resource Utilization, Capacity, Appropriateness Recognisability, Learnability, Operability, User Interface Aesthethics, Modularity, Analysability, Modifability, Confidentiality, Integrity, NonRepudiation, Authenticity, CoExistence, Interoperability, Adaptability, Installability, and Replaceability. I.e. information in this SRS is insufficient for assessing 24 quality subcharacteristics out of 31 subcharacteristics (according to ISO 25010).

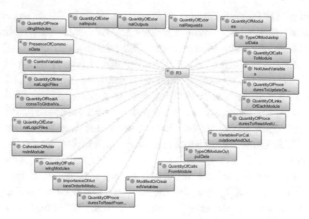

Fig. 3. Component of the ontology of the "Software Engineering" subject domain ("SRS (Quality Indicators)" part)—indicators present in Sect. 3 of the SRS according to ISO 29148:2011 (template of Sect. 3 of the SRS in terms of availability of quality indicators).

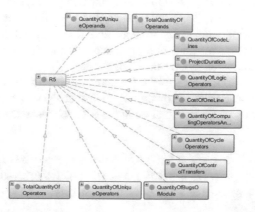

Fig. 4. Component of the ontology of the "Software Engineering" subject domain ("SRS (Quality Indicators)" part)—indicators present in Sect. 5 of the SRS according to ISO 29148:2011 (template of Sect. 5 of the SRS in terms of availability of quality indicators).

On the basis of conducted analysis, the OBIA concluded that the information in the SRS of the transport logistics decision support system is insufficient for assessing the software quality according to ISO 25010. Numerical evaluation of the level of sufficiency of the information in the SRS for assessing software quality according to ISO 25010 is

$$
\begin{aligned}
D_{Quality} = (31 - (&1/4 + 1/5 + 1/6 + 1/14 + 0/4 + 1/5 + 0/7 + 1/7 + 2/14 + 2/5 \\
&+ 1/6 + 1/8 + 2/13 + 0/11 + 1/6 + 0/5 + 2/7 + 0/6 + 1/6 + 1/8 + 0/6 + 3/10 \\
&+ 3/8 + 1/2 + 0/2 + 1/1 + 2/4 + 1/5 + 3/11 + 2/4 + 2/3))/31 = 0.77.
\end{aligned}
$$

$$(3)$$

Consequently, OBIA concludes: "Measures of the SRS are insufficient for assessing software quality. The level of sufficiency of information of the SRS is 77%. There is a need to supplement the measures (list of measures is given) in this SRS".

Similarly, collation and comparison of information in the sections of the SRS of the transport logistics decision support system with the base ontologies of the sections of the SRS (Figs. 2, 3 and 4) was made (1st, 3rd, and 5th sections of the SRS contain indicators necessary for assessing software quality by metric analysis). As a result of this collation and comparison, it has been found that the following indicators are missing in this SRS: Quantity of Modules, Project Duration, Total Quantity of Operators, and Project Type.

As a result of analysis of the influence of each element in the set of missing indicators on software quality and its metrics, conducted on the basis of the base ontology of "Software Engineering" subject domain ("Software Quality. Metric Analysis" part), developed in [18], numbers of missing indicators for metrics of software quality are calculated. In addition, this analysis provided the conclusion that on the basis of indicators available in the SRS, it is impossible to assess such quality metrics as Chepin's metric, Jilb's absolute metric, McClure's metric, Kafur's metric, Quantity of found bugs in models, Expected estimate of interfaces complexity, Time of models modification, Software design total time, Design stage time, Realization productivity, Halstead's metric, McCabe's metric, Jilb's relative metric, Quantity of program statements, Effort applied by Boehm, and Development time by Boehm. I.e. information in this SRS is insufficient for assessing 16 quality metrics out of 24 metrics (which can be calculated already at the design stage).

Thus, information in the SRS of the transport logistics decision support system is insufficient for assessing software quality by metric analysis. Numerical evaluation of the level of sufficiency of the information in the SRS for assessing the software quality by metric analysis is

$$
\begin{aligned}
D_{Quality\_metr} = (24 - (&1/5 + 1/2 + 1/5 + 1/4 + 0/2 + 0/5 + 0/2 + 1/3 + 1/2 + 0/1 \\
&+ 1/5 + 1/3 + 1/3 + 1/2 + 1/3 + 1/2 + 1/3 + 0/2 + 0/4 \\
&+ 1/2 + 0/3 + 0/5 + 1/2 + 1/2))/24 = 0.75
\end{aligned}
$$

$$(4)$$

OBIA concludes: "Indicators of the SRS are insufficient for assessing the software quality by metric analysis. The level of sufficiency of information of the SRS is 75%. There is a need to supplement the indicators (list of indicators is given) in this SRS".

Consequently, nowadays OBIA assesses sufficiency of information in the SRS for the software quality assessment. It provides the conclusion about sufficiency or insufficiency of information, numerical evaluation of the sufficiency level, the list of measures and/or indicators that should be supplemented in the SRS to increase the level of its information sufficiency.

During this experiment, OBIA cannot calculate numerical evaluation of the level of sufficiency of information in the SRS for assessing all non-functional software features, because nowadays knowledge base of OBIA doesn't contain ontologies for other non-functional features of software.

5 Conclusions

In this paper, the method of activity of the ontology-based intelligent agent for evaluating initial stages of the software lifecycle is developed for the first time. Equations for numerical evaluations of the level of sufficiency of information in the SRS for assessing each non-functional software feature and for assessing all non-functional software features are proposed for the first time.

The intelligent agent, which is working on the basis of the developed method, assesses sufficiency of information in the SRS for assessing non-functional software features. It makes conclusions about sufficiency or insufficiency of information, evaluates the level of sufficiency of SRS information for assessing each non-functional feature in particular and all non-functional characteristics in general, provides the list of attributes (measures) and/or indicators, which it is necessary to supplement into the SRS to increase the level of sufficiency of its information.

For the experiment, the intelligent agent used the ontology of the "Software Engineering" subject domain (with the following parts: "Software Quality," "Software Quality. Metric Analysis," "Software Requirements Specification (Quality Measures)," "Software Requirements Specification (Quality Indicators)"). The information in the SRS for the transport logistics decision support system was evaluated in terms of its sufficiency. During the experiments, the intelligent agent determined that information in this SRS is insufficient for assessing software quality according to ISO 25010 (level of sufficiency is 77%) and for assessing software quality by metric analysis (level of sufficiency is 75%). Therefore, OBIA recommended adding to this SRS measures and indicators necessary for assessing software quality (with the list of missing measures and indicators).

Obviously, for evaluating sufficiency of information in the SRS for assessing all non-functional features of software, it is necessary to develop base ontologies of SRS sections (that contain necessary attributes (measures) and/or indicators for assessing all non-functional features), and base ontologies for all non-functional software features. After downloading developed ontologies into the knowledge base of the agent, it will be able to draw conclusions about sufficiency or insufficiency of information in the SRS for assessing all non-functional features. In addition, it is necessary to develop a separate agent that will conduct semantic analysis of natural-language specifications for the search of attributes (measures) and/or indicators. Further efforts of the authors will be directed to solving such tasks.

References

1. Sivarajah, U., Kamal, M.M., Irani, Z., Weerakkody, V.: Critical analysis of big data challenges and analytical methods. J. Bus. Res. **70**, 263–286 (2017)
2. Mauerhoefer, T., Strese, S., Brettel, M.: The impact of information technology on new product development performance. J. Prod. Innov. Manag. **34**(6), 719–738 (2017)
3. Kinch, M.W., Melis, W.J.C., Keates, S.: Reviewing the current state of machine learning for artificial intelligence with regards to the use of contextual information. In: The Second Medway Engineering Conference on Systems: Efficiency, Sustainability and Modelling Proceedings. University of Greenwich (2017)
4. Noor, A.K.: Potential of cognitive computing and cognitive systems. Open Eng. **5**(1), 75–88 (2015)
5. LeCun, Y., Bengio, Y., Hinton, G.: Deep learning. Nature **521**(7553), 436–444 (2015)
6. Ristoski, P., Paulheim, H.: Semantic web in data mining and knowledge discovery: a comprehensive survey. J. WEB Seman. **36**, 1–22 (2016)
7. Golub, K.: Subject Access to Information: An Interdisciplinary Approach. Libraries Unlimited, Westport (2015)
8. Hastie, S., Wojewoda, S.: Standish Group 2015 Chaos Report – Q&A with Jennifer Lynch. http://www.infoq.com/articles/standish-chaos-2015. Accessed 13 Mar 2018
9. A Look at 25 Years of Software Projects. What Can We Learn? https://speedandfunction.com/look-25-years-software-projects-can-learn/. Accessed 13 Mar 2018
10. McConnell, S.: Code Complete. Microsoft Press, Redmond (2013)
11. Wooldridge, M., Jennings, N.R.: Intelligent agents – theory and practice. Knowl. Eng. Rev. **10**(2), 115–152 (1995)
12. Freitas, A., Bordini, R.H., Vieira, R.: Model-driven engineering of multi-agent systems based on ontologies. Appl. Ontol. **12**(2), 157–188 (2017)
13. Ossowska, K., Szewc, L., Weichbroth, P., Garnik, I., Sikorski, M.: Exploring an ontological approach for user requirements elicitation in the design of online virtual agents. In: Information Systems: Development, Research, Applications, Education, vol. 264, pp. 40–55 (2017)
14. Lezcano-Rodriguez, L.A., Guzman-Luna, J.A.: Ontological characterization of basics of KAOS chart from natural language. Rev. Iteckne **13**(2), 157–168 (2016)
15. Garcia-Magariño, I., Gomez-Sanz, JJ.: An ontological and agent-oriented modeling approach for the specification of intelligent ambient assisted living systems for parkinson patients. In: Hybrid Artificial Intelligent Systems, vol. 8073, pp. 11–20 (2013)
16. Wilk, S., Michalowski, W., O'Sullivan, D., Farion, K., Sayyad-Shirabad, J., Kuziemsky, C., Kukawka, B.: A task-based support architecture for developing point-of-care clinical decision support systems for the emergency department. Methods Inf. Med. **52**(1), 18–32 (2013)
17. Rakib, A., Faruqui, R.U.: A formal approach to modelling and verifying resource-bounded context-aware agents. Lecture Notes of the Institute for Computer Sciences Social Informatics and Telecommunications Engineering, vol. 109, pp. 86–96 (2013)
18. Hovorushchenko, T.: Information technology for assurance of veracity of quality information in the software requirements specification. Advances in Intelligent Systems and Computing II, vol. 689, pp. 166–185 (2018)

Ontological Representation of Legal Information and an Idea of Crowdsourcing for Its Filling

Anatolii Getman[1] , Volodymyr Karasiuk[1](✉) ,
Yevhen Hetman[2] , and Oleg Shynkarov[1]

[1] Yaroslav Mudryi National Law University, Kharkiv, Ukraine
karasiuk@yahoo.com
[2] National Academy of Law Science of Ukraine, Kharkiv, Ukraine

Abstract. This article represents consideration of the creation process of legal knowledge ontology for study purposes. The peculiarities of legal information and experience of legal knowledge formalization have been scrutinized. The peculiarities of complex systems self-organization have been considered and application of these principles to legal information on the basis of four features of self-organization has been proved. It has been determined that the most reasonable way of legal knowledge description is ontology, as a basis for forming of knowledge structure. The review of existing ontologies that are used in the field of law has been carried out. Mathematical description of the knowledge base structure has been introduced. The software package has been developed for working with legal knowledge ontology. This package of programs is used by students at the Yaroslav Mudryi National Law University. The method of collective filling and editing of the knowledge base is proposed to be used as the basis of methodology for working with the knowledge base. The ontology of legal knowledge at the University has been created not only by experts but by all the users. Principles of crowdsourcing are considered as a basic technique of technological process of the ontology filling. Results of filling of this ontology by a number of users have been briefly reviewed. The legal knowledge ontology that is being created is proposed to be used for forming an individual learning style of students.

Keywords: Legal information · Ontology
Self-organization of legal information · Crowdsourcing
Individual learning style

1 Introduction

Legal information is a set of documentary or publicly disseminated information about law, its system, sources, implementation, juridical facts, legal relations, law enforcement, law violation, prevention of law violation etc. [1]. However, this set is not homogeneous. Informational resources are stored in different allocated storages, are structured and described in a different way, using diverse terminology. As for available

O. Chertov et al. (Eds.): ICDSIAI 2018, AISC 836, pp. 179–188, 2019.
https://doi.org/10.1007/978-3-319-97885-7_18

electronic resources, they also have versatile formats that are incompatible, with no ways of codifying their content.

There is an urgent need to structure legal information in a single well-organized information system, navigation in which should be simple and available for an average user. This is especially important from the point of view of improving educational process of teaching law, when a significant part of resources is transferred to electronic realm, and the primary users are often not very skilled students.

One of the modern ways of knowledge integration is use of ontological description. Many conceptual interpretations of ontology are widely known. Therefore, here we can use a working definition (formulated by Valkman in [2] on the basis of definitions of Gruber et al.) oriented at provision of intellectual modeling processes. Ontology is a specialized knowledge base, in which all concepts and relationships between them, necessary and sufficient to solve the problems of a particular problem area, are revealed.

However, the process of ontologies creation is quite complicated. To solve the task of filling of the modern information system by actual content is possible only due to collaborative work of a large number of users, as it cannot be done only by experts in the domain. Therefore, for entry of new information, it is reasonable to attract many users and provide them with tools for this work.

2 Self-organization of Legal Information

In social sciences, the concept of self-organization was established long ago, due to the presence of obvious interaction between subjects. The idea of self-organization is also admitted in the humanities, although this paradigm also needs detailed study. For example, the development of humanity was also explained in terms of self-organization idea from the perspective of explanation of historical development processes [3]. In philosophy, the concept of self-organization is used for substantiation of development of science and epistemology. Yet in 1920s, there was raised an idea of philosophy development based on holism principles, which are actually prerequisites of self-organization [4]. At present time, self-organization principles are actively used in educational process [5].

As for legal information, it is stated in [6, 7] that legal informational structures have a special feature of self-organization. This statement is based on several factors. For example, organizational structure of judicial power has strict hierarchy (courts in Justitia: Local—Appeal—Supreme). Structure of law-enforcement authorities complies with the structure of local self-government authorities. A similar hierarchy exists in the structure of legal information. For example, laws and other regulatory documents should not contradict the Constitution. Laws and other normative documents are to be on the higher level than instructions, provisions, etc. Also, legal information structures and the dynamics of legal systems are examined in [8]. Here, any new notion (concept), loaded into the knowledge structure, is uniquely associated with its level of hierarchy. However, clarifications and restrictions that complicate this scheme occur in practice. Garbarino simulates self-organization in accordance with extraction rules of prece-dential decisions. In the hierarchy that is defined by binding power of precedent—

delivery rule that connects the precedent and decision. In [6], self-organization principles in court-oriented sectors of legal information with focus on bankruptcy cases were analysed. Trujillo concludes that self-organization is the central process in the content of legal information and establishes the fact that the order emerges due to variations.

Systems that have a feature of self-organization exhibit distinctions [9]. These peculiarities were first discovered in thermodynamic systems and determined in social and informational systems [10]. These features are basic, and include openness, availability of sufficient number of interactive elements, disequilibrium, dissipativity. We are going to analyze if legal information has these features.

Legal information is located in the local informational network of the higher educational establishment, and all the students have access to it for reading and editing. In the described project, legal knowledge ontology is formed with the help of all interested students of the University.

The network information system, in which every user can receive access and change information, should be considered as open. The potential number of the network users (and there are significantly more than 1,000 registered users in the network at the University) allows to consider it as sufficient in measures for beginning of self-organization processes. That means properties of openness and presence of sufficient number of elements are met [11].

The fundamental principle of self-organization is emergence and strengthening of order via fluctuations. In the case of their emergence, the system tends to weaken them and liquidate. Although in open systems these processes acquire other scopes and due to strengthening of disequilibrium these deviations increase with time and finally cause "smearing" of order exciting in the system and emergence of new order. Fluctuation of state in the information system is a change of content (informational filling) of system. At the same time, if there a possibility of liquidation of last changes (i.e. a return to the previous state), then it will be considered as suppression of incorrect content fluctuations. Correspondingly, the huge accumulation of informational filling changes should be considered as a creation of the order due to fluctuations. The disequilibrium causes selectivity of the system and unusual reactions to external displays. That means informational filling of the system will not always be correct in separate fragments, but with the passage of time and rising of change processes, these cases of incorrectness will be eliminated.

The open disequilibrium systems acquire a property of dissipativity during the process of interaction with external environment. Dissipativity is considered as macroscopic emergence of the processes that occur on microlevel. Due to dissipativity in disequilibrium systems, new structures may spontaneously occur, transfer from chaos and mess to organization, and order may emerge. In the context of information systems, this is tantamount to uncoordinated work of multiple users that finally converges to creation of correct knowledge structures. Due to that, the presence of the third and fourth peculiarities of the self-organizing systems [12] can be considered to be proved. Therefore, laws of self-organization of information occur in the legal information system by analogy with the general laws of self-organization that take place in living nature [10]. In fact, based on the results of the analysis, it can be considered proven that in legal systems, there is a principle of self-organization of information.

It should be noted that at the present moment, there is no mathematical model of self-organization process for complex, widespread information systems. However, as shown above, these principles operate in practice in legal information systems and can be used.

3 Ontology Legal Knowledge as a Tool for Building a Training System

Ontological structures (ontologies) are widely used for description of resources in semantic Web applications, corporate databases, systems for document processing [13]. Peculiarities that are typical for ontological resources are presented in [14]. In [15], it is proved why ontologies are attractive in the legal domain. There is a number of software systems that have proved their effectiveness in the field of law. Reviews of characteristics of some systems and methodologies of their building can be found in [16–19]. The most interesting systems are FOLaw (Functional Ontology of Law) [20], LRI Core [21], Frame-based Ontology, CLO (Core Legal Ontology), Jurwordnet [22]. Legal case-based reasoning (LCBR) systems should also be mentioned [23, 24]. To resolve online disputes, the Ontomedia platform is used that works with Core Mediation Ontology and the ontological subdomain is used to mediate with individual subdomains (for example, the consumer's problem area, the family model problem area, etc.) [25].

With regard to the methodology for the considered ontology, all these projects involve compulsory participation of experts in the problem domain knowledge (domain experts), in knowledge engineering (knowledge engineers), and ontology engineers. This leads to high costs, limited "bandwidth" of the system of ontology building and updating its content, which eventually leads to the viability of the relatively small amount of knowledge projects. In largescale projects, this scheme is too costly, and it is necessary to organize the work and the interaction between a number of experts.

Thus, the existing legal ontology has developed structure and interfaces, however, to replenish them, experts are needed, and their scales are within the local branches of law. At present, the principle of ontological description of the data is the most appropriate mechanism for modeling the conclusions in the legal systems and, consequently, to educate students and interested users.

4 Formalization of Legal Knowledge Ontology Description

For completeness of the presentation of study domain, we give a formal description of the ontology of legal knowledge. As a basis, we take the description given in [11]. At a set-theoretic level, the ontology can be presented as:

$$O = \langle P, R, F \rangle, \tag{1}$$

where P is the final set of concepts (terms) in the domain law; R is a finite set of relations between concepts (terms) of a given domain; F is a final set of interpretation functions (axiomatization) given on the concepts and/or relationships of ontology O. The only restriction, which is imposed on P, is its extremity and non-emptiness:

$$P = \{P_i\}, \quad i = \overline{1, m},$$
$$P \neq \{\emptyset\}, \quad m = \{1, 2, 3 \ldots, M\}, \tag{2}$$

where P_i is a single concept that has its own semantic representation, which is connected with a specific set of facts and the set of admissible syntactic constructions; i is a number of concepts in the structure of knowledge; m is a number of elements in P, M is a positive finite integer. Thus, ontology is a dictionary for representing and sharing knowledge about a certain subject area and the set of links established between terms in this dictionary. In this case, the concept P_i can be formally represented as a set of phrases W_j^i. Each phrase may consist of a group of concepts. Each concept can have synonyms S_q^i:

$$P_i = \left(W_1^i, \ldots, W_n^i\right),$$
$$W_j^i = \left(S_1^i, \ldots, S_k^i\right), \quad i = \overline{1, m}; \quad j = \overline{1, n}; \quad q = \overline{1, x}; \tag{3}$$
$$S_q^i = \left(Q_{j1}^i, \ldots, Q_{jx}^i\right).$$

That is, the concept P_i can be intricately described and have many variants of representation (n), each of which also can have a number of components (1 to k). These components, in turn, can have synonyms (from 1 to x). An element of the ontology is also a connection R_r between concepts or a group of concepts:

$$(P_l, \ldots, P_t) R_r (P_v, \ldots, P_w). \tag{4}$$

In (4), the set of concepts $(P_l, \ldots, P_t)$ has a connection with the type R_r with a group of concepts $(P_v, \ldots, P_w)$. Also, each version of the concept presentation W_j^i has communication with the legislative (accurate) definitions of this concept in the regulations:

$$\left(W_1^i, \ldots, W_n^i\right) R_y \left(Z_1^i, \ldots, Z_n^i\right), \tag{5}$$

where $\left(Z_1^i, \ldots, Z_n^i\right)$ is a set of legal definitions or explanations, with the indication of the period of the legitimacy of this definition. In general, the number of elements in W_j^i need not coincide with the number of elements in Z_j^i. The case when Z_j^i will be empty is possible; legislative definition of a concept is absent [26].

5   Crowdsourcing to Form an Ontology of Legal Knowledge

A distinctive principle of crowdsourcing is the possibility of dividing the work into small parts (fragments) that can be carried out by different performers. In principle, any performers who wish to work on conditions of crowdsourcing are allowed. However, in a large number of practical systems, we are talking about additional conditions (procedures) for allowing performers to work with information systems. Such filtering of performers is considered as a prerequisite for the application of crowdsourcing to more complex information structures having responsible content. In our case, students of the University are encouraged to work with legal knowledge ontology, that is itself some sort of qualification condition for access.

For technical support of this work, a software package called JURONT was developed. Software implementation of the package is in the form of four sub-systems, using Java object-oriented programming language. The software package uses a web-based user interface and an automated mode of working with the knowledge base. The system interface is written in three languages. The basic subsystem that defines practical significance of the package is application of a user. The global tasks of the user's application are: (1) navigation in ontology; (2) selection of fragments of text-sources complying with ontology elements; (3) browsing complete text sources; (4) browsing texts divided into sections; (5) browsing marked text and others [27]. The package is installed in the local network of the University. Users were assigned to (voluntarily) to fill ontology concepts that cover the material studied in the "Criminal Law of Ukraine" and "Legal information" courses. During the period from 2014 to 2017, the software package was improved, the experience of using the package by students of different specializations was accumulated.

As it was already mentioned, for principle actuation of self-organization in the case of ontology filling, there is a need in emergence and strengthening of order in the information structure via fluctuations. For example, the considered principle of legal information self-organization emerges as follows. In legal ontology, there are branches of high level: criminal law, civil law, administrative law, and others. If any user has mistakenly included a concept from area of criminal law in a branch of civil law, then during the process of operation and filling of the system, there will appear the other user who will edit a position of the mentioned concept and include it in the branch of criminal law. Every user of the University network (a student, teacher, research associate) has a possibility to supplement and correct the knowledge basis. The mathematical model of the given process for similar social systems does not exist yet.

The study of self-organizing systems is important for artificially created structures, because their prototypes in physical life clearly demonstrate self-organization, and this property should be used in created intelligent information systems.

As a result of the practical filling of the ontology, according to experts, the completeness of extracting concepts from the texts of legal sources in the analyzed ontology branches amounted to more than 90%. In general, more than 6,000 definitions (concepts) were included in the ontology. At the same time, concepts that had complex formulation or multivalued definitions remained unreached in ontology. Students highly appreciated the use of ontology. When doing research work, students also turned to the created ontology.

6 Individual Learning Style with the Help of Ontology

Modern organization of education in universities can be characterized as follows. The problematic area of knowledge and its specific formulations as a rule are designated by a teacher. Students then designate their personal profiles, as a rule only in the social term, little connected with a process of study. Purpose of study is defined by an educational establishment, but not by a student. On the other hand, the target settings for the student should be the basis for the formation of students' learning profiles in terms of the subject area, the definition of learning objectives, and the selection of social subjects that may be involved in the learning process. And from the point of view of the structure of knowledge, these settings should become the basis for the creation of some "semiotic social space" [28], capable of supporting the student community in the process of information exchange while mastering new knowledge. Therefore, the problem of creating an individual educational information space for students of law is an urgent problem, the solution to which will increase the quality of education and adapt the future professional lawyer to using the information sphere of modern society.

The postulate for building individual education systems should be considered as an adaptation to the level of initial knowledge of students. Based on existing knowledge of the student in the information space, it is necessary to generate individual trajectories of learning, which is most expedient to do on the basis of ontologies of subject areas. After all, modern ontologies contain many concepts, they have a format that can be conveniently processed by computer programs, and they also have a strict logical structure.

The study [29] considers four scenarios for the application of ontologies in the learning process: application of existing ontologies to execution of studying activities; application of ontologies to systematization and description of new objects and facts; joint development of ontology as a part of general structure of knowledge; joint research based on many ontologies.

At present, a number of ontologies have been implemented and used in problem areas, which allow to organize an individual trajectory of the learning process for the student [30–34]. In accordance with the developed ontology of legal knowledge, tools are currently being developed to implement an individual learning style within the curriculum of the "Law" major.

7 Conclusions

In the paper, the process of creating the legal knowledge ontology has been analyzed. The peculiarities of legal information and experience of legal knowledge formalization have been studied. The conclusion is drawn that ontology is the most modern mechanism for describing legal knowledge. The principle of self-organization of legal information is grounded. A software application has been developed that ensures the construction of an ontology of legal knowledge. To fill the ontology with current knowledge, the method of crowdsourcing is used. The structure of knowledge on the basis of ontology makes it possible to implement the next step in the formalization of education, i.e. to automate the construction of an individual trajectory of students' training in the "Law" major. In general, the paper suggests the structure of the knowledge base of legal information, the methodology for filling and maintaining the knowledge base in an up-to-date state.

The perspective directions of the research include the following ideas. The supposed improvement of the methodology and software should automatically solve the problem of finding a place in the ontology (the existing node in the structure), to which the new definition (concept) is closest. This problem can be solved due to development of metric for evaluation of ontologies similarity. As a result, the software complex will automatically offer the user the best possible location for the new concept, which is added to the knowledge structure. Also, a promising problem is the development of software to solve the following problems: informational and referential; information and searching; arrangement of the knowledge base structure; building and extension of the individual knowledge base for every user.

References

1. Law of Ukraine: "On information" No. 2657-XII. In: Summaries of the Verkhovna Rada of Ukraine (VRU), 48 (1992)
2. Valkman, Yu., Stepashko, P.: Principles of constructing the ontology of intelligent modeling. In: Valkman, Yu., et al. (eds.) Proceedings of the XVI International Conference «Intellectual Information Analysis» IIA-2016, pp. 25–34, Prosvita, Kiev (2016)
3. Laszlo, E.: A Strategy for the Future: The Systems Approach to World Order. Braziller, New York (1974)
4. Smuts, J.: Holism and Evolution. The Macmillan Company, New York (1926)
5. Sun, Z., Xie, K., Anderman, L.: The role of self-regulated learning in students' success in flipped undergraduate math courses. Internet High. Educ. **36**, 41–53 (2018). https://doi.org/10.1016/j.iheduc.2017.09.003
6. Trujillo, B.: Self-organizing legal systems: precedent and variation in bankruptcy. In: Farell, N. (ed.) Utah Law Review, vol. 2, pp. 483–562 (2004). http://papers.ssrn.com/sol3/papers.cfm?abstract_id=924673. Accessed 18 Feb 2018
7. Kawaguchi, K.H.: A Social Theory of International Law: International Relations As a Complex System. Brill Academic Publishers, The Netherlands (2003). https://doi.org/10.1007/978-94-017-4978-7
8. Garbarino, C.: A model of legal systems as evolutionary networks: normative complexity and self-organization of clusters of rules. Bocconi Legal Studies Research Paper No. 1601338, Milan (2010). https://papers.ssrn.com/sol3/papers.cfm?abstract_id=1601338. Accessed 15 Sept 2017
9. Nicolis, G., Prigogine, I.: Self-organization in Nonequilibrium Systems. Wiley, New York (1977)
10. Hacken, G.: Information and Self-organization: A Macroscopic Approach to Complex Systems. World, Moskow (1991)
11. Getman, A., Karasiuk, V.: A crowdsourcing approach to building a legal ontology from text. Artif. Intell. Law **22**(3), 313–335 (2014)
12. Naidysh, V.: Concepts of Modern Natural Science. Alpha-M, Infra-M, Moskow (2004)
13. Tuzovsky, A., Chirikov, S., Yampolsky, V.: Knowledge Management Systems (Methods and Technologies). Publishing House of Scientific and Technical Literature, Tomsk (2005)

14. Soloviev, V., Dobrov, B., Ivanov, V., Lukashevich, N.: Ontologies and thesauri. Kazan State University, Kazan (2006). http://window.edu.ru/resource/722/41722. Accessed 25 May 2016

15. Bench-Capon, T., Visser, P.: Ontologies in legal information systems; the need for explicit specifications of domain conceptualisations. In: Proceedings of the 6th International Conference on Artificial Intelligence and Law, pp. 132–141, Melbourne, Australia (1997)

16. Saravanan, M., Ravindran, B., Raman, S.: Improving legal information retrieval using an ontological framework. Artif. Intell. Law 17(2), 101–124 (2009). https://doi.org/10.1007/s10506-009-9075-y

17. Nardi, J., Falbo, R., Almeida, J.: Foundational ontologies for semantic integration in EAI: a systematic literature review. In: Douligeris, C., Polemi N., Karantjias, A., Lamersdorf, W. (eds.) Proceedings I3E 2013, pp. 238–249. (2013). https://hal.inria.fr/hal-01470537/document. Accessed 22 Mar 2018

18. Nguyen, V.: Ontologies and information systems: a literature survey. DSTO-TN-1002. Defence Science and Technology Organization, Edinburg, South Australia, (2011). http://www.dtic.mil/get-tr-doc/pdf?AD=ADA546186. Accessed 24 Apr 2017

19. Ding, Y., Foo, S.: Ontology research and development, Part 1 - A review of ontology generation. J. Inf. Sci. 28(2), (2002). http://www.ntuedu.sg/home/sfoo/publications/2002/02jis01_fmt.pdf. Accessed 07 Mar 2014

20. Breuker, J., Hoekstra R.: Epistemology and ontology in core ontologies: FOLaw and LRI-Core, two core ontologies for law. In: Proceedings of the EKAW04 Workshop on Core Ontologies in Ontology Engineering, Northamptonshire, UK, pp. 15–27. http://dare.uva.nl/document/8751. Accessed 01 Mar 2014

21. Gangemi, A., Prisco, A., Sagri, M.T., Steve, G., Tiscornia, D.: Some ontological tools to support legal regulatory compliance, with a case study. In: Meersman, R., Tari, Z. (eds.) On the Move to Meaningful Internet Systems 2003, OTM 2003 Workshops. LNCS, 2889, pp. 607–620. Springer, Heidelberg (2003). http://www.loa.istc.cnr.it/Papers/WORM-CORE.pdf. Accessed 10 Sept 2015

22. Sagri, M., Tiscornia, D., Bertagna, F.: Jur-WordNet. In: Second International Wordnet Conference, pp. 305–310. Masaryk University, Brno (2004). https://www.fi.muni.cz/gwc2004/proc/111.pdf. Accessed 10 Mar 2018

23. Henderson, J., Bench-Capon, T.: Dynamic arguments in a case law domain. In.: Loui, R. P. (ed.) ICAIL 2001, Proceedings of the 8th International Conference on Artificial Intelligence and Law, pp. 60–69. ACM Press, New York (2001). https://doi.org/10.1145/383535.383542

24. Zeng, Y., Wang, R., Zeleznikow, J., Kemp, E.: Knowledge representation for the intelligent legal case retrieval. In: Khosla, R., Howlett, R.J., Jain, L.C. (eds.) Knowledge-Based Intelligent Information and Engineering Systems. KES 2005. LNCS (Part 1), vol. 3681, pp. 339–345. Springer, Berlin, Heidelberg (2005). https://doi.org/10.1007/11552413_49

25. Poblet, M., Casanovas, P., López-Cobo, J.-M., Castellas, N.: ODR, ontologies, and Web 2.0. J. Univ. Comput. Sci. 17(4), 618–634 (2011). http://wwwresearchgate.net/publication/258046605_ODR_Ontologies_and_Web_2.0. Accessed 21 Jan 2018

26. Karasiuk, V.: Ontological paradigm of process of content for education purposes. Bull. V. Karazin Kharkiv Natl. Univ. 19(1015), 148–154 (2012)

27. Tatsyi, V., Getman, A., Ivanov, S., Karasiuk, V., Lugoviy, O., Sokolov, O.: Semantic network of knowledge in science of law. In: Shokin, Yu., Bychkov, I., Potaturkin, O. (eds.) Proceedings of the IASTED International Conference on Automation, Control, and Information Technology (ACIT 2010), pp. 218–222. ACTA Press, Anaheim, USA (2010)

28. Gee, J.: Semiotic social spaces and affinity spaces: from the age of mythology to today's schools. In: Barton, D., Tusting, K., (eds.) Beyond Communities of Practice. Language Power and Social Context. Cambridge University Press, Cambridge, MA, pp. 214–232 (2005).   http://www.bendevane.com/RDC2012/wp-content/uploads/2012/08/Gee-Social-Semiotic-Spaces.pdf. Accessed 2018/02/23

29. Wang, S.: Ontology of learning objects repository for pedagogical knowledge sharing. Interdiscip. J. E-Learn. Learn. Obj. **4**, 1–12 (2008). http://www.ijello.org/Volume4/IJELLOv4p001-012Wang200.pdf. Accessed 18 Mar 2018

30. Lendyuk, T., Vasylkiv, N.: Fuzzy model of individual learning path forming and ontology design on its basis. In: Oborsky G.A. (ed.) Informatics and Mathematical Methods in Simulation, vol. 7, no. (1–2), pp. 103–112 (2017)

31. Sokolov, A., Radivonenko, O., Morozova, O., Molchanov, O.: Use of the ontological test in the system of assessing the quality of training. In: Sokol, E. (ed.) Bulletin of the National Technical University "KhPI", vol. 2, pp. 79–85 (2011)

32. Morozova, O.: Information technology of the training process organization based on the identification of individual parameters. Sci. Techn. J. Sci. Technol. the Air Force. Ukraine **3** (36), 265–268 (2013)

33. Telnov, Yu., Kazakov, V., Kozlova, O.: Dynamic intellectual system of process management in information and education environment of higher educational institutions. J. Open Educ. **1**(96), 40–49 (2013). https://doi.org/10.21686/1818-4243-2013-1(96)-40-49

34. Valaski, J., Malucelli, A., Reinehr, S.: Recommending learning materials according to ontology-based learning styles. In: CAPES (2011). https://www.ppgia.pucpr.br/pt/arquivos/pesquisa/engsoft/2011/recommending_learning_materials_according_to_ontology-based_learning_styles.pdf. Accessed 25 Feb 2018

Methods for Automated Generation of Scripts Hierarchies from Examples and Diagnosis of Behavior

Viktoriia Ruvinskaya[1] and Alexandra Moldavskaya[2(✉)]

[1] Odessa National Polytechnic University, Odessa, Ukraine
iolnlen@te.net.ua
[2] Odessa College of Computer Technologies, Odessa, Ukraine
am.poly@ya.ru

Abstract. The aim of the research is to increase the reliability of the behavior diagnostics by developing new models and methods based on scripts automatically extracted from data. An improved model of script hierarchies is proposed by adding concepts of role, forest of hierarchies, as well as the support function that connects them. An improved model of multilevel behavior pattern construction is proposed. That, unlike existing models, enabled using methods based on machine learning, along with an expert, to formulate scripts. The 2-staged method for diagnosing the objects behavior based on script hierarchies is developed: at the first stage, identification of the tested behavior to one or several script hierarchies is made; in the second stage, based on the naive Bayesian classifier, it is detected if the object belongs to one or more classes. Approbation of models and methods for the subject area of detecting malicious programs is carried out. The results show an increase in detection reliability.

Keywords: Knowledge-oriented systems · Sequential patterns
Behavior analysis · Scripts · Sequential pattern mining · Malware

1 Introduction

The most appropriate models for describing behavior are models of knowledge representation, in particular, scripts first proposed by Schank and Abelson [1], and also by Minsky [2]. Then there were various attempts to formalize scripts, including on the basis of regular expressions [3]. In [4] it is suggested to use sequential patterns, both one- and multi-level for behavior description on the basis of examples.

However, it's natural to describe object behavior with at least two-level constructions. The top level represents semantic actions. For example, in program lifecycle it's creating a file, requesting server data and so on. The bottom level is their implementation in the form of elementary actions which can be recorded. The task of the behavior analyzer is recovering the top level having the bottom one, and thus discovering typical behaviors. So for the further development of such systems, actual tasks are, on the one hand, the improvement of the formal script model and the script hierarchy, as well as sequential patterns, and on the other hand, the development of methods for analyzing and diagnosing behavior based on scripts.

O. Chertov et al. (Eds.): ICDSIAI 2018, AISC 836, pp. 189–198, 2019.
https://doi.org/10.1007/978-3-319-97885-7_19

2 Review of Existing Solutions

The definition of the behavior concept depends on the scientific field, where this concept is applied. From the point of view of system analysis, when modeling the behavior of investigated object, it is not a static structure but a process of development of a certain system that reflects its basic properties. For this purpose, simulation models are created that treat an object of arbitrary complexity as an independent system and allow one to explore its features of functioning and development [5]. The behavior of the analyzed object is studied through the interactions occurring in the modeled system [6]. In system analysis, as well as in knowledge engineering, behavior is understood as a chain of actions ensuring the development of a system (object) [5]. The final result of the execution of such chain is called the goal [1, 7].

To describe and analyze the behavior, various models are used, in particular, finite automata that simulate the execution of actions [8], but they are not sufficiently intuitive. Knowledge representation models are more suitable for describing behavior. Growing pyramidal networks [9] have been applied in robotic systems, for design and others, realizing the hypothesis about structuring the information during the perception process. However, knowledge representations, especially specialized for the description of behavior, are also expedient. The feature vectors are often used, but describing the behavior in the form of a sequence of actions is more natural. To do this, the most applicable models of knowledge representation in the form of scripts. The Minsky script [2] is a typical structure describing some event while taking its context into account. The unit of the script corresponding to the frame slot is in this case the answer to the characteristic question related to the situation. In contrast to Minsky approach, the Schank-Abelson scripts [1, 10] represent successively performed typical actions. They also defined actors (roles) and introduced the concept of the goal – the main reason to make the script actions. Both scripts involve the use of this knowledge representation model in artificial intelligence when creating recognition systems, planning, and others. Minsky frames and scripts were used for robotics, linking the tasks of environmental recognition and adequate response of the machine. In addition, scripts have found application in psychology and linguistics [11], because they are based on human perception of the world and the transmission of meaning.

For the further development of such systems, it is necessary to create knowledge bases based on scripts for different subject areas. However, their manual development by experts requires a lot of effort. In addition, a large amount of data (examples) has been accumulated, on the basis of which it is expedient to build knowledge. Sequential pattern mining [12] is used for automatic construction of behavioral patterns [4]. This is the appropriate approach to develop with the goal of creating scripts based on patterns and analysis on their basis.

Currently, the tasks of behavior analysis are relevant, in particular, for diagnosing the behavior of information systems components in such subject areas as user behavior on web resources, consumer behavior [13], program behavior, in particular malicious [14], in order to ensure the safety of computer systems, detection of spam messages. Some threats, for example, polymorphic malware, can change their code, but their behavior in the system will remain similar. Intrusion detection is carried out by detecting mismatches with the norm [3]. The diagnostic process includes the following steps: (1) identification: recognition or assessment of the object state or behavior; (2) classification: assessment of the accordance of the object state or behavior to a certain class; (3) localization: the determination of specific objects showing signs of violating the functionality of the system being diagnosed [15].

It is concluded that existing models that describe behavior in the form of scripts need improvement in order to obtain these scripts from examples and to use them afterwards in behavior diagnosing.

3 Advanced Model of Script, Script Hierarchy, and Forest of Script Hierarchies

The following requirements should be presented to the script-based model of knowledge representation, which could be constructed in two ways, based on data by machine learning, and on an expert's experience:

1. The knowledge representation model in the form of scripts should contain layers:

- scripts describing the behavior in a human-readable manner and suitable for analysis; such knowledge answers the questions "What?" and "Why?" for the expert;
- sequential patterns, that is, typical sequences of actions, with which learning algorithms that operate with low-level input data can work.

2. It is necessary to convert gained sequential patterns into scripts.

Let's construct a formal model of the script. Let G be the set of goals. For example, the goals for malicious programs are: file infection, calling API functions and the like. $At \subseteq G$ is the set of atomic goals (or atomic actions), those that are not detailed in the model, for example, the API function call. $G \backslash At$ is the set of non-atomic goals, which are detailed in the model. The script is the triple $\Lambda = (g, S, R)$, where $g \in G$ is the goal of the script, $S \subseteq G$ is the set of script sub-goals that are necessary for the goal g realization, R is the regular language over the alphabet S, described by regular expression. A regular expression includes sub-goals $g_i \in S$ and the following basic operations: concatenation is denoted by "–", OR is denoted by the symbol "|", the iteration is denoted by the "*" symbol; and additional: ? – an optional sub-goal (the name ends with a "?" symbol), + similar to *, but repetitions must happen more than 0 times.

Let's formally define the script hierarchy. The script hierarchy is the triple $H = (V, E, \Lambda_0)$, which is the root tree, where V is the set of vertices that are scripts, E is the edges, $\Lambda_0 \in V$ is the root node (the root script in the hierarchy). Let $\Lambda = (g, S, R) \in V$ is a vertex of this tree. Then for each non-atomic sub-goal $g_i \in S \backslash At$ in the tree there is an edge $e_i \in E$ joining the vertex Λ with the vertex Λ_i such that the goal of Λ_i is g_i.

The peculiarity of the proposed model is that hierarchical relationships are not generated by vertices (scripts), but by vertex elements, that is, by sub-goals that are combined into a regular expression. Figure 1 presents an example of script hierarchy describing the work of a parasitic virus.

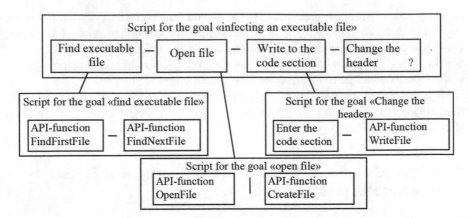

Fig. 1. An example of a script hierarchy for describing a parasitic virus

The script hierarchy corresponds to the criteria of the hierarchical model: there are no loops, hammocks, hierarchical links between elements (scripts) of the same level.

A forest model of script hierarchies $F = \{H_1, H_2, ..., H_k\}$ is proposed. This is the set of all constructed script hierarchies for some subject area. Let's introduce the set of roles (classes), or the set of typical variants of the behavior, Roles = $\{C_1, ..., C_r\}$. Let's also introduce the support function Sup: $F \times Roles \rightarrow [0,1]$. $Sup(H_i, C_j)$ is the probability of encountering a behavior corresponding to the script hierarchy Hi for the class C_j (Fig. 2).

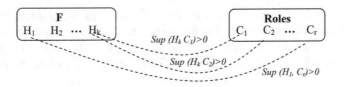

Fig. 2. Forest model script hierarchies

So, the class is described as its own forest of script hierarchies and the behavior of some objects can correspond to several roles $C_{i1}, C_{i2}, ..., C_{in}$. This is typical for modern malicious programs: the behavior of several roles can, for example, manifest a Worm program that also contains the behavior similar to that of the Trojan.

A model is proposed on the basis of sequential patterns for describing the behavior of objects, which can be obtained by machine learning, namely sequential analysis based on sequences of actions, formally representable as follows: $Sq = <a_1, ..., a_n>$, where n is the length of the sequence, $a_i \in At$ is the action, $i = (1, ..., n)$.

A one-level pattern $p \in At^*$ is a sequence of atomic actions. The patterns are extracted from the data (the set of sequences for each role) D: Roles $\rightarrow 2^{At^*}$ using sequential pattern mining. This means that each role is associated with a set of sequences (for Trojans its input set of samples in the form of its own sequence of API functions, own one for viruses, and so on). As a result, let's obtain a set of patterns of the first level $P_1 \subseteq At^*$ with their support for different roles Sup_1: $P_1 \times \text{Roles} \rightarrow [0,1]$.

Since the behavior is advisable to represent hierarchically, a model of a multilevel behavior pattern is proposed. In it, elements of patterns of a higher level are lower-level patterns. From the first level patterns in the samples $D_1 \subseteq P_1^*$ let's obtain the patterns of the second level $P_2 \subseteq P_1^*$ and support Sup2: $P_2 \times \text{Roles} \rightarrow [0,1]$ and so on, from the pattern sequences of the $h-1$ level in the samples $D_{h-1} \subseteq P_{h-1}^*$ let's obtain a set of patterns $P_h \subseteq P_{h-1}^*$ of level h and support Sup_h: $P_h \times \text{Roles} \rightarrow [0,1]$.

The developed model based on sequential patterns (one-level and multi-level) is a transitional between the input data and the knowledge representation model using script hierarchies. It allows to obtain knowledge about typical sequences of actions with the help of sequential pattern mining, level-by-level forming more and more generalized knowledge.

Let's consider Example 1, illustrating the models of behavior. Input sequences with allocated frequently repeated subsequences and their belonging to two classes C_1 and C_2 are shown in Fig. 3. All six patterns of the 1st and 2nd level are revealed in class C_1 with support of 100%; in class C_2 three one-level patterns are identified with support for 100%, and the rest with support of 50% (Fig. 4).

The corresponding script hierarchies are shown in Figs. 5 and 6.

Sq₁: **27 28** 31 40 41 *1 2 3*44 *10 20 30*41 41 51 55 *15 16 17 18*40 1045 | C1
Sq₂: **27 28**44 44 *1 2 3*40 *10 20 30*7 49 5 6 7 8 50 60 7044 12 14 *15 16 17 18* | C1
Sq₃: *15 16 17 18*1130 3 30 *1 2 3*10 20 30*3290 91 90 *5 6 7 8*50 60 70 | C2
Sq₄: **27 28** 12 15*1 2 3*44 45 46*10 20 30*11 *15 16 17 18* 4032 7 18 49 *5 6 7 8* | C2

Fig. 3. Example: input sequences for classes

p^1_1=(27,28) p^1_2=(1,2,3) p^1_3=(10,20,30) p^1_4=(15,16,17,18) – Single-level Patterns
p^2_1= (p^1_1, p^1_2, p^1_3) = ((27, 28), (1, 2, 3), (10, 20, 30)),
p^2_2= (p^1_1, p^1_2, p^1_3, p^1_4) = ((27, 28), (1, 2, 3), (10, 20, 30), (15,16,17,18)) – Double-level patterns

Fig. 4. Example: behavioral patterns

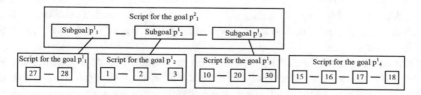

Fig. 5. Example: first script hierarchy

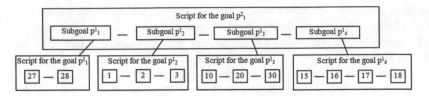

Fig. 6. Example: second script hierarchy

4 Method for Diagnosing the Behavior of Objects Based on Scripts

To classify an object behavior, it is necessary to determine how it matches to the previously learned script hierarchies. However, the same script hierarchy may occur in several different classes, but with different support. In addition, several script hierarchies may be detected in the diagnosed sample, and each of them may have different support in different classes. How, then, should the diagnosed sample be classified? It is proposed to classify using a simple probabilistic classifier, the so-called naive Bayesian classifier, since the support for script hierarchies is of probabilistic nature.

The proposed method for two-stage diagnosis of behavior involves two steps:

1. Matching the diagnosed Sq^T behavior to one or more script hierarchies obtained during training.
2. The classification of behavior by roles reveals to which roles the behavior of Sq^T corresponds, or shows that there is no correspondence.

Let's describe the first stage. The diagnosed behavior Sq^T is checked against the script hierarchies obtained during training. To do this, let's minimize each script hierarchy containing more than one level, from the top level to the lower level. As a result, at the lower level of each hierarchy, let's obtain a regular expression consisting of atomic sub-goals. Then let's determine whether the Sq^T sample matches the regular expressions for each script hierarchy. On the output we'll obtain a set of script hierarchies, for which correspondence with Sq^T is revealed.

Let's describe the second stage. Classification of the behavioral sample by roles stands for discovering to which class(es) the sample belongs on the basis of the behavioral scripts found in the sample, obtained at the training stage. A naive binary Bayesian classifier is used: in the case of not intersecting classes, it is determined to which one class the sample belongs; in the case of intersecting classes, a binary classification is carried out, that is, it is determined whether or not the sample belongs to each of the classes.

Further, to classify behavior samples on the basis of script hierarchies, it is advisable to use the support of script hierarchies in classes ($Sup(H_i, C_j)$ — the probability for samples of class C_j to meet the behavior corresponding to the script hierarchy Hi). However, it is obvious that some support values will be zero, since not all script hierarchies are present in all classes, and there will be issues during calculations. This kind of problem is solved using additive smoothing, namely adding one, yet not to relative probability values, but to all absolute values. These are the number of class samples in which the script hierarchies were met during learning stage [16]. Further, let's introduce the notation for additional information necessary for calculating the values of a naive Bayesian classifier.

Let Tr_{All} be the set of samples of all the classes on which the learning was conducted. During the training $F = \{H_1, H_2, \ldots, H_n\}$ was obtained, which is a forest of behavioral script hierarchies. Let $B(H_i)$ be the set of samples in which the behavior corresponding to the script hierarchy H_i is identified during the training.

Let's suppose the sample is $x \in Tr_{\text{All}}$, then x: C means that x is from class C. Let's denote $Tr_C = \{x \in Tr_{\text{All}} \,|\, x : C\}$, $Tr_C \subseteq Tr_{\text{All}}$ is the set of samples of class C. $Tr_{\overline{C}} = Tr_{\text{All}} \backslash Tr_C$ is the set of samples not of class C.

Let's now describe in more detail the use of a binary classification for intersecting classes. For each script hierarchy H_i, let's determine how many samples from class C it corresponds to: $Q_C(H_i)$: $H \to N$, where N is the set of non-negative integers; $Q_C(H_i) = |\{x \in Tr_c \,|\, x \in B(H_i)\}| = |Tr_c \cap B(H_i)|$. Similarly, we determine the number of samples of all classes except C, corresponding to H_i: $H \to N$; $Q_{\overline{C}}(H_i) = |Tr_{\overline{C}} \cap B(H_i)|$. $Q_{\overline{C}}(H_i)$ is the number of samples not from class C, which corresponds to the script hierarchy Hi. The value of $Q_C(H_i)$ can be zero if there is no behavior corresponding to H_i in C. Similarly, $Q_{\overline{C}}(H_i)$ can be zero if the behavior for Hi is absent in all classes except C. The solution is additive Laplace smoothing, namely adding one to each value of the function $Q_C(H_i)$ and $Q_{\overline{C}}(H_i)$.

Classification of the sample behavior for each class C, in which several script hierarchies $H_{i1}, \ldots, H_{ik}$ are encountered, is carried out according to the following formula:

$$C_{res} = \arg\max_{c \in \{C, \overline{C}\}} \left[\log P(C) + \sum_{i=1}^{k} \log P(H_i | C)\right]$$

where C_{res} is a class or not a class, $P(C) = |Tr_C|/|Tr_{\text{All}}|$ and $P(\overline{C}) = |Tr_{\overline{C}}|/|Tr_{\text{All}}|$, $P(H_i|C) = Q_C/|Tr_C|$ and $P(H_i|\overline{C}) = Q_{\overline{C}}/|Tr_{\overline{C}}|$.

Let's consider Example 2 of the script-based method of diagnosing behavior. As an input set of script hierarchies for a role, there are six script hierarchies presented in Example 1 of Sect. 3. Additional information about the training is given in Table 1.

Table 1. Detailed data about training

No.	Script hierarchies	Q_{C1}	$Q_{\overline{C1}}$	Σ	Sup_{C1}	Q_{C2}	$Q_{\overline{C2}}$	Σ	Sup_{C2}
1	H_1^1	2	1	3	100%	1	2	3	50%
2	H_2^1	2	2	4	100%	2	2	4	100%
3	H_3^1	2	2	4	100%	2	2	4	100%
4	H_4^1	2	2	4	100%	2	2	4	100%
5	H_1^2	2	1	3	100%	1	2	2	50%
6	H_2^2	2	2	4	100%	1	2	3	50%

The sample of behavior is set in the form of a sequence of actions: Sq^T = <**27, 28,** 43, 44, **1, 2, 3, 10, 20, 30,** 36, 37, 33, 12, 10, 11, 20> with allocated frequently repeated subsequences during training. The results of the first stage, determining whether the sample contains scripts inherent in the roles, are as follows: Sq^T match $H_1^1, H_2^1, H_3^1, H_1^2$; Sq^T doesn't match H_4^1, H_2^2 – that is, the sample corresponds to the four script hierarchies found during training. The second stage defines the values of the naive Bayesian classifier:

$$\text{Cres}_1 = \text{argmax}(-0.7, -2.08) = C_1, \text{Cres}_2 = \text{argmax}(-2.08, -0.7) = \overline{C}_2.$$

The maximum values indicate that the sample belongs only to class C_1.

5 Approbation of Models and Methods for the Subject Area of Malware Detection

At the training stage, from the set of real data (dataset) in the form of sequences of executable API functions for common samples of pre-classified malware, as well as a control class of non-malicious programs (Benign), a set of patterns, including multi-level ones, were obtained. Afterwards, script hierarchies were built on their basis [4]. Since in the dataset each sample has a single label that determines which class the sample belongs to, and malwares often exhibit behavior of several classes, information about sample classes has been supplemented by expert judgment based on reports of antivirus companies, marking in the tested samples the presence of features of various classes. Fragments of the results are presented in Table 2 on the left. Due to the fact that cross-classes are typical for malicious programs, a binary classification was chosen. This allowed to determine the belonging of the tested sample to each of the classes separately. The fragments of the results are shown in the right part of Table 2, the columns correspond to the classes for which detection is performed, the rows stand for the test samples. At the intersection, the test sample corresponds to the class. "+" means existence and "−" means absence of the features of each of the given classes in sample's behavior.

Table 2. Data for comparing the results of the experiment with the sample's labels (fragments)

Test samples by classes	Real marks					Classifier estimation				
	Backdoor	Trojan	Worm	Virus	Benign	Backdoor	Trojan	Worm	Virus	Benign
Trojan. Win32. Delf.a	−	+	+	−	−	−	+	−	−	−
Trojan. Win32. Delf.b	−	+	−	−	−	−	+	−	−	−
Trojan. Win32. Delf.bv	−	+	−	−	−	−	+	−	−	−
Trojan. Win32. Delf.ch	−	+	+	−	−	−	−	+	−	−

Table 3. Reliability and errors in detection

Class	Number of + in the left part	Number of coincidences in the left and right parts	Validity	Number of mismatches ± (FN)	Percentage of false negatives	Number of mismatches ∓ (FP)	Percentage of false positives
Backdoor	158	395	0.992	2	0.005	1	0.003
Trojan	103	350	0.879	36	0.090	12	0.030
Worm	146	339	0.852	44	0.111	15	0.038
Virus	37	387	0.972	10	0.025	1	0.003
Benign	15	396	0.995	2	0.005	0	0.000
Total/ average	459	1867	**0.938**	94	**0.047**	29	**0.015**

Based on a comparison of real and experimental data from Table 2, integral estimates are obtained, namely, the reliability of the detection results is calculated (Table 3). Reliability is the number of coincidences in the left and right parts of the Table 2, divided by the total number of test samples. The FN (False Negatives) is the number of mismatches between the "+" on the left and "−" in the right in the table, divided by the total number of samples. The FP (False Positives) is the number of mismatches between "−" on the left and "+" in the right in the table, divided by the total number of samples. The reliability of detection was 93.8%, the average error of the FP was 4.7%, and the FN was 1.5%.

A FN for the considered subject domain is the case when a test sample, according to an expert assessment belonging to a checked class, is classified as not belonging. Such errors are especially dangerous, since a malware can be "missed". A FP is a case when a sample that does not belong to the class being tested is wrongly attributed to the class. A large number of FP, when a safe program is identified as malware, adversely affects the credibility of the classifier.

6 Conclusions

In the paper, new models and methods for describing and analyzing the objects behavior have been created. The main results are as follows:

1. Developed formal models of scripts, script hierarchies, forests of script hierarchies, single-level and multi-level sequential patterns allow to conduct an analysis of the objects behavior.
2. A method for diagnosing whether a given behavior belongs to one or more roles using a naive Bayesian classifier with smoothing is proposed. An experiment is conducted on the binary classification of programs separately for each of the five classes. On average, the reliability of detection is 93.8%, which is 5.5% higher than the average figure obtained by Mohaisen et al. [17]. In the classification of safe and malicious programs, the FP is 1.5%, the FN is 4.7%.

Further research areas should be directed to the development of methods for the formation of script hierarchies from patterns in an automated mode with the opportunity for the expert to influence the process of obtaining knowledge.

References

1. Schank, R.C., Abelson, R.P.: Scripts, plans, and knowledge. Yale University, New Haven, Connecticut USA (1975)
2. Minsky, M.: Freymy dlya predstavleniya znaniy. Energiya, Moscow (1979)
3. Ruvinskaya, V.M., Berkovich, E.L., Lotockiy, A.A.: Heuristic method of malware detection on the basis of scripts. Iskusstvenniy intellekt **3**, 197–207 (2008)
4. Moldavskaya, A.V., Ruvinskaya, V.M., Berkovich, E.L.: Method of learning malware behavior scripts by sequential pattern mining. In: Gammerman, A., Luo, Z., Vega, J., Vovk, V. (eds.) Conformal and Probabilistic Prediction with Applications. COPA 2016. Lecture Notes in Computer Science, Vol. 9653, pp. 196–207. Springer, Cham (2016)
5. Surmin, Yu.P.: Teoriya sistem i sistemniy analiz. MAUP, Kyiv (2003)
6. Chernyshov, V.N., Chernyshov, A.V.: Teoriya sistem i sistemnyy analiz. Izdatel'stvo Tambovskogo gosudarstvennogo tekhnicheskogo universiteta, Tambov (2008)
7. Tocenko, V.G.: Metody i sistemy podderzhki prinyatiya resheniy. Algoritmicheskiy aspekt. Naukova dumka, Kyiv (2002)
8. Polikarpova, N.I., Shalyto, A.A.: Avtomatnoe programmirovanie, 2nd edn. Piter, Saint Petersburg (2010)
9. Gladun, V.P.: Obnaruzhenie znaniy na osnove setevyh struktur. Int. J. Inf. Technol. Knowl. **4**(4), 303–328 (2010)
10. Schank, R.C., Abelson, R.P.: Scripts, plans and goals. In: IJCAI 1975, Proceedings of the 4th International Joint Conference on Artificial intelligence, vol. 1, pp. 151–157, San Francisco, CA, USA (1975)
11. Polatovskaya, O.S.: Freym-scenariy kak tip konceptov. Vestnik IGLU **4**(25), 161–163 (2013)
12. Gupta, M., Han, J.: Approaches for pattern discovery using sequential data mining. In: Pattern Discovery Using Sequence Data Mining: Applications and Studies. IGI Global, pp. 137–154 (2012)
13. Il'in, V.I.: Povedenie potrebiteley. Piter, Saint Petersburg (2000)
14. Rieck, K., Holz, T., Willems, C., Düssel, P., Laskov, P.: Learning and Classification of Malware Behavior. In: Zamboni, D. (ed.) Detection of Intrusions and Malware, and Vulnerability Assessment. DIMVA 2008. Lecture Notes in Computer Science, vol. 5137, pp. 108–125. Springer, Heidelberg (2008)
15. Kovalenko, A.S., Smirnov, O.A., Kovalenko, O.V.: Subsystem technical diagnostics for automation of processes control in integrated information systems. Systemy ozbroiennia i viyskova tekhnika **1**, 126–129 (2014)
16. Yuan, Q., Cong, G., Thalmann, N.M.: Enhancing naive bayes with various smoothing methods for short text classification. In: Proceedings of the 21st International Conference on World Wide Web, pp. 645–646. Lyon, France (2012)
17. Mohaisen, A., Alrawi, O., Mohaisen, M.: AMAL: high-fidelity, behavior-based automated malware analysis and classification. Comput. Secur. **52**, 251–266 (2015)

Development of the Mathematical Model of the Informational Resource of a Distance Learning System

Vyacheslav Shebanin[1] , Igor Atamanyuk[1] ,
Yuriy Kondratenko[2]([⊠]) , and Yuriy Volosyuk[1]

[1] Mykolayiv National Agrarian University, Mykolaiv, Ukraine
volosyuk@mnau.edu.ua
[2] Petro Mohyla Black Sea State University, Mykolaiv, Ukraine
y_kondrat2002@yahoo.com

Abstract. Analysis of the existing models of knowledge representation in informational systems allowed to make a conclusion about considerable advantages of combined network models, which are able to take into account indistinct content of some information. That's why the semantic network that differs from the well-known ones by the peculiarities stated in this article is accepted as a mathematical model of informational resource of a distance learning system.

Keywords: Information technology · Informational resource
Semantic network · Predicate · Fuzzy logic

1 Introduction

Analysis of existing models of knowledge representation in informational systems allowed the authors to make a conclusion about considerable advantages of combined network models, which are able to take into account indistinct content of some information. That's why the semantic network that differs from the well-known ones by the peculiarities stated in this article is accepted as a mathematical model of informational resource of a distance learning system.

A semantic network is one of the ways of knowledge representation. From time immemorial, a semantic network was planned as a model of long-term memory structure presentation in psychology but with time it became one of the main ways of knowledge presentation in knowledge engineering. In many dictionaries, the word "semantics" is defined as meaning, contents of a word, an art work, an action, a circumstance etc., conveyed with the help of some presentation and expression. But it is impossible to give rather precise definition of the word "semantics" as a psychological notion even by some explanations. In spite of this, of course, concepts and images associated with some object are taken into consideration, and depending on the case, it is taken as an independent essence.

Let us be given an N-ary heterogeneous semantic network $S = (V, D, \Gamma)$, where V is a vertex set (concepts) of the network $V = \{v_i\}$, $i = \overline{1, n}$; D is a set of

© Springer Nature Switzerland AG 2019
O. Chertov et al. (Eds.): ICDSIAI 2018, AISC 836, pp. 199–205, 2019.
https://doi.org/10.1007/978-3-319-97885-7_20

arcs (relations between the concepts) of the network $D = \{d_j\}$, $j = \overline{1, m}$; $\Gamma = (\Gamma_V, \Gamma_D)$ is a set of weights of vertices and arcs of the network correspondingly by power $|\Gamma| = |\Gamma_V \cup \Gamma_D| = n + m$, $n, m \in N$, $\gamma_i \in \Gamma_V$, $i = \overline{1, n}$; $\gamma_j \in \Gamma_D$, $j = \overline{n+1, m}$ (Fig. 1).

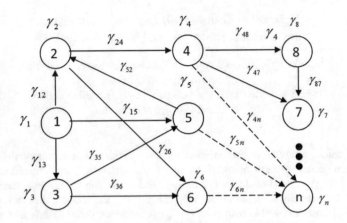

Fig. 1. *N*-ary heterogeneous semantic network $S = (V, D, \Gamma)$

Analysis of the literature showed that a modern theory of loaded directed graphs is developed for the case when $|\Gamma| = |\Gamma_D| = m$, $m \in N$, that is only the weights of arcs are taken into account, and $\Gamma_V = \emptyset$. Thus, the theoretical problem of mathematical formalization of *N*-ary heterogeneous semantic network $S = (V, D, \Gamma)$ is pressing [1].

2 Solution

The following approach to solving the given problem is offered [2]. We will introduce the concept of "elementary semantic first-order network" as a network consisting of two vertices and an arc between them with corresponding weights (Fig. 2).

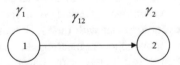

Fig. 2. Elementary semantic first-order network

Then, it is logical to introduce the concept of "induced elementary semantic second-order network" with the weight of the arc γ, which corresponds to the elementary semantic first-order network (Fig. 3):

Fig. 3. Induced elementary semantic second-order network

$$\gamma = \Theta_k(\gamma_1, \gamma_2, \gamma_{12}) \equiv m \quad m^T = (m_0, m_1), \tag{1}$$

where $\Theta_k(\gamma_1, \gamma_2, \gamma_{12}) \equiv m$ is a predicate that takes the value of fuzzy logic vector.

It is obvious that the network from Fig. 3 is obtained after transforming the network in Fig. 2. Range of values of the predicate is expanded from traditional 0 and 1 to fuzzy two-component logical vector [3] $m_{ijk}^T = (m_{0ijk}, m_{1ijk})$. Besides, if $m_{ijk}^T = (0, 1)$, the vector takes the value "true" and if $m_{ijk}^T = (1, 0)$, it takes the value "false". Moreover, the following conditions must be met:

$$0 \geq m_{0ilk}, m_{1ilk} \geq 1, m_{0ijk} + m_{0ilk} = 1. \tag{2}$$

Negation of a vector corresponds to interchanging its elements $\overline{m}_{ilk}^T = (m_{1ijk}, m_{0ilk})$.

Entropy $S(m_{ijk}) = -m_{0ijk} \log_2 m_{0ijk} - m_{1ilk} \log_2 m_{1ijk}$ serves as the measure of indistinctness of the logic vector m_{ilk}. The measure of indistinctness of a semantic network is the value $S(M) = \sum_{i,j} S(m_{ijk})$ where $m_{ijk} \in M$, $i = \overline{1, l}$, $j = \overline{1, c}$, $i, j \subset N$.

During each logical operation between vector variables, a tensor of rank 3 is compared. During this, tensors keep the form, which they had in vector representation of traditional crisp logic. This allows to describe unequivocally the operations over indistinct logical variables. Besides, the same relations, which were present in crisp logic, are kept between operations. Essential convenience of vector representation is in that operations over logical variables can be represented in matrix form [4]. The introduced approach differs from the existing theory of indistinct predicates by that instead of the value of a fuzzy logical vector of a predicate, indistinct semantic network is placed in line. This permitted to combine advantages of the theory of predicates, vector representation of a logical variable, and matrix theory under logical conclusions [5]. Let's introduce the concept of "elementary semantic second-order network" (Fig. 3) $S = (V, D, \Gamma)$, $|V| = n$, $|D| = m$, $|\Gamma| = |\Gamma_V \cup \Gamma_D| = n + m$, $n, m \in N$, as a network that as a result of transformations can become so called "induced elementary semantic second-order network" $S^* = (V^*, D^*, \Gamma^*)$, for which

$$|V| > |V^*| > 2, |D| > |D^*| > 1, |\Gamma| > |\Gamma^*| > 3 \tag{3}$$

Thus the introduced approach allowed to solve the theoretical problem of mathematical formalization of N-ary heterogeneous semantic network.

The developed model that combines advantages of the theories of predicates, fuzzy logic, and semantic networks is the most prospective in our view [6].

Let variables $\gamma_1, \gamma_2, \ldots, \gamma_{n+m}$ take the values, which belong to random sets: $\gamma_i \in \Gamma_V$, $i = 1, 2, \ldots, n$, $\gamma_j \in \Gamma_D$, $j = n + 1, n + 2, \ldots, m$. Then, the function

$y = \Theta(\gamma_1, \gamma_2, \ldots, \gamma_{n+m})$, to which indirect semantic network $S = (V, D, \Gamma)$ can be put in line, such that $\Theta(\gamma_1, \gamma_2, \ldots, \gamma_{n+m}) \equiv S$, where V is a set of vertices (concepts) of the network of the power $|V| = n$; D is a set of arcs (relations between the concepts) of a network of the power $|D| = m$; $\Gamma = (\Gamma_V, \Gamma_D)$ is a set of weights of vertices and arcs of a network correspondingly, the power $|\Gamma| = |\Gamma_V \cup \Gamma_D| = n + m$, $n, m \in N$, $\gamma_i \in \Gamma_V$, $i = \overline{1,n}$; $\gamma_j \in \Gamma_D$, $j = \overline{n+1,m}$ is called $n + m$-local predicate on an indistinct semantic network [7].

That's why if any indistinct semantic network $S = (V, D, \Gamma)$ can be represented as a loaded directed graph in the form of an induced elementary semantic second-order network, above mentioned network can be described by one and only one matrix of the contiguity M. Thus mathematical formalization of the given statement is of the form $\forall \gamma_i \in V$, $i = \overline{1,n}$, $\gamma_j \in D$, $j = \overline{n+1,m} \Rightarrow \exists$.

$$\Theta_k(x_1, x_1, \ldots, x_m) \equiv \begin{Vmatrix} m_{11k} & m_{12k} & \cdots & m_{1ck} \\ m_{21k} & m_{22k} & \cdots & m_{2ck} \\ \vdots & & \cdots & \vdots \\ m_{l1k} & m_{l2k} & \cdots & m_{lck} \end{Vmatrix} \tag{4}$$

Range of values of matrix elements of the contiguity is expanded from traditional 0 or 1 to fuzzy two-component logic vector $m_{ijk}^T = (m_{0ijk}, m_{1ijk})$.

So the logic of indistinct predicates is developed in vector-matrix representation. We will represent a predicate as a vector field of indistinct variables over given set of terms. We will study operations over predicates, develop the variant of construction of indistinct conclusion on the basis of rules formulated in the form of relations between predicates. We will give the definition and determine the method of indistinct quantifiers $\forall$ and $\exists$ calculation.

In works [7, 8], matrix representation of fuzzy logic, naturally generalizing apparatus of ordinary "crisp" logic, was developed. Tensor representation of logic introduced in the work of Mizraji [9] was chosen as a starting point. Logic variables are represented by 2D vectors $m^T = (m_0, m_1)$ whose components satisfy the conditions $0 \geq m_0$, $m_1 \geq 1$, $m_0 + m_1 = 1$. Negation $\overline{m}$ of the vector m is equivalent to the permutation of its components: $\overline{m}^T = (m_1, m_0)$. Space of fuzzy vectors is designated with the symbol F. Entropy $S(m) = -m_0 \log_2 m_0 - m_1 \log_2 m_1$ serves as the measure of indistinctness of a logic vector $m \in F$.

Tensor of rank 3 $T^{(P)}$ that realizes the representation $m \otimes y \xrightarrow{P} z$ (or $F \otimes F \xrightarrow{P} F$) is compared during each logic operation P between vector variables m, y. During this tensors $T^{(P)}$ keep the form which they had in vector representation of "crisp" logic. This allows to interpret unambiguously the operations over indistinct logic variables. Besides, between operations the same relations which were in "crisp" logic, for example, laws of De Morgan are kept.

We will work out a matrix model of indistinct predicates. "Crisp" predicates, that is, classical predicates are determined as the functions on the set of "terms" M, which take value in Boolean space $B = \{0, 1\}$. So if $M = \{\gamma_1, \gamma_2, \gamma_3, \gamma_4, \gamma_5\}$, the function

$$\begin{array}{cccccc} m & \gamma_1 & \gamma_2 & \gamma_3 & \gamma_4 & \gamma_5 \\ P(m) & 1 & 0 & 1 & 0 & 0 \end{array} \tag{5}$$

can serve as the example of monadic predicate P(m) where $m \in M$.

Two-place, three-place and etc. predicates are determined by analogy. For example, two-place predicate $P(m, y)$, $m \in M$, $y \in N$ is determined on the set $M \otimes N$.

We determine indistinct predicate $P(m)$ as the function given on the set M and taking the value in the space of vector indistinct variables F, which was determined above. So, indistinct logic vectors $P(m) \in F$ or $[P(m)]^T = (P_0(m), P_1(m))$ are the range of values of the predicate, besides for all m it is right

$$0 \leq P_0(m), \quad P_1(m) \leq 1, \quad P_0(m) + P_1(m) = 1 \tag{6}$$

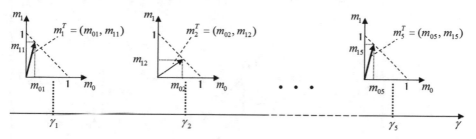

Fig. 4. The example of an indistinct predicate $P(m)$ as a vector field of indistinct variables given on the set $M = \{\gamma_1, \gamma_2, \gamma_3, \gamma_4, \gamma_5\}$.

So indirect predicate $P(m)$ sets on M some vector field as it is shown in Fig. 4.

As predicates are logic variables, all indirect logic operations introduced in [9] and studied shortly above can be applied for them. This allows to build new predicates, more complex ones from some those which are set on M, which gives the possibility to expand on the predicate domain the rules of logical inference.

In standard statement of fuzzy logic, the concept of linguistic variable L is used. The same as a predicate a linguistic variable L is determined on some set M but has "grade of membership" $\mu_L(m)$, $m \in M$ points of set M of the given linguistic variable as a range of values. Depending on the context the grade of membership is interpreted either as a true value of indistinct logic variable or as indistinct values of characteristic function. Indistinct logical variables are interpreted as "one-dimensional" logical rules and are deduced as some empirical laws. In the same way, operations over indistinct sets are introduced.

The considered above scheme of application of indistinct predicates in vector-matrix representation allows to introduce logical operations without random assumptions. Logical operations over indistinct variables are described with the same tensors as in "crisp" logic. As a result, flexible and well-grounded system of calculations that comprises empirical expert estimations only at "the entry" of the algorithms comes out.

Based on the aforementioned, the model of knowledge representation is developed (Fig. 5).

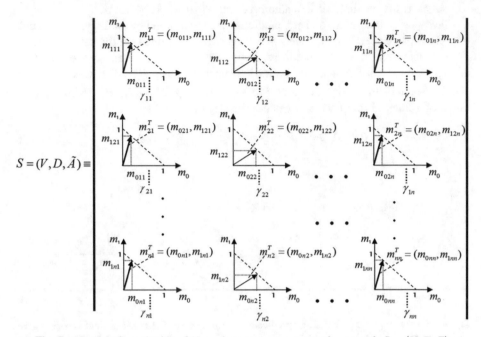

Fig. 5. Matrix of contiguity of N-ary heterogeneous semantic network $S = (V, D, \Gamma)$

The efficiency of the developed model of informational resource is reasonable to estimate by the criterion of complexity.

3   Conclusion

Advanced mathematical model of knowledge representation in the form of indistinct semantic network, which differs from the known ones by that it combines advantages of semantic networks and vector-matrix representations of fuzzy logic that gives possibility to introduce logical operations without random assumptions. Great convenience of vector representation is in that operations over logical variables can be represented in a matrix form. Introduced approach of the transformation of indistinct semantic network on the basis of the application of elementary semantic networks allowed to complete the problem of mathematical formalization of N-ary heterogeneous indistinct semantic network.

References

1. Russell, S., Norvig, P.: Artificial Intelligence: A Modern Approach. Pearson (2007)
2. Volosiuk, Y.: Suchasni metody praktychnoho vyluchennia znan v intelektual'nykh systemakh navchannia. Materialy IV Mizhnarodnoi naukovo-praktychnoi konferentsii «Informatsijni tekhnolohii v ekonomitsi, menedzhmenti i biznesi. Problemy nauky, praktyky i osvity». Yevropejs'kyj universytet, T.1, pp. 58–61 (2009)
3. Pospelov, D.: Iskusstvennyj intellekt. V 3-h kn. Kn. 2. Modeli i metody: Spravochnik. Radio i svjaz', 304 p. (1990)
4. Asai, K., Vatada, D.: Prikladnye nechetkie sistemy. Mir, 368 p. (2008)
5. Oksiiuk, O.: Metodyka otsinky strukturnoi skladnosti N-arnoi neodnoridnoi semantychnoi merezhi. Suchasni informatsijni tekhnolohii u sferi bezpeky ta oborony, №2(9), pp. 49–50 (2010)
6. Marcenjuk, M.: Matrichnoe predstavlenie nechetkoj logiki. Nechetkie sistemy i mjagkie vychislenija, T. 2. № 3, pp. 7–36 (2007)
7. Oksiiuk, O.: Kompleksna model' predstavlennia znan na osnovi predykativ i nechitkykh semantychnykh merezh. In: Kozheduba, I. (ed.) Materialy shostoi naukovoi konferentsii Kharkivs'koho universytetu Povitrianykh Syl im. «Novitni tekhnolohii – dlia zakhystu povitrianoho prostoru». (14–15 kvitnia 2010 r.). KhUPS, Kharkiv, p. 127 (2010)
8. Shebanin, V., Atamanyuk, I., Kondratenko, Y., Volosyuk, Y.: Application of fuzzy predicates and quantifiers by matrix presentation in informational resources modeling. In: Perspective Technologies and Methods in MEMS Design, Proceedings of the International Conference MEMSTECH-2016, Lviv-Poljana, Ukraine, 20–24 April 2016, pp. 146–149 (2016)
9. Mizraji, E.: Vector logics: "the matrix-vector representation of logical calculus". Fuzzy Sets Syst. 50, 179–185 (1992). Author, F.: Contribution title. In: 9th International Proceedings on Proceedings, pp. 1–2
10. Atamanyuk, I.P.: Algorithm of extrapolation of a nonlinear random process on the basis of its canonical decomposition. J. Cybern. Syst. Anal. 41, 267–273 (2005). Mathematics and Statistics. Kluwer Academic Publishers Hingham
11. Kondratenko, Y., Simon, D., Atamanyuk, I.: University curricula modification based on advancements in information and communication technologies. In: Ermolayev, V., et al. (eds.) Proceedings of 12th International Conference on Information and Communication Technologies in Education, Research, and Industrial Application. Integration, Harmonization and Knowledge Transfer, 21–24 June 2016, Kyiv, Ukraine, ICTERI 2016, CEUR-WS, vol. 1614, pp. 184–199 (2016)
12. Kondratenko, Y., Khademi, G., Azimi, V., Ebeigbe, D., Abdelhady, M., Fakoorian, S.A., Barto, T., Roshanineshat, A.Y., Atamanyuk, I., Simon, D.: Robotics and prosthetics at Cleveland State University: modern information, communication, and modeling technologies. In: Ginige, A., et al. (eds.) ICTERI 2016, CCIS 783. Springer, Cham, pp. 133–155 (2017). https://doi.org/10.1007/978-3-319-69965-3_8
13. Shebanin, V., Atamanyuk, I., Kondratenko, Y., Volosyuk, Y.: Canonical mathematical model and information technology for cardio-vascular diseases diagnostics. In: 2017 14th International Conference the Experience of Designing and Application of CAD Systems in Microelectronics, CADSM 2017 - Proceedings Open Access, pp. 438–440 (2017). https://doi.org/10.1109/cadsm.2017.7916170
14. Poltorak, A.: Assessment of Ukrainian food security state within the system of its economic security, Actual problems of economics, № 11(173), pp. 120–126 (2015)
15. LNCS Homepage. http://www.springer.com/lncs. Accessed 30 Mar 2018

Use of Information Technologies to Improve Access to Information in E-Learning Systems

Tetiana Hryhorova$^{(\boxtimes)}$ and Oleksandr Moskalenko

Kremenchuk Mykhailo Ostohradskyi National University, Kremenchuk, Ukraine
grital0403@gmail.com

Abstract. Various forms of training use e-learning systems. Existing e-learning systems do not have the ability to find information quickly in all the training courses that have been downloaded, which limits access to the information provided. Several commonly used search engines were researched to access the possibility of using them for efficient access to all materials presented in the e-learning system. Two e-learning systems EFront and Moodle were considered in terms of organization of the database structure. The EFront system is based on the database with clustered indices, and it was chosen for testing search engines. Use of an additional programming module connected to EFront system is proposed. This module would allow a full-text search of the information needed. During the research, two modules of full-text searching were considered. The Sphinx technology was selected as the most efficient one. Thus, the proposed solutions made it possible to expand the students' ability to access the necessary educational information.

Keywords: e-learning systems · Database structure · Full-text search

1 Information Technologies in Different Forms of Education

1.1 Innovative Forms of Learning

In a modern life, more and more attention is paid to educational content. Special influence on change of contents and forms of education was made by development of information technologies. Innovative forms of learning that are built on information technology such as STEM education, peer-2-peer education, distance education, and blended education are increasingly used in learning process. The main task for universities is search for new forms of organization of studying process, which will keep up with the times.

The form of education that is now developing and spreading is STEM-education. Main task of STEM-education is to teach future specialists for different industries of a human activity. STEM workers use their knowledge to try to understand how the world works and to solve problems [1].

General requirements for future STEM specialist are creative thinking and ability to solve problems and different tasks. Main task for STEM education is to teach communication and pursuit of continuous improvement in professional activity. In September 2016 in Ukraine, STEM Education Coalition was created. Among founders

© Springer Nature Switzerland AG 2019
O. Chertov et al. (Eds.): ICDSIAI 2018, AISC 836, pp. 206–215, 2019.
https://doi.org/10.1007/978-3-319-97885-7_21

of this Coalition are such famous companies as Ukrainian Nuclear Society, Samsung, Ericsson, Kyivstar, Syngenta, United Minerals Group, Microsoft Ukraine, and Energoatom. Urgent task of this Coalition is development of recommendations for the Ministry of Education and Science of Ukraine in teaching STEM disciplines, organization of professional oriented projects for youth, teacher training of innovative approaches in teaching. The Coalition is planning to create conditions for experimental research at schools, to conduct scientific and technical contests, Olympiads, quests, hackathons, etc.

STEM combines science, mathematics, technology, and engineering in the literal sense of the word. Combining of disciplines could be provided by means of:

1. Creation of new experimental base (laboratory with modern equipment).
2. Using virtual laboratories and simulators.
3. Development of courses in mathematical and natural sciences, in which examples will be the tasks taken directly from professional disciplines. In classic universities, there is a division into math, natural sciences, and professional disciplines.

The next form of education is peer-2-peer education. In Ukraine, it is represented by UNIT Factory School based on the model of the French programming school "42" [2]. The basic principle of this school is collaboration. Students use self-education and help each other. They have access to a database, in which projects are stored, and choose tasks for themselves to find solutions. They work in computer rooms at a time convenient for them. Administrators monitor the equipment and record who uses it, but have no right to help students. Students can use online resources and mutual assistance. Leading professionals periodically lecture for students of different levels of education. The course is for three years.

The classical school has the opportunity to promote implementation of group student projects in every way possible. If the team works smoothly, the interest and responsibility of its members is high, and weaker students pull up to a higher level. Such groups can be created by uniting students of different departments and majors. This will allow students to immerse themselves in solving a wide range of problems that exist in society and help them understand how professionals collaborate in different fields. Virtual communities are used for this form of training such as Google educational communities. The Google Educator Group provides a platform for collaboration, as well as an exchange of experiences and ideas. Google Apps for Education[1] is a system of free tools that provides the necessary education technologies based on cloud-based services in educational institutions.

Distance learning is a form of training, which uses computer and telecommunication technologies that provide interactive communication between teachers and students at different stages of learning, and also support independent work with information network materials [3]. To organize this process, e-learning systems are used that provide:

- educational materials for students;
- control of students' progress;

---

[1] https://www.google.com/landing/geg/.

- interactive teacher-student collaboration;
- teaching materials toolkit.

Another form of education that combines traditional methods of education and information technologies is blended education. This form of education is suitable for employing in higher education and for advanced training courses. Its basis is the organization of independent work of students.

There are many varieties of this form of education [4], including "Face-to-Face Driver," "Rotation," "Flex," "OnlineLab," "Self-blend," "OnlineDriver," "Program Flow," and "Core-and-Spoke." All these forms are distinguished by the time division between traditional (in-person) and computer-oriented forms of learning organization. Typically, blended learning consists of the following stages:

- study of theoretical material using technologies of electronic, distance, or mobile learning;
- practice in the form of traditional classroom lessons;
- discussion of problem situations online and/or offline;
- control and evaluation of students using computer testing;
- final control over the discipline (examination, credit) and/or defense of the term project in the traditional (in-person) form.

Main stages of blended learning use some e-learning system. Such a system may be Moodle, eFront, ATutor, Blackboard, and others, or Open Learning Portal. Such portals are intended for creating communication space and information field of educational professional community by means of the Internet. Educational portals have the ability to automatically update their content regularly and include all the latest communications and training services.

1.2 Motivation for Research

All new forms of education use different e-learning systems that help organize the learning process by providing interaction between its participants: those who create training courses and those who get access to created courses. However, any course has limited amount of information, and very often does not have enough answers for students' research activities. Also, some courses require additional knowledge in other related subjects. Course developers can supplement their material with links to other resources, or students can independently search for the necessary information on the Web. It is time consuming for both types of users, and does not always give the necessary result.

Systems of e-learning and the system of open educational portals are constantly expanding and being filled with knowledge, but do not have an internal search engine. Some systems would provide the opportunity to search for additional information within the system and, if necessary, expand this search beyond the system to other portals.

The aim of this work is to research the possibility of a quick access to all course materials presented in the e-learning system using quick search on a specified set of keywords. To do this, it is necessary to investigate the organization of the database in an e-learning system and various modules for full-text search of information in the database.

2 Results of the Research

2.1 Organization of the Database in the E-Learning System

At the Department of Computer Science and Higher Mathematics of the Kremenchuk Mykhailo Ostohradskyi National University, two e-learning systems [4] based on EFront and Moodle for introduction of blended learning are used. They differ in organizational approach and are intended to increase opportunities for independent work of students through access to educational material and self-esteem during the tests. In these systems, from various disciplines in accordance with the syllabus, an electronic version of lectures and teaching materials was prepared, which includes test assignments. For preparation of control or term project for each student or a group thereof, a system of flexible control in the form of control points is introduced, which is accompanied by a detailed description of the results to be achieved before that date. Student work results are checked directly in the information system environment, or clearly in consultation with the instructor. Systems contain glossary, hyperlinks to additional literature from the Internet [5].

Consider the database organization in these systems. The database in EFront uses clustered indices, which allows us to create clusters. The cluster is used to physically co-store one or more tables that are joined together in an SQL query [6]. The EFront distance learning system has the structure for storage of objects given in Fig. 1. The courses table contains all the subjects; each subject stores lectures of the given subject ("lessons" table). To store the content of lectures, the "content" table is used. To link lectures and subjects, the "lessons_to_courses" table is used, which stores the unique ID of objects and lectures.

The Moodle distance learning system, unlike the EFront system, uses foreign keys to link the tables and has the structure for storing items given in Fig. 2. The "course_categories" table contains categories of items and has a one-to-many relationship with the "courses" table. The "courses" table contains data and a list of courses and has a one-to-many relationship with the "lesson" table. The "lesson" table contains data and a list of course lectures and has a one-to-many relationship with the "lesson_pages" table. To store the content of lectures, the "lesson_pages" table is used.

After constructing the queries, it was found that query for tables with cluster links are performed faster than query for tables with usual links.

In addition to the higher speed of database queries, these links can be used for tree-like structures. With time, cluster indices grow in the process of adding data, and the tables start to branch. The specific node of such a tree includes indices of all child nodes or the final data if the node is the last. Nodes can be referred to each other only in one direction from the parent to the child. If the amount of data increases, the tree will

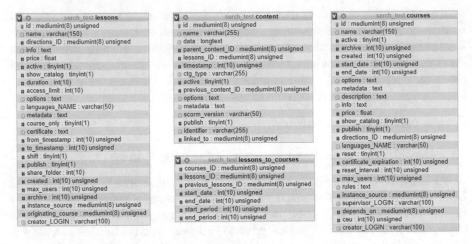

Fig. 1. Database structures for storing subjects in EFront

Fig. 2. Database structures for storing subjects in Moodle

become more complicated and branched. Each branch of tree-like structure is a higher level abstraction. Such a structure should be organized for convenient and accurate search of data according to the specified parameters and criteria. If one creates such a structure with a usual link, then searching will be very complex and slow.

Most full-text search solutions work on the principle of index search [7]. Index is built based on text data and keeps all search results along with the information where these data were found in the document. This approach increases the search speed more than scanning the entire content of all (or multiple) database tables.

The search service consists of two components: a search engine and an indexer. The indexer receives the text on the input, does the processing of the text (cutting out endings, insignificant words, etc.), and stores everything in the index. The structure of the index allows us to conduct a very fast search on it. Search engine works by searching interface by index: it receives a query from the client, parses the phrase, and looks for it in the index.

Search technologies with a full-text index are used to create a full-text search on the website. These include built-in full-text search in Sub, Sphinx, Elastic Search, Apache Solr, etc. Technologies Sphinx and Apache Solr are most suitable for full-text search in the distance learning system EFront.

Sphinx technology [8] is a server for full-text search with open source. To work with it on the server, a Sphinx search daemon (server) should be installed, which will index search data in the database after a specified period of time.

For a full-text search on the content of lessons, it needs to be indexed (Fig. 1). First, a configuration file is created. In the configuration file, source settings are specified where the content is stored. Then, a query that makes a selection of all records is written, which must be indexed, and text attributes are defined. Also, index building settings, settings for storing values of document attributes, indexer settings, and search daemon settings should be set. The query that makes a selection of all the records that should be indexed is as follows:

```
sql_query = SELECT content.id AS contid, content.name AS
contname, content.data AS cdata, lessons.name AS lname,
courses.name AS cousesname FROM `content` INNER JOIN les-
sons ON content.lessons_ID=lessons.id INNER JOIN courses
INNER JOIN lessons_to_courses ON lessons.id = les-
sons_to_courses.lessons_ID AND cours-
es.id=lessons_to_courses.courses_ID
```

After creating the configuration file, the data need to be indexed, and the configuration daemon (the server, Search Daemon) needs to be run.

After successful indexing, several search queries are executed in the console. For example, for searching for all subjects and lectures that contain the word "zend," the following command should be executed:

```
/usr/local/sphinx/bin/search --config /usr/local/sphinx/
etc/sphinx.conf zend.
```

The full-text search uses PHP language. To do this, a class is connected to the Sphinx API project. The result of a full-text search using PHP for searching subjects and lectures that contain the word "zend" is shown in Fig. 3.

Apache Solr technology [9], in contrast to Sphinx technology, creates not only a search index, but is also used as a data storage, that is, the data are stored entirely and do not need to be duplicated in the database.

For full-text search, one needs to index the content of lectures given names of lectures and the course associated with it.

In the db-data-config.xml file, the data for connecting to the database and the query for data indexing are described, and numerical and text attributes are defined. These attributes must be added to the search schema (schema.xml).

A fragment of the Apache Solr schema looks as follows:

```
<field name="contname" type="text_general" indexed="true"
stored="true" termVectors="true" termPositions="true"
termOffsets="true"/>
<field name="cdata" type="text_general" indexed="true"
stored="true" termVectors="true" termPositions="true"
termOffsets="true"/>
<field name="lname" type="text_general" indexed="true"
stored="true" termVectors="true" termPositions="true"
termOffsets="true"/>
<field name="cousesname" type="text_general" in-
dexed="true" stored="true"/>
```

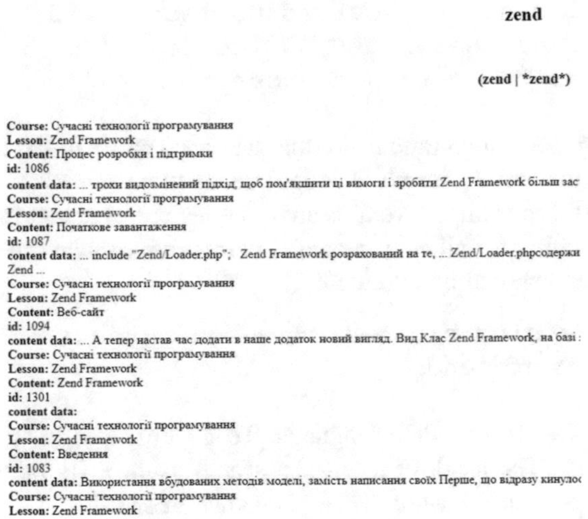

Fig. 3. Indexing data and starting the search server

After that, we index the data from MySQL. After successful indexing the data in the search engine, we can make several search queries, and find all the subjects and lectures that contain the word "zend" (Fig. 4).

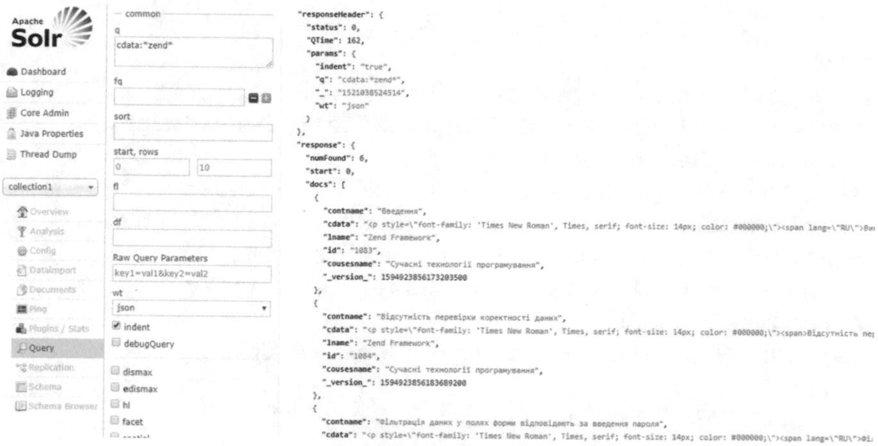

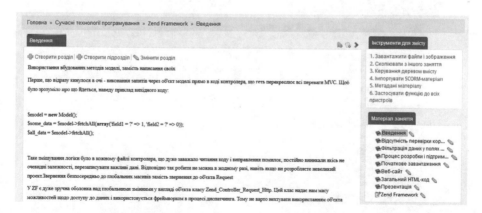

Fig. 4. Search results

The results of full-text search using engines Sphinx (Fig. 3) and search results Apache Solr (Fig. 4) are the same. The search query containing the word "zend" was found in the course, which is called "Modern programming technologies," in the content of the lesson "Zend Framework," in the contents of the following lesson sections: "Initial boot," "Development and support process," "Introduction," "Web site," "Zend Framework," "Filtering data in form fields," "Lack of data validation." Results of the search correspond to information stored in the distance learning system (Fig. 5).

Fig. 5. Content of the lesson Zend Framework

Consider the full-text search using PHP (Fig. 6). We need to download and connect a client to the PHP project `solr`. Example of PHP code for full-text search is as follows:

```php
<?php
header("Content-Type: text/html; charset=utf8");
include 'solr-php-client/Apache/Solr/Service.php';
//connect Apache Solr api
$solr = new Apache_Solr_Service( 'localhost', '8983',
'/solr/collection1' ); // connect to search server
$query="contname:Debug*"; // search query
$results = $solr->search($query, 0, 10); // search
//results output
foreach ($results->response->docs as $p)
  {
    echo "$p->id</br>";
    echo "$p->cousesname </br>";
    echo "$p->lname </br>";
    echo "$p->contname </br>";
    echo "$p->cdata </br>";
  }
?>
```

Based on our research, it is thus advisable to use Sphinx technology for e-learning system EFront since Apache Solr is designed for more complex data structures.


```
0 =>
  object(Apache_Solr_Document)[12]
    protected '_documentBoost' => boolean false
    protected '_fields' =>
      array (size=6)
        'contname' => string '1. Learn to Debug' (length=17)
        'cdata' => string '<table border="0" cellspacing="2" cellpadding="2" width="98%" align="center">
      <tbody>
              <tr>
                      <td align="justify">
                      <h1>Learn to Debug</h1>
                      Debugging is the cornerstone of being a programmer. The first meaning of the verb to debug is to
                      remove errors, but the meaning that really matters is to see into the execution of a program by examining
                      it. A programmer that cannot debug effectively is blind.<br />
                      <br />
                      Idealists that think design, or analysis, or complexity theory, or wha'... (length=3987)
        'lname' => string 'How to be a programmer - Beginner' (length=33)
        'id' => string '31' (length=2)
        'cousesname' => string 'How to be a programmer' (length=22)
        '_version_' => int 1594923854769160192
    protected '_fieldBoosts' =>
      array (size=6)
        'contname' => boolean false
        'cdata' => boolean false
        'lname' => boolean false
        'id' => boolean false
        'cousesname' => boolean false
        '_version_' => boolean false
1 =>
  object(Apache_Solr_Document)[13]
    protected '_documentBoost' => boolean false
    protected '_fields' =>
      array (size=6)
        'contname' => string '2. How to Debug by Splitting the Problem Space' (length=46)
        'cdata' => string '<table border="0" cellspacing="2" cellpadding="2" width="98%" align="center">
      <tbody>
              <tr>
                      <td align="justify">
                      <h1>How to Debug by Splitting the Problem Space</h1>
                      Debugging is fun, because it begins with a mystery. You think it should do something, but instead
                      it does something else. It is not always quite so simple---any examples I can give will be contrived
                      compared to what sometimes happens in practice. Debugging requires creativity and ingenuity. If
                      there is a single key t'... (length=3537)
```


Fig. 6. Search results

3 Conclusions

In the paper, innovative forms of training based on information technologies were discussed. The main component of such forms of education is an e-learning system. Results of the discussion show that these systems do not have the ability to quickly find information in all training courses loaded into the system. This limits access to knowledge and the possibility of obtaining more complete information.

The aim of the work was to study the possibility of quick access to all the materials presented in the e-learning system through quick search in any course on the specified set of keywords. EFront was chosen as the system that is built on the basis of a database with clustered indices. In order to search for information within the same e-learning system, it was suggested to connect an additional programming module that will allow organizing a full-text search in the system's database. During the research, two modules of full-text search were considered, and for the selected e-learning system, it was suggested to use the Sphinx technology. Thus, proposed solutions made it possible to expand the students' ability to access the necessary educational information through the use of full-text search.

References

1. Vilorio, D.: STEM 101: Intri to tomorrow's jobs. Bureau of Labor Statistics (BLS), US (2014)
2. Andrews, M., Manning, N.: Peer learning in public sector reforms. Effective Institutions Platform (EIP), Paris (2015)
3. Hryhorova, T., Lyashenko, V., Kalinichenko, D.: Approaches for implementation of distance learning in higher education. In: Digests 38 International Convention on Information and Communication Technology, Electronics and Microelectronics "MIPRO 2015", 25–29 May, Opatija, Croatia, pp. 1070–1074 (2015)
4. Staker, H., Horn, M.B.: Classifying K 12 Blended learning. Innosight Institute, CA (2012)
5. Shamsi, K.N., Khan, Z.I.: Development of an e-learning system incorporating semantic web. Int. J. Res. Comput. Sci. 2(5), 11–14 (2012)
6. Cioloca, C., Georgescu, M.: Increasing database performance using indexes. Database Syst. J. II(2/2011), 13–22 (2011)
7. Lincheng, S.: A large-scale full-text search engine using DotLuence. In: Proceedings of the 3rd International Conference on Communication Software and Networks, IEEE Conferences, pp. 793–795 (2011)
8. Arjun Atreya, V., Chaudhari, S., Bhattacharyya, P., Ramakrishnan, G.: Building multilingual search index using open source framework. In: Proceedings of the 3rd Workshop on South and Southeast Asian Natural Language Processing (SANLP), Mumbai, India, pp. 201–210, December 2012
9. Kuć, R.: Solr Cookbook, 3rd edn. Packt Publishing LTD, Birmingham (2015)

Multi-agent Approach Towards Creating an Adaptive Learning Environment

Maksim Korovin$^{(\boxtimes)}$ ⓘ and Nikolay Borgest ⓘ

Samara University, Samara, Russian Federation
maks.korovin@gmail.com, borgest@yandex.ru

Abstract. The paper describes a concept of an intelligent agent-based tutoring system to guide students throughout the course material. The concept of adaptivity is applied to an adaptive education system, based on the profiling of students using the Felder and Silverman model of learning styles. Benefits that the multi-agent organization structure provides for an adaptive tutoring system are outlined. In this paper, a conceptual framework for adaptive learning systems is given. The framework is based on the idea that adaptivity is finding the best match between the learner's profile and the course content's profile. Learning styles of learners and content type of learning material are used to match the learner to the most suitable content.

Keywords: Adaptive · Tutoring system · Multi-agent

1 Introduction

Adaptivity is a crucial issue in today's online learning environments. In [1], it is argued that virtual learning environments are best at achieving learning effectiveness when they adapt to the needs of individual learners. In service-job-training, it is especially necessary to identify learning needs and customize solutions that foster successful learning and performance, with or without an instructor to supplement instruction. Learning management systems provide educational services to a wide range of students and they can help students to achieve their learning goals by delivering knowledge in an adaptive or individualized way. In [2], it is stated that as long as the competition on the market of Web-based educational system increases, "being adaptive" will become an important factor for winning the customers. Intelligence of a learning management system is largely attributed to its ability to adapt to specific student needs during the learning process. Each student may have different backgrounds, learning styles, individual preferences, and knowledge levels. This raises the need for adaptivity of learning environments. Learning systems must be able to adapt to be suitable for different kinds of students.

Brusilovsky in [2] gives a detailed review of these systems with a focus on their supporting adaptive technologies are provided. Adaptive technologies include tutoring systems utilizing curriculum sequencing, intelligent analysis of student's solutions, interactive problem solving support; adaptive hypermedia technologies involving

© Springer Nature Switzerland AG 2019
O. Chertov et al. (Eds.): ICDSIAI 2018, AISC 836, pp. 216–224, 2019.
https://doi.org/10.1007/978-3-319-97885-7_22

adaptive presentation and adaptive navigation support; Web-inspired technologies like student model matching.

Pedagogical agents are autonomous agents that support human learning by interacting with students in the context of the learning environment. They extend and improve upon previous work on intelligent tutoring systems in a number of ways. They adapt their behavior to the dynamic state of learning environment, taking advantage of learning opportunities as they arise. They can support collaborative learning as well as individualized learning, because multiple students and agents can interact in a shared environment. Adaptive learning system is a type of e-learning technology, which behaves intelligently to understand the learner's needs and characteristics before delivering the content for them. However, the ability to adapt to user's needs is crucial in a wide variety of applications where the user has to interact with large volumes of data [3]. The purpose of this paper is to develop a multi-agent system that provides adaptivity for learning management systems. The secondary purpose is to construct a conceptual framework defining learner modeling, content modeling, and adaptation strategies.

2 Approach

2.1 Adaptive Learning Environments

Intelligent tutoring systems (ITS) are adaptive instructional systems developed with the application of artificial intelligence (AI) methods and techniques. ITS provides learner oriented design and much more pedagogical knowledge implemented in the system [4]. Benefits of individualized instruction are the essence of ITS, which uses artificial intelligence to tailor multimedia learning.

All adaptive systems are based on a conceptual framework defining the adaptation mechanism. In this paper, a conceptual framework for adaptive learning management systems is defined. This framework describes the ways to model the learner and content, and also to find the best match between the learner and content. We model the learner according to three factors, namely, behavioral factors, knowledge factors, and personality factors.

The adaptation strategy in the framework is to find the best match between the learner and the instruction set. Using the classification information, we find the best match between the learner profile and the course profile by applying a normalized Euclidian distance function.

2.2 Learning Styles

Learners have different ways of perception, construction, and retention of knowledge. Each learner has a unique learning process, because each has different prior knowledge, mental abilities, and personality factors. Individuals perceive and process knowledge in different ways. This leads to the theory defined as "Learning Styles Theory."

The learning styles theory began with Jung [5] who underlined major differences between individuals in terms of perception, decision, and interaction. Authors in [6] have also followed this paper and focused on understanding the differences in learning.

According to the learning styles theory, each learner has different ways of perception, and one style does not address all individuals. Therefore, instruction must be presented in different ways according to these differences.

2.3 Adaptation Techniques

Brusilovsky [7] explains the adaptive techniques under two main categories: adaptive hypermedia technologies and intelligent tutoring techniques.

In [8], it is stressed that a learning system is considered adaptive if it is capable of monitoring activities of its users, interpreting these on the basis of domain-specific models inferring user requirements and preferences out of the interpreted activities, appropriately representing these in associated models, and acting upon available knowledge of its users and the subject matter at hand, to dynamically facilitate the learning process.

3 Agent Approach

Intelligent Software Agent (ISA) is a software agent that uses AI in the pursuit of the goals of its clients [9]. AI is the imitation of human intelligence by mechanical means. Clients, then, can reduce human workload by delegating tasks to ISAs that normally would require human-like intelligence.

The agents in education are defined as pedagogical agents. Pedagogical agents are autonomous agents that support human learning, by interacting with learner in the context of interactive learning environments. They extend and improve upon previous work on intelligent tutoring systems in a number of ways. They adapt their behavior to the dynamic state of the learning environment, taking advantage of learning opportunities as they arise. They can support collaborative learning as well as individualized learning, because multiple students and agents can interact in a shared environment [10].

If a system is called adaptive, it should be able to:

1. Get an initial model of a particular student, i.e. a student model. Some misunderstanding, misconception, or miscalibration may arise.
2. Give some initial educational material from the knowledge base, revise the initial model indicating what content to skip, what to present, and in what order, i.e. the current model.
3. Deliver educational material and revise the current model.
4. Repeat until all content has been mastered by the student, and the system has a good model of the student for future use, i.e. a student's profile.

4 A System's Framework

In this paper, a conceptual framework for adaptive learning systems is given. The framework is based on the idea that the adaptivity is finding the best match between the learner profile and the course content profile. Learning styles of learners and content type of learning material are used to match the learner to the most suitable content.

The conceptual framework involves learner profile, course content profile, matching strategies between learner and course content profiles, and the initialization and update strategies of the profiles. The framework is based on the idea that the effectiveness of the adaptivity is highly dependent on how much we know about the learner and how much the available content fits the learner profile. Therefore, we need to model the learner and the course content. The matching is between the style of the learner and the type of the content. The style of the learner is configured using learning style theories. The content can be a diagram, a question, an exercise, an experiment, a figure, a graph, text, a table, or a slide retrieved from type of learning resources defined by IEEE LOM Metadata specification [11]. Howard et al. used hypertext, slides, audio, and video content types in [12]. Concept, example, activity are the content types referred to in [13]. Diagram, fact, procedure, innovation, theory content types are mentioned as types of learning resources in [14]. The rest of the content types are derived from the descriptions provided by the learning style model [15]. According to the learner profile, we try to find the appropriate content using these content types.

4.1   Learner Profile

Learner modeling is crucial to provide adaptive instruction. Each learner requires an individualized student model. The better we model our learners, the more we know their personalization needs. To know the learner, we need to keep a variety of information about the learner, such as learning styles, domain knowledge, progress, preferences, goals, interests, etc. In this paper, as in [15], we model the learner according to three factors, namely, behavioral factors, knowledge factors, and personality factors (Fig. 1).

Behavioral attributes	Knowledge attributes	Personality attributes
ID of the action	ID of the content block	According to Felder-Silverman model
ID of the learner	course concepts IDs	IMS LIP standard
owner of the action	content media type	
type of the action	learner's knowledge level	
action starting time	testing results	
action ending time	last modification date	
description of the action		
preferred media type		

Fig. 1. Description of the learner profile

4.2   Learner-Course Content Matching

According to the framework, the adapted content is the content that is best matched with the learner profile where the required bits of information are extracted from the knowledge base with regard to the user's level of competence and personal preferences [16]. Therefore, we need to define a matching mechanism between the learner profile and the course content profile. There might be several approaches for calculating the matching rates of the course contents. In this paper, we simply use Euclidian distance and find a matching score based on the normalized distance.

We keep the learner style information in eight dimensions, $x = [x_1, x_2, ..., x_8]$. We have course content profile information classified into eight dimensions of learning style, $y = [y_1, y_2, ..., y_8]$. We normalize the dimension values by substituting the maximum value of the dimension.

The Euclidian distance between these two dimensions is computed as

$$D(x, y) = \|x - y\| = \sqrt{(x_{1n} - y_{1n})^2 + (x_{2n} - y_{2n})^2 + \ldots (x_{8n} - y_{8n})^2}. \tag{1}$$

The matching score is defined as: $S(x, y) = -D(x, y)$. $S(x, y)$ gives the matching score for the learner and the course content profiles. The matching score is calculated for each course content profile. Resulting scores are sorted, and the course content with the highest score is accepted as the best fitted course content regarding the learner profile. After finding and sorting the scores, a filtering might be applied depending on the application choices. One might prefer to display the first three best matched contents or two, or may accept only the best one.

Figure 2 shows an overview of the interface of an educational system in the field of airplane design. The bottom left section of the image shows the prototype hardware.

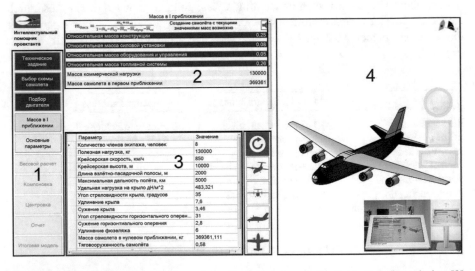

Fig. 2. An overview of the interface of an adaptive system in the field of airplane design [3]

The interface consists of several major parts: timeline (1), interactive section (2), collapsible general information window (3), and the model preview window (4).

The timeline represents the scenario of creating an aircraft at a high level of abstraction. Therefore, subsections of the timeline represent design processes of a lower level of abstraction. The interactive section consists of collapsible separate blocks that facilitate particular design decisions both with calculation tools and necessary reference material, presented in the form most suitable to the student's profile. Consequences of the implemented design decisions are demonstrated live on the model preview window.

Choosing optimal behaviors within this problem domain is challenging to accomplish in acceptable timeframes and can easily be intractable. Generating plans via symbolic planning can typically be accomplished within reasonable timeframes, but will not account for emerging complexities, and may not provide realistic estimates of generated course durations.

The goal of the tutoring system is the autonomous generation of supportive behaviors, with the explicit goal of improving a collaborating agent's performance through the simplification of various aspects of a task via calculated educational environmental manipulations. The ability to optimize an agent's behaviors for a supportive role broadens its utility, especially within problem domains, in which the agent cannot fully perform a specified task either due to liability or lack of capability. Further, the potential reduction in cognitive load attained by incorporating a supportive agent into a problem domain can serve to increase student's productivity.

A permissible action sequence is generated that results in an educational environment, which minimizes the expected execution time of the learners working on the task. Conceptually, this is accomplished by finding the most desirable environmental configurations for the set of likely execution policies being carried out by the lead agents (learners), limited by the supportive agent's ability to effect the environment in a way that achieves this ideal educational environmental configuration.

The supportive behavior problem could be defined as the tuple $\sum$ such that

$$\sum = (\Pi_T, a_S, C_S, S_C, P, T), \tag{2}$$

where Π_T is a set of symbolic plan solutions for T, a_S is a supportive agent, C_S is a mapping function indicating blocks from T that are available for a_S, S_C is the current educational environment state, P is a set of predicates that indicate undesirable environmental states, and $T = \{A, O, C, s_O, s_G\}$ is a problem definition, where:

- A is a set of agents representing students;
- O is a set of course material blocks competing for the student's available time;
- C is a capabilities mapping function between learners and material blocks that indicates what blocks are "available" for the learners from a comprehension standpoint;
- s_O and s_G are sets of predicates that specify the starting and desirable states of the learners.

A solution to T is a policy $\pi \in \Pi_T$ that achieves state s_G through a specified sequence of information blocks in O, executed by a subset of agents in A. A solution to the supportive behavior problem $\sum$ is a plan π_s, which, when executed by a_S, reduces the expected duration for an agent in A of physically executing a plan in Π_T, or reduces the expected search complexity required for agents in A to find solutions $\Pi \subseteq \Pi_T$.

To solve the supportive behavior problem described by $\sum$, alternative environmental configurations are sampled, evaluating them based on metrics such as the estimated cognitive (planning complexity) or physical (expected comprehension time requirements) demands imposed on the agents in the original problem. This optimization must account for the anticipated time costs associated with achieving the hypothesized improved educational environment states, as well as relative likelihoods of a learner following particular plans in Π_T.

At the current environment state S_C, a set of parameterized information blocks usable by a_S is built, which is denoted by O_s.

With O_s constructed, different environmental configurations ("hypothetical educational environments") obtainable by changing individual aspects of the environment are sampled. The sampling method used involves choosing a single available information block. For each hypothesized environment, a plan is determined using actions in O_s enabling the support agent to reconfigure the current environment into the hypothesized state. An estimate is computed for the execution duration of each plan associated with a hypothetical environment (unobtainable configurations are discarded). These duration estimates are later used to evaluate the inconvenience associated with the support agent creating this environment (and tying up the associated resources where necessary).

Each attainable hypothetical environment is finally encoded as the tuple $\xi = \{\pi, d, s\}$ indicating a plan π composed of allowable parameterized information blocks, an estimate of the duration d required to achieve the desired environment, and a set of predicates s describing the resulting educational environment state. A set of plan weights is determined to influence the type of support provided. Selection of this weighting function can have strong effects on the plans generated. Three types of weighting schemes are characterized that can be used to direct a supportive agent toward particular outcomes. A plan execution duration approximation can be defined as

$$m = \min_{\pi \in \Pi_T} duration(T, \pi, s_0, f(t) = 1) \tag{3}$$

to be the duration of the shortest (temporally optimal) known plan, where $f(t) = 1$ indicates that the main task of the system is accomplished, meaning that the set of competences that the student have indicated as desirable is contained within the set of competences obtained through interacting with the adapted educational environment. Three possible cases of useful weighting functions and their resulting behaviors are:

1. A conservative optimization function that weights plans relative to their estimated optimality of execution duration. Any positive function that monotonically decreases as the sub-optimality of a given plan increases can be used.
2. A greedy optimization function that optimizes for what is estimated to be the best known plan and ignoring consequences for all other possible policies.
3. An aggressive optimization function that not only prioritizes making the best plans better, but also making the worst plans even worse, with the intention of driving a rational agent away from selecting poorly. Functionally, this can be accomplished by providing undesirable plans with negative plan weights, for example, for optional information blocks that the student have indicated to be undesirable during the initial testing for the course. It is important to keep the magnitude of negative weights less than the positive weights (using a normalization factor α), or else the support agent may perpetually block task progress in the case of partial plan overlap

between "good" and "bad" plans. A functional example of such a weighting scheme can be achieved by modifying the results from the conservative weighting approach:

$$w_\pi = \begin{cases} w_\pi, & duration(T, \pi, s_0, f(t) = 1) \leq \varepsilon \\ -\alpha w_\pi, & \text{otherwise} \end{cases}, \tag{4}$$

where ε is a value equal to or slightly greater than m.

Combining the information gathered thus far, a supportive agent will choose the best supportive action plan according to

$$\min \sum_{\pi \in \Pi_T} w_\pi \cdot duration(T, \pi, s_c, \varepsilon, \gamma), \tag{5}$$

where w_π is a plan's associated weight, and the duration function computes an estimate of the solution accounting for the cost of the supportive behavior. The final argument of the duration function $\gamma : Z^+ \rightarrow [0, 1]$ is a decay function used to modulate the prioritization of supportive actions causing near-term effects over those that cause long-term consequences.

5 Conclusion

In this paper, a multi-agent approach in the context of educational domain was presented. We provided a description of intellectual educational agents. Benefits that the multi-agent organization structure provides for an adaptive tutoring system were outlined. The main negative aspect of adaptivity is the necessity of content redundancy in the system, compared to a traditional approach to course composition. However, in the established education fields, there is generally enough content to facilitate different kinds of learners. A conceptual framework for adaptive learning systems was presented. The framework is based on the idea that the adaptivity is finding the best match between the learner profile and the course content profile. Learning styles of learners and content type of learning material are used to match the learner to the most suitable content.

References

1. Park, I., Hannafin, M.J.: Empirically based guidelines for the design of interactive multimedia. Educ. Technol. Res. Dev. **41**(3), 63–85 (1993)
2. Brusilovsky, P.: Adaptive and intelligent technologies for web-based education. KI – Kunstl. Intell. **13**(4), 19–25 (1999)
3. Borgest, N.M., Vlasov, S.A., Gromov, Al.A., Gromov, An.A., Korovin, M.D., Shustova, D. V.: Robot-designer: on the road to reality. Ontol. Des. **5**(4(18)), 429–449 (2015)
4. Ong, J., Ramachandran, S.: Intelligent tutoring systems: the what and the how. In: Learning Circuits (2000)

5. Jung, C.: Psychological Types. Rascher Verlag, Zurich (1921)
6. Briggs, K.C., Myers, I.B.: Gifts Differing: Understanding Personality Type. Davies-Black Publishing, Palo Alto (1980)
7. Brusilovsky, P.: Methods and techniques of adaptive hypermedia. User Model. User-Adapt. Interact. 6(2–3), 87–129 (1996)
8. Paramythis, A., Loidl-Reisinger, S.: Adaptive learning environments and e-learinng standards. Electron. J. e-Learn. 2(1), 181–194 (2004). ISSN 1479-4403
9. Croft, D.W.: Intelligent software agents: definitions and applications (1997). http://alumnus. caltech.edu/ ~ croft/research/agent/definition/
10. Shaw, E., Johnson, W.L., Ganeshan, R.: Pedagogical agents on the web. In: AGENTS 1999: Proceedings of the Third Annual Conference on Autonomous Agents, pp. 283–290. ACM, New York (1999)
11. IEEE information technology, learning technology: Learning objects metadata from the World Wide Web. http://ltsc.ieee.org
12. Carver, C.A., Richard, A.H.: Enhancing student learning through hypermedia courseware and incorporation of student learning styles. IEEE Trans. Educ. 42, 33–38 (1999)
13. Stash, N.V., Cristea, A.I., Bra, P.M.: Authoring of learning styles in adaptive hypermedia: problems and solutions. In: WWW Alt. 2004 Proceedings of the 13th International World Wide Web Conference on Alternate Track Papers & Posters, pp. 114–123. ACM Press, New York (2004)
14. Parvez, S.M., Blank, G.D.: A pedagogical framework to integrate learning style into intelligent tutoring systems. J. Comput. Small Coll. 22(3), 183–189 (2007)
15. Felder, R.M., Silverman, L.K.: Learning and teaching styles in engineering education. Eng. Educ. 78(7), 674–681 (1998)
16. Borgest, N., Korovin, M.: Ontological approach towards semantic data filtering in the interface design applied to the interface design and dialogue creation for the "robot-aircraft designer" informational system. In: Advances in Intelligent Systems and Computing, vol. 534, pp. 93–101. Springer (2017)

Intelligent Search and Information
Analysis in Local and Global Networks

Usage of Decision Support Systems in Information Operations Recognition

Oleh Andriichuk[✉], Dmitry Lande, and Anastasiya Hraivoronska

Institute for Information Recording of National Academy of Sciences of Ukraine, Kyiv, Ukraine
andriichuk@ipri.kiev.ua, dwlande@gmail.com, nastya_graiv@ukr.net

Abstract. In this paper, usage of decision support systems in information operations recognition is described. Based on information of experts, the knowledge engineer constructs a knowledge base of subject domain, using decision support system tools. The knowledge base provides the basis for specification of queries for analysis of dynamics of respective informational scenarios using text analytics means. Based on the analysis and the knowledge base structure, decision support system tools calculate the achievement degree of the main goal of the information operation as a complex system, consisting of specific informational activities. Using the results of these calculations, decision makers can develop strategic and tactical steps to counter-act the information operation, evaluate the operation's efficiency, as well as efficiencies of its separate components. Also, decision support system tools are used for decomposition of information operation topics and evaluation of efficiency rating of these topics in dynamics.

Keywords: Information operation · Decision support system · Expert estimate Content-monitoring system

1 Introduction

Given the present level of development of information technologies, it is hard to overestimate their impact on human life. Information media, which involve every man, any social group or population, form their worldview, affect their behavior and decision making [1, 2]. Therefore, the problems related to formation and modification of these information media are extremely important nowadays.

Under information operation (IO) [3], we assume the complex of information activities (news articles in the Internet and the papers, news on TV, comments in social networks, forums, etc.) aimed at changing public opinion about a definite object (person, organization, institute, country, etc.). For example, spreading the rumors about problems in a bank can provoke its clients to take back their deposits, which in turn can cause its bankruptcy. Mainly, this activity has a disinformation character. The information operation belongs to so called ill-defined subject domains [4, 5], because it possesses several such characteristic: uniqueness, inability to formalize the objective of its function and, as a consequence, inability to construct the analytical model,

O. Chertov et al. (Eds.): ICDSIAI 2018, AISC 836, pp. 227–237, 2019.
https://doi.org/10.1007/978-3-319-97885-7_23

dynamics, incompleteness of description, presence of human factor, absence of standards. These domains are treated using expert decision support systems (DSS) [6].

In [3], techniques of IO identification are presented, which are based on analysis of time series built on the basis of thematic information stream monitoring. The following problematic situations, which can appear in IO identification due to drawbacks of current methods and techniques, can be noted:

1. Given a sufficiently large number of publications about the IO object, the number of publications (information stove-piping) about its definite component can be very small and, as a consequence, respective system distortions of typical dynamics of information plots will not be revealed (such as, for example, "Mexican Hat" and Morlet wavelets [7] discovered on the respective wavelet scalogram). Some IOs may be complex and respective information stove-pipings may be staged, related to different components of the IO object in different time periods. If their number is blurred at the background of the total number of publications about the IO object ("information noise"), and the respective information attacks are not identified, then the beginning of the information campaign on the object discredit can be missed, and some information damage to its image will not be taken into account.
2. Content-monitoring tools control queries consisting of keywords to search for respective publications. Keywords are formed based on the IO object title. But the complex IO object can have a great number of components with respective titles, which are not accounted for in queries and, as a consequence, not all the publications on the issue will be found.
3. Queries related to IO object have different degrees of importance according to IO components, to which they are related. Absence of information about values of these importance degrees (i.e. their equivalence) leads to reduction of the IO model relevance.

To overcome the above-mentioned drawbacks, we suggest using the following techniques for application of decision support systems in IO recognition.

2 Methodology for Usage of Decision Support Systems in Information Operations Recognition

The core of the methodology for usage of expert decision-making support tools in information operations recognition [3, 8] proceedings along the following lines:

1. Preliminary investigation of the IO object is carried out, and its target parameters (indexes) are being selected. Then it is suggested that formerly, in retrospective, IOs against the object were taking place, and, thus, its condition (respective target indices) has been deteriorated.
2. The group expertise is conducted to determine and decompose information operation purposes and to estimate its degree of influence. Thus, the IO object is being decomposed as a complex weakly structured system. For this purpose, means of the system for distributed collection and processing of expert information (SDCPEI) are

used. For obtaining expert information in a full range and without distortions, the expert estimation system is used.

3. The respective knowledge base (KB) is constructed using DSS tools, based on the results of the group expertise performed by means of SDCPEI and using available objective information.
4. Analysis of dynamics of the thematic information stream by means of content-monitoring system (CMS) is carried out. KB of DSS is complemented.
5. Recommendations are calculated by means of DSS based on the constructed KB. For this, the IO target achievement degrees are calculated in retrospective and are compared with the respective changes of the IO object condition. The mean value of IO target achievement degrees is calculated, at which degrees the deterioration of the IO object target indices occurs. Thus, monitoring of the IO object condition for the current period of time allows to predict the deterioration of the IO object target values based on the comparison of the calculated value of IO target achievement degree with the above mentioned mean value. In the case of availability of statistically sufficient sample size, and given sufficient correlation between values of IO target achievement degrees and deterioration of IO object target index values, one can even predict the quantitative value of the IO object target index for the current period.

Advantages of the suggested methodology are as follows:

1. Great specification of the model. At the background of numerous publications about IO object in general, the change in publication number dynamics due to the stove-piping about one of IO components will be insignificant and, therefore, will not be revealed.
2. The number of found thematic publications will increase, because of a larger number of queries and keywords.
3. Weighing of IO components allows to avoid the situation when all components are of equal importance. The IO model constructed in such a way will be more relevant.
4. The constructed KB can be used again later during a long period without the necessity to carry out a new expertise.
5. Application of SDCPEI tools makes it possible for experts to work through the global network, thus saving time and resources.

Drawbacks of the suggested methodology are as follows:

1. Application of expert techniques requires time and financial efforts for implementation of the group expertise. Besides, the timely actualization of KB should be done for its second use in the future.
2. Complexity and sometimes ambiguity of the presentation of some sufficiently complex statements of IO components in the form of queries in the content-monitoring system.

Let us show in details the suggested methodology at the example of information operation against National Academy of Sciences (NAS) of Ukraine. It is known that presently NAS of Ukraine is going through hard times. In recent years, the funding of NAS is getting worse. The total funding of NAS of Ukraine decreases, and the share of the budget of NAS of Ukraine in the total budget of Ukraine decreases as well. This is well seen from the information about distribution of expenses for the State Budget of Ukraine for 2014–2016 [9–11]. Suppose that this cut of funding is the result of information operation against NAS of Ukraine.

As SDCPEI for group expert decomposition, we use Consensus-2 system [12] aimed at performing estimations by territorially distributed expert groups.

Based on interpretation of the knowledge base generated after operation of Consensus-2 SDCPEI, Solon-3 DSS [13] generates the respective KB. In Solon-3 DSS, a novel method is used based on decomposition of main objective of the program, construction of knowledge base (hierarchy of targets), and dynamic target estimation of alternatives [6]. Expert's pair comparisons were performed in the Level system [14].

Let's consider some aspects of the above mentioned method. The KB has a structure of goal hierarchy graph. Nodes (vertices) of the graph represent goals or KB objects. Edges reflect the impact of one set of goals on achievement of other goals: they connect sub-goals to their immediate "ancestors" in the graph (super-goals).

To build the goal hierarchy, the method of hierarchic target-oriented evaluation of alternatives is used [6]. A hierarchy of goals is built, after which respective partial impact coefficients (PIC) are set, and the relative efficiency of projects is calculated. First, the main goal of the problem solution is formulated, as well as potential options of its achievement (projects) that are to be estimated at further steps. After that, a two-phase procedure of goal hierarchy graph construction takes place: "top-to-bottom" and "bottom-to-top" [6]. "Top-to-bottom" phase envisions step-by-step decomposition of every goal into sub-goals or projects, that influence the achievement of the given goal. The main goal is to decompose into more specific components, sub-goals that influence it. Then these lower-level goals are further decomposed into even more specific sub-components that are, in their own turn, decomposed as well. When a goal is decomposed, the list of its sub-goals may include (beside the just formulated ones) goals (already present in the hierarchy) that were formulated in the process of decomposition of other goals. Decomposition process stops when the sets of sub-goals that influence higher-level goals include already decomposed goals and decision variants being evaluated. Thus, when decomposition process is finished, there are no goals left unclear. "Bottom-to-top" phase envisions definition of all upper-level goals (super-goals, "ancestors") for each sub-goal or project (i.e., the goals this project influences).

As mentioned, experts build a hierarchy of goals represented by an oriented network-type graph. Its nodes are marked by goal formulations. Presence of an edge connecting one node (goal) to another indicates impact of one goal upon achievement of the other one. As a result of the above-mentioned process of goal hierarchy construction, we get a graph that is unilaterally connected, because from each node, there is a way to the node marking the main goal. Each goal is assigned an indicator from 0 to 1 of achievement level. This indicator equals 0 if there is absolutely no progress in achievement of the goal, whereas if the goal is completely achieved, it equals 1. Impact of one goal upon the other can be positive or negative. Its degree is reflected by the respective value, i.e. a PIC. In the method of target-oriented dynamic evaluation of alternatives, the delay of impact is also taken into consideration [6]. In case of projects, their implementation time is taken into account as well. PIC are defined by experts, and, in order to improve credibility of expert estimation process, pairwise comparison based methods are used.

In the result of the group expertise, 15 expert statements were obtained presenting the components of IO against NAS of Ukraine, namely, "Bureaucracy in NAS of Ukraine," "Inefficient personnel policy of NASU," "Corruption in NAS of Ukraine," "Underestimation of the level of scientific results of NAS of Ukraine," "Lack of introduction of scientific developments into manufacture," "Underestimation of the level of international collaboration," "Misuse and inefficient use of the realty of NASU," "Misuse and inefficient use of land resources of NASU," "Discredit of President of NAS of Ukraine," "Discredit of Executive secretary of NAS of Ukraine," "Discredit of other well-known persons of NAS of Ukraine," "Juxtaposition of scientific results of Ministry of Education and Science (MES) and NAS," "Juxtaposition of scientific results of other academic organizations to NAS," "Juxtaposition of developments of Ukrainian companies to NAS of Ukraine," "Juxtaposition of scientific results of foreign organizations to NAS."

By means of InfoStream CMS [15], analysis of thematic information stream dynamics was made. For this, in accordance with each of the above listed IO component, queries were formulated in a special language, using which the above mentioned process (analysis of publication dynamics on target issues) takes place.

Figure 1 presented results of the express analysis [3] of thematic information stream corresponding to the IO object, i.e. NAS of Ukraine. In the result of the analysis by means of InfoStream CMS, the respective information stream from the Ukrainian Internet segment was obtained. To reveal information stove-piping, publication dynamics was analyzed on the target issue using available analytical tools. In Fig. 1, one characteristic fragment of the dynamics is shown (for the period from July 1, 2015 to December 31, 2015).

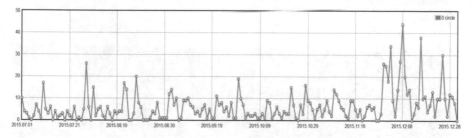

Fig. 1. Publication dynamics on target issue

To reveal the degree of similarity of fragments of respective time series to IO diagram in different scales, one can use wavelet analysis [3, 16]. A wavelet is a function which is well localized in time. In practice, we often use the Mexican Hat wavelet):

$$\psi(t) = C(1 - t^2) \exp\left\{-\frac{t^2}{2}\right\},\tag{1}$$

and the Morlet wavelet:

$$\psi(t) = \exp\left\{ikt - \frac{t^2}{2}\right\}.\tag{2}$$

The essence of the wavelet transform is to identify regions of time series, which are similar to the wavelet. To explore different parts of the original signal with various degrees of detail, the wavelet is transformed by stretching/squeezing and moving along the time axis. Therefore, the continuous wavelet-transform has the location parameter (l) and the scale parameter (s). The continuous wavelet transform of function $x \in L^2(R)$ is determined as follows:

$$W(l, s) = \frac{1}{\sqrt{s}} \int_{-\infty}^{\infty} x(t)\psi^*\left(\frac{t - l}{s}\right)dt,\tag{3}$$

where $l, s \in R, s > 0$, ψ^* is complex conjugate of ψ, values $\{W(s, l)\}$ are called coefficients of wavelet transform, or wavelet-coefficients. The wavelet-coefficients are visualized in the plot with a location axis and a scale axis.

The reason to use the Mexican hat wavelet and the Morlet wavelet is a possibility to detect spikes in time series. Wavelet coefficients show, to which extent the behavior of the process in a definite point is similar to the wavelet in a definite scale. In the respective wavelet spectrogram (Fig. 2), all the characteristic features of the initial series can be seen: scale and intensity of periodic variations, direction and value of trends, presence, position, and duration of local features.

Fig. 2. Wavelet spectrogram (Morlet wavelet) of the information stream

IO dynamics is represented by most exactly the "Mexican Hat" and Morlet wavelets [3, 7, 16]. Therefore, the time series according to each of 15 IO components are analyzed during four periods (January 1, 2013–December 31, 2013; January 1, 2014–December 31, 2014; January 1, 2015–December 31, 2015; and January 1, 2016–December 15, 2016), and the presence of the above-mentioned wavelets is identified.

Based on revealed mentioned above information stove-pipings and their parameters (position, duration and intensity), the knowledge engineer complements KB of Solon-3

DMSS. In particular, the stove-piping was identified on the IO component "Underestimation of the scientific results of NAS of Ukraine" situated at November 30, 2015 with the duration of 14 days. Correspondingly, as a characteristic of the project "Underestimation of the scientific results of NAS of Ukraine," the parameter of duration of the project execution of 14 days is introduced, and as a characteristic of the project "Underestimation of the scientific results of NAS of Ukraine" influence on the objective "Discredit of scientific results of NAS of Ukraine," the parameter of delay in influence distribution for 10 months term is introduced. For other revealed information stove-pipings, characteristics of projects and influences are introduced in a similar fashion.

Thus, for the time period from January 1 2015 to December 31, 2015, KB is complemented and has the structure shown in Fig. 3. Correspondingly, the list of numbers and statements of all targets and projects of KB is as follows: 0 — "Information operation against National Academy of sciences of Ukraine", 1 — "Discredit of scientific results of NAS of Ukraine," 2 — "Discredit of the structure of NAS of Ukraine," 3 — "Discredit of well-known persons of NAS of Ukraine," 4 — "Overestimation of scientific results of competing with NASU organizations," 5 — "Lack of introductions of scientific developments into production," 6 — "Underestimation of the level of international collaboration," 7 — "Discredit of the organization structure of NASU," 8 — "Juxtaposition of scientific results of MES and NAS," 9 — "Discredit of President of NAS of Ukraine," 10 — "Discredit of other well-known persons of NAS of Ukraine," 11 — "Juxtaposition of scientific results of other academic organizations to NAS of Ukraine," 12 — "Juxtaposition of scientific results of foreign organizations to NAS of Ukraine," 13 — "Juxtaposition of developments of Ukrainian companies to NAS of Ukraine," 14 — "Underestimation of the level of scientific results of NAS of Ukraine," 15 — "Corruption in NAS of Ukraine 2," 16 — "Bureaucracy in NAS of Ukraine 2," 17 — "Inefficient personnel policy of NASU 2," 18 — "Misuse and inefficient use of the realty of NASU 1," 19 — "Misuse and inefficient use of the realty of NASU 2," 20 — "Misuse and inefficient use of land resources of NASU 1," 21 — "Misuse and inefficient use of land resources of NASU 2," 22 — "Discredit of the actions of the Case Management department of NASU," 23 — "Discredit of Executive secretary of NAS of Ukraine," 24 — "Corruption in NAS of Ukraine 1," 25 — "Bureaucracy in NAS of Ukraine 1," 26 — "Inefficient personnel policy of NASU 1."

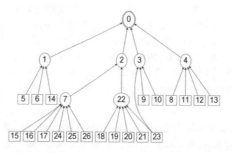

Fig. 3. Structure of goal hierarchy of knowledge base for information operation against National Academy of Sciences of Ukraine

It should be noted that for some IO components, namely, "Corruption in NAS of Ukraine," "Bureaucracy in NAS of Ukraine," "Inefficient personnel policy of NASU," "Misuse and inefficient use of land resources of NASU," and "Misuse and inefficient use of the realty of NASU," stove-pipings were revealed twice during 2015, therefore, respective projects were entered into KB twice. For example, for the IO component "Bureaucracy in NAS of Ukraine," projects "Bureaucracy in NAS of Ukraine 1" and "Bureaucracy in NAS of Ukraine 2," but each of them has different characteristics of the execution duration (9 and 15 days) and respective influences have different characteristics of delay in distribution (9 and 11 months).

Then, in the Solon-3 DSS, degrees of project implementation are introduced. If for some IO components no stove-piping was found, as, in particular, for "Juxtaposition of developments of Ukrainian companies to NAS of Ukraine" and "Discredit of the actions of the Case Management department of NASU," then for respective projects, implementation degrees are set equal to 0%. For all other projects they equal 100%.

Next, the results are obtained of the calculation of recommendations, namely, the degree of achievement of the main IO target and of project efficiency (relative contribution into achievement of the main target).

For the periods January 1, 2013–December 31, 2013; January 1, 2014–December 31, 2014; January 1, 2015–December 31, 2015; and January 1, 2016–December 15, 2016, the degrees of achievements of the main target have the following values: 0.380492, 0.404188, 0.570779 and 0.438703, respectively.

In the retrospective, the mean value of the degree of achievements of the main target is equal: to $(0.380492 + 0.404188 + 0.570779)/3.0 \approx 0.45182$.

Thus, since the mean retrospective and current value of the degree of achievements of the IO main target are sufficiently close (differ less than by 3%), then the conclusion can be drawn that IO during the current period can provoke the deterioration of the target indexes of the object with high probability.

3   Methodology for Constructing Knowledge Bases of Decision Support Systems in Information Operations Recognition

Application of the approach and the methodology described in the previous section calls for availability of a group of experts. Labor of experts is rather costly and requires considerable time. Also, for small expert groups, it is necessary to take into account competence of experts [17], which entails the need for additional information and time for its collection and processing. So, reduction of expert information usage during construction of DSS KB during IO detection represents a relevant issue.

The essence of the methodology of DSS KB building during IO detection [3, 16] is as follows:

1. Group expert estimation is conducted in order to define and decompose the goals of the information operation. Thus, the IO is decomposed as a weakly structured system. For this purpose, the means of SDCPEI are used.

2. Using DSS, the respective KB is constructed based on the results of expert examinations conducted by SDCPEI as well as using available objective information.
3. Analysis of dynamics of thematic information flow is performed by means of CMS. KB of DSS is supplemented.
4. Recommendations of the decision-maker are calculated by means of DSS, based on the KB already built.
5. As part of the "goal hierarchy graph" model, a hierarchy of goals, or KB, represented in the form of an oriented network-type graph (an example for Brexit is shown in Fig. 4), is built by experts.

The methodology is illustrated by the example of Brexit. Currently, Brexit is a topical issue that is widely researched by the scientific community [18].

The Consensus-2 system [12], intended for evaluation of alternatives by a group of territorially distributed experts, was used as SDEICP for group decomposition. Based on interpretation of the data base, formed in the Consensus-2 SDCPEI, the knowledge engineer created the respective KB of the Solon-3 DSS" [13]. The structure of goal hierarchy of this KB is provided in Fig. 4.

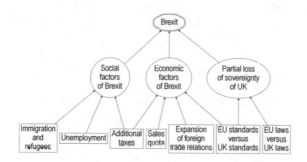

Fig. 4. Structure of goals hierarchy of knowledge base for Brexit

After constructing the goal hierarchy, analysis of dynamics of thematic information flow was conducted by means of InfoStream CMS [15]. For this purpose, in accordance with each component of the IO, queries were formulated in the specialized language. Based on these queries, dynamics analysis of publications on the target topic was performed. During formulation of queries in accordance to the goal hierarchy structure (Fig. 4), the following rules were used:

1. When moving from top to bottom, respective queries of lower-level components of the IO were supplemented by queries of higher-level components (using '&' symbol), for clarification.
2. In cases of abstract, non-specific character of IO components, movement from bottom to top took place, whereas respective queries were supplemented by queries of lower-level components (using '|' symbol).
3. In cases of specific IO components, the query was made unique.

Based on results of fulfillment of queries to the InfoStream system, particularly, on the number of documents retrieved, respective PICs were calculated for each of them. PICs were calculated under assumption that the degree of impact of IO component was proportional to the quantity of the respective documents retrieved. Obtained PIC values were input into the KB. Thus, we managed to refrain from addressing the experts for evaluation of impact degrees of IO components.

Recommendations (in the form of dynamic efficiency ratings of information operation topics) produced in the way describe above are used to evaluate the damage caused by the information operation, as well as to form the information counteractions [19] with information sources impact [20] taken into consideration. Based on the KB for Brexit, the Solon-3 DSS provided the following recommendations (rating of information impact of the publications topics): "Immigration and refugees" (0.364), "EU standards versus UK standards" (0.177), "EU laws versus UK laws" (0.143), "Expansion of foreign trade relations" (0.109), "Additional taxes" (0.084), "Sales quota" (0.08), "Unemployment" (0.043).

4   Conclusions

In the paper, feasibility of application of decision support systems is demonstrated in the process of recognition of information operations. The methodology was suggested for usage of decision support systems in information operations recognition, which allows to predict changes in values of target indices of the object for the current period based on the analysis of retrospective data. Also, the methodology was suggested for constructing knowledge bases of decision support systems in information operations recognition, which allows to provide the decomposition of the topics of the information operations and assess rating of the effectiveness of these topics.

References

1. Lewandowsky, S., Ecker, U.K.H., Seifert, C.M., Schwarz, N., Cook, J.: Misinformation and its correction continued influence and successful debiasing. Psychol. Sci. Public Interest **13** (3), 106–131 (2012). Department of Human Development, Cornell University, USA
2. Berinsky, A.J.: Rumors and health care reform: experiments in political misinformation. Br. J. Polit. Sci. **47**(2), 241–262 (2017)
3. Dodonov, A.G., Lande, D.V., Tsyganok, V.V., Andriichuk, O.V., Kadenko, S.V., Hraivoronska, A.M.: Raspoznavaniye informatsionnykh operatsiy (Recognition of information operations) – K., Engineering, p. 282 (2017). (in Russian)
4. Taran, T.A., Zubov, D.A.: Iskusstvennyy intelekt. Teoriya i prilozheniya (Artificial intelligence. Theory and applications). Lugansk: V. Dal VNU, p. 239 (2006). (in Russian)
5. Glybovets, M.M., Oletsikiy, O.V.: Shtuchnyy intelekt (Artificial intelligence). – K.: "KM Akademiya" publishers, p. 366 (2002). (in Ukrainian)
6. Totsenko, V.G.: Metody i sistemy podderzhki prinyatiya resheniy. Algoritmicheskiy aspect (Methods and systems for decision-making support. Algorithmic aspect). – K.: Naukova Dumka, p. 382 (2002). (in Russian)
7. Addison, P.S.: The Illustrated Wavelet Transform Handbook: Introductory Theory and Applications in Science, Engineering, Medicine and Finance, p. 446. CRC Press/Taylor & Francis Group, Boca Raton (2016)

8. Andriichuk, O.V., Kachanov, P.T.: Metodika primeneniya instrumentariya ekspertnoy podderzhki prinyatiya resheniy pri identifikatsii informatsionnykh operatsiy (A methodology for application of expert data-based decision support tools while identifying informational operations). In: CEUR Workshop Proceedings (ceur-ws.org). Vol-1813 urn:nbn:de:0074-1813-0. Selected Papers of the XVI International Scientific and Practical Conference "Information Technologies and Security" (ITS 2016), Kyiv, Ukraine, pp. 40–47, 1 December 2016. http://ceur-ws.org/Vol-1813/paper6.pdf. (in Russian)

9. Law of Ukraine On State budget of Ukraine for 2014. http://zakon4.rada.gov.ua/laws/show/719-18. (in Ukrainian)

10. Law of Ukraine On State budget of Ukraine for 2015. http://zakon4.rada.gov.ua/laws/80-19. (in Ukrainian)

11. Law of Ukraine On State budget of Ukraine for 2016. http://zakon4.rada.gov.ua/laws/show/928-19. (in Ukrainian)

12. Tsyganok, V.V., Roik, P.D., Andriichuk, O.V., Kadenko, S.V.: Certificate of registration of copyright on the work № 75023 of Ministry of Economic Development and Trade of Ukraine. Software "System for distributed acquisition and processing of expert information for decision support systems – "Consensus-2"", 27 November 2017. (in Ukrainian)

13. Totsenko, V.G., Kachanov, P.T., Tsyganok, V.V.: Certificate of registration of copyright on the work № 8669. Ministry of education and Science of Ukraine. State Department for Intellectual Property. Software "Decision support system SOLON-3" (DSS SOLON-3), 31 October 2003. (in Ukrainian)

14. Tsyganok, V.V., Andriichuk, O.V., Kachanov, P.T., Kadenko, S.V.: Certificate of registration of copyright on the work № 44521 of State Intellectual Property Service of Ukraine. Software "Software tools for expert estimations using pair comparisons "Riven"("Level")", 03 July 2012. (in Ukrainian)

15. Grigoriev, A.N., Lande, D.V., Borodenkov, S.A., Mazurkevich, R.V., Patsyora, V.N.: InfoStream. Monitoring novostey iz Internet: tekhnologiya, sistema, servis (InfoStream. Monitoring of Internet news: Technology, system, service) – K: Start-98, p. 40 (2007). (in Russian)

16. Lande, D.V., Andriichuk, O.V., Hraivoronska, A.M., Guliakina, N.A.: Application of decision-making support, nonlinear dynamics, and computational linguistics methods during detection of information operations. In: CEUR Workshop Proceedings (ceur-ws.org). Vol-2067 urn:nbn:de:0074-2067-8. Selected Papers of the XVII International Scientific and Practical Conference on Information Technologies and Security (ITS 2017) Kyiv, Ukraine, pp. 76–85, 30 November 2017. http://ceur-ws.org/Vol-2067/paper11.pdf

17. Tsyganok, V.V., Kadenko, S.V., Andriichuk, O.V.: Simulation of expert judgements for testing the methods of information processing in decision-making support systems. J. Autom. Inf. Sci. 43(12), 21–32 (2011)

18. Bachmann, V., Sidaway, J.D.: Brexit geopolitics. Geoforum 77, 47–50 (2016)

19. Tsyganok, V.: Decision-making support for strategic planning. In: Proceedings of the International Symposium for the Open Semantic Technologies for Intelligent Systems (OSTIS 2017), Minsk, Republic of Belarus, 16–18 February, pp. 347–352 (2017)

20. Dodonov, A.G., Lande, D.V.: Issledovaniye istochnikov informatsionnogo vliyaniya veb-resursov seti Internet (A study of informational impact sources of the world wide web). In: CEUR Workshop Proceedings (ceur-ws.org). Vol-1813 urn:nbn:de:0074-1813-0. Selected Papers of the XVI International Scientific and Practical Conference "Information Technologies and Security" (ITS 2016) Kyiv, Ukraine, pp. 1–7, 1 December 2016. http://ceur-ws.org/Vol-1813/paper1.pdf. (in Russian)

How Click-Fraud Shapes Traffic: A Case Study

Dmytro Pavlov and Oleg Chertov[✉]

Igor Sikorsky Kyiv Polytechnic Institute, Kyiv, Ukraine
uctpavlov@gmail.com, chertov@i.ua

Abstract. This paper provides a real-life case-study of click-fraud. We aim to investigate the influence of invalid clicks on the time series of advertising parameters, such as the number of clicks and click-through-rate. Our results show that it can be challenging to visually identify click-fraud in real traffic. However, using powerful methods of signal analysis such as 'Caterpillar'-SSA allows efficiently discovering fraudulent components. Finally, our findings confirm the hypothesis from previous works that attacks can be discovered via behavioral modeling of an attacker.

Keywords: Internet advertising · Click-fraud · 'Caterpillar'-SSA

1 Introduction

Internet became an important part of our day-to-day life and continues to invade new fields of human activities. The advertising industry is not an exception. Nowadays, Internet is the main advertising media. For instance, Google and Facebook receive 18% and 7% of global ad revenues, concentrating a quarter of the world ad spend within these two companies [1]. Internet advertising can be based on three payment models: *pay-per-impression* (advertiser pays certain amount every time a web page with an ad is demonstrated to a user), *pay-per-click* (a payment is made every time a user clicks on the ad), and *pay-per-conversion* (a payment is made after a user commits a predefined action, for example, a purchase or account registration) [2]. Pay-per-click model is the most common. However, it is also vulnerable to so-called *'click-fraud'*, when an attacker generates additional clicks with the aim to exhaust the advertiser's budget [3]. We will refer to such clicks as *invalid clicks*. The Internet advertising service providers constantly improve their protection systems, see, for example, Google's Protection against Invalid Clicks[1]). However, as it is mentioned in the report of PPC Protect Company[2], the level of fraud in Internet advertising remains high.

---

[1] Google's Protection against Invalid Clicks, https://www.google.com/intl/en/ads/adtrafficquality/invalid-click-protection.html.

[2] List of Fraud Statistics 2018, https://ppcprotect.com/ad-fraud-statistics/.

ⓒ Springer Nature Switzerland AG 2019
O. Chertov et al. (Eds.): ICDSIAI 2018, AISC 836, pp. 238–248, 2019.
https://doi.org/10.1007/978-3-319-97885-7_24

In this paper, we aim to study the effects of click-fraud on real advertising campaigns. The obtained insights and ideas can be further used for the development of the system of automatic or semi-automatic click-fraud detection.

2   'Caterpillar'-SSA for Time Series Analysis

Following [4], we aim to discover how click-fraud can re-shape the time distribution of Internet advertising parameters. In the general case, a time series model consists of the following components [5]: stable long-term component or *trend*, cyclic fluctuations or *periodic component*, and random component or *noise*. Various statistical approaches are widely used to identify the mentioned components, for example, criteria for testing the trend existence hypothesis (series criterion, squares of successive differences criterion), methods for trend detection (ARIMA regression models, nonparametric models), methods for periodic components detection (detection of components with known oscillation periods) [6].

Unlike statistical approaches, 'Caterpillar-SSA' does not make any statistical assumptions regarding the structure of the studied series. This also allows identifying unforeseen components and non-obvious trends. 'Caterpillar'-SSA became popular in numerous applications, for example, geodynamics [7], meteorology [8] and analysis of network traffic [9]. Researchers also developed special cases of this method for the analysis of time series with missing values [10] and multi-dimensional time series analysis [11]. The advantages of 'Caterpillar-SSA' over other methods were demonstrated in [12].

The basic 'Caterpillar'-SSA algorithm consists of two phases, and each of the phases consists of two steps: (1) decomposition with steps (a) embedding and (b) singular decomposition, (2) reconstruction with steps (a) grouping and (b) diagonal averaging. The algorithm has only one parameter, the length of the window L, which is used in the step 1a. The cyclic components of the time series can be identified more efficiently if L is multiple of their oscillation periods. Also, using the values of L close to the half of the analyzed signal length usually results in better performance.

Let us consider time series $F = (f_0, \ldots, f_{N-1})$ of length N. During the embedding step of the decomposition phase (1a), the algorithm forms $K = N - L + 1$ embedded vectors of the form $X_i = (f_{i-1}, \ldots, f_{i+L-2})$, $1 \le i \le K$. These vectors are used to construct a trajectory matrix $\mathbf{X}$ $(dim\,(\mathbf{X}) = L \times K)$ according to the following formula:

$$\mathbf{X} = \begin{bmatrix} f_0 & f_1 & f_2 & \cdots & f_{K-1} \\ f_1 & f_2 & f_3 & \cdots & f_K \\ f_2 & f_3 & f_4 & \cdots & f_{K+1} \\ \cdots & \cdots & \cdots & \cdots & \cdots \\ f_{L-1} & f_L & f_{L+1} & \cdots & f_{N-1} \end{bmatrix} \tag{1}$$

From (1), we can see that matrix $\mathbf{X}$ has the same elements on the diagonals perpendicular to its main diagonal (diagonals composed of elements whose indices satisfy the following condition $i + j = const$). It is also obvious that there is a one-to-one correspondence between the trajectory matrix $\mathbf{X}$ and the time series F.

In the step 1b the trajectory matrix is decomposed into a sum of elementary matrices of the unit rank as follows: $\mathbf{X} = \mathbf{X}_1 + \ldots + \mathbf{X}_d$. Next, the grouping of elementary terms in this sum is performed in the step 2a. In order to achieve this, the set of indices $\{1, \ldots, d\}$ is divided into m non-intersecting subsets $I_1, \ldots, I_m$. For example, if $I = \{i_1, \ldots, i_p\}$ then the matrix $\mathbf{X}_I$ corresponding to the group I is determined as follows: $\mathbf{X_I} = \mathbf{X}_{i_1} + \ldots + \mathbf{X}_{i_p}$. Such matrices are calculated for each $I_1, I_2, \ldots, I_m$. In the end of this step, the result of the decomposition can be written as follows: $\mathbf{X} = \mathbf{X}_{I_1} + \ldots + \mathbf{X}_{I_m}$.

This formula represents the decomposition of the initial time series F into m components. However, these components have to be converted into numerical sequences. This can be done using the diagonal averaging procedure (step 2b of the algorithm). During this procedure, each matrix $\mathbf{X}_I$ is approximated with a Hankel matrix which corresponds uniquely to a particular time series. The procedure of the diagonal averaging that transforms an arbitrary matrix into a Hankel matrix is determined by the following formula $y^*_{i,j} = mean\,(y_{m,n}|_{m+n=i+j})$, where $y_{m,n}$ stand for the elements of the input matrix Y and $y^*_{i,j}$ stand for the elements of the corresponding Hankel matrix $\mathbf{Y}^*$. As a result of the diagonal averaging procedure, each matrix $\mathbf{X}_I$ is matched with a time series $\widetilde{F}_k$, and the initial input series F is decomposed into a sum of m time series as follows: $F = \widetilde{F}_1 + \ldots + \widetilde{F}_m$.

Performing successful grouping of elementary terms and selecting appropriate length of the window L, it is possible to efficiently discover the trend, fluctuations and the noise components of the original series F. Based on the results described in the dissertation [13], 'Caterpillar'-SSA method can be automated. This allows using this powerful technique for automated analysis of Internet traffic.

3   A Case-Study: Analysis of Advertising Parameters

In this section, we present a case study of an advertising traffic analysis. We ran an advertising campaign in Google AdWords[3]. We collected the real-time information about the number of clicks and impressions during a period of 438 days from December 1, 2015 to February 10, 2017. It is known that Google automatically filters out non-valid clicks in both on-line and off-line modes[4]. Invalid clicks filtered out in on-line mode are not registered by Google and thereby cannot be captured. However, those invalid clicks that were filtered out in off-line mode disappear from the statistical reports provided by Google after some time. Thereby, by comparing our collected data with the data present in Google reports we can identify cases of click-fraud that were not immediately discovered by the protection systems of Google. From the collected data and the data obtained from Google reports we calculated the value of CTR for each day. CTR

---

[3] Google AdWords https://adwords.google.com/home/.

[4] AdWords Help. Invalid clicks https://support.google.com/adwords/answer/42995.

(click-through-rate) is a ratio of the number of clicks to the number of impressions. We choose to further analyse the distribution of clicks and CTR, as the latter one captures not only the distribution of impressions but also a relation between the number of impressions and the number of clicks.

Thereby, we perform our analysis on four time series: collected distribution of clicks (backup-clicks) and corresponding CTR values (backup-CTR) and distribution of clicks and CTR obtained from Google reports (Google-clicks and Google-CTR). These time series are presented in Fig. 1. As we can see from the figure, invalid clicks are present during the whole period of the campaign activity. This fact also has an impact on the distribution of CTR. Note that some values are missing due to the fact that the considered campaign is not active during weekends and public holidays. Such settings are very popular in real cases. By means of visual inspection of the presented distributions, we can also conclude that it is difficult to suspect click-fraud presence.

In order to analyze the strategy of the attacker, we subtract Google-clicks from backup-clicks. In this way, we obtain the distribution of invalid clicks (click-fraude rate). We further perform the analysis of the obtained distribution with the 'Caterpillar'-SSA method. By grouping the first principal components that correspond to eigenvectors 1, 2 and 3, we obtain the trend of this distribution presented in Fig. 2. Note that in the discovered trend we can identify intervals corresponding to 'probe' – first try to perform the fraud (from day 50 to day 190), 'slack' – system normalization (from day 190 to day 210), and 'attack' - actual attack (from day 210 to day 393). These intervals form the 'information attack' pattern [14]. This patter was discussed in the literature before as a typical behavioral strategy of an attacker committing click-fraud [15].

To continue our analysis, we further decompose the distributions of clicks (see Fig. 3 for backup-clicks and Fig. 4 for Google-clicks) and CTR (see Fig. 5 for backup-CTR and Fig. 6 for Google-CTR) into main components. Note, that the length of the main components is equal to the width of the window L, in our case $L = 217$.

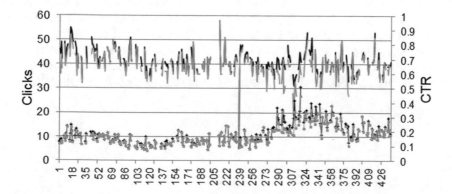

Fig. 1. Distribution of the number of clicks and CTR (grey without markers – Google-clicks, black without markers – backup-clicks, grey with markers – Google-CTR, black with markers – backup-CTR)

Let us analyze the decomposition of backup-clicks and Google-clicks (Figs. 3 and 4). We can see that the components of both distributions are almost identical, except the component corresponding to the eigenvector #4 in Fig. 3. As the only difference between the two analyzed time series is the absence of invalid clicks, we can conclude that this component actually corresponds to click-fraud distribution presented in Fig. 2.

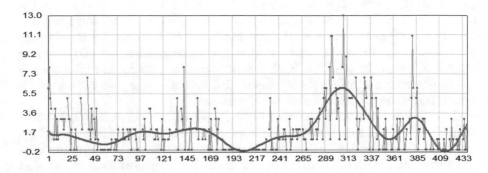

Fig. 2. Click-fraude rate and its trend component

Let us now analyse the decomposition of backup-CTR and Google-CTR (Figs. 5 and 6). As in the previous case, we can see that the main components are almost identical. The difference is visible only in the second component. This component is largely stable, thereby it corresponds to an element of the trend. We can see that this component has inverse positioning of maximum and minimum values for CTR-backup and CTR-Google. In Fig. 5 it has a maximum towards the end of the series (closer to the 'attack' phase), while in Fig. 6 the maximum is situated in the middle of the series (closer to the 'slack' phase). This means that, as compared with a usual interval, during the actual attack more clicks with *higher density* (less number of impressions per each click) were produced. Such strategy is also typical for an attacker. Thus the second component of the decomposition given in Fig. 5 reflects the activities of an attacker.

Thereby, using 'Caterpillar'-SSA method we can successfully identify those components of the distributions, that are caused by attacker activities. This proves the high potential of this method in the task of click-fraud identification.

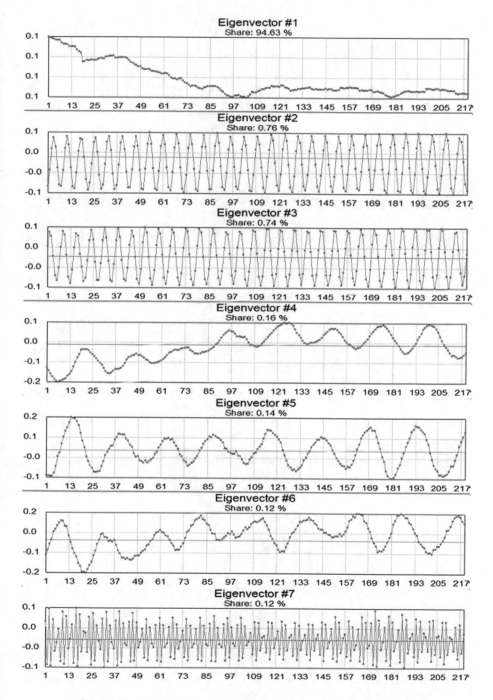

Fig. 3. First 8 main components of backup-clicks distribution

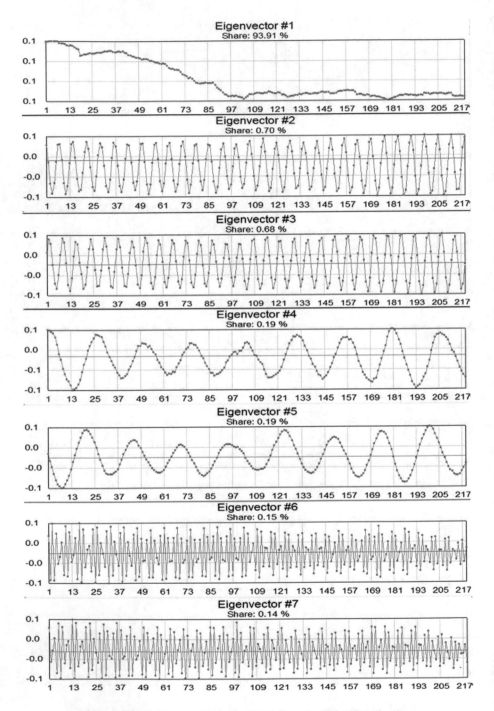

Fig. 4. First 8 main components of Google-clicks distribution

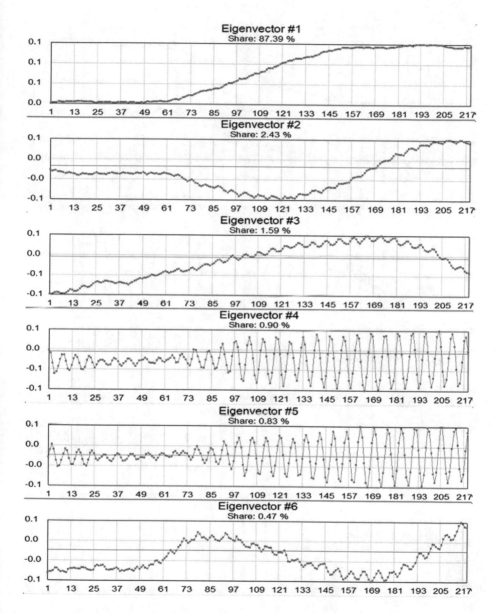

Fig. 5. First 6 main components of the backup-CTR distribution

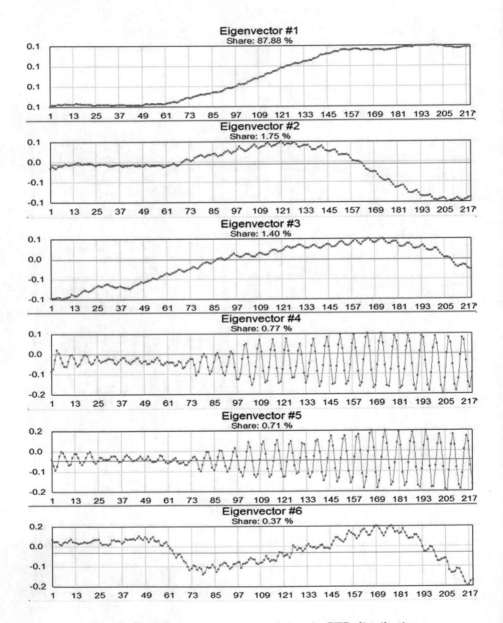

Fig. 6. First 6 main components of Google-CTR distribution

4   Conclusions

In this paper, we presented a real-life case study of how click-fraud can shape the distribution of Internet advertising parameters. After analyzing the distribution of clicks and CTR from a real advertising campaign we can formulate several conclusions. Although, it can be difficult to visually identify the appearance of invalid clicks, the appropriate methods of signal processing, for example, 'Caterpillar'-SSA, can efficiently discover fraudulent components. Thereby, such techniques can be potentially successful in automated or semi-automated click-fraud identification. While analyzing the time series of invalid clicks we also discovered the presence of a behavioral pattern known as 'information attack'. This pattern was previously studied in the literature as a factor for click-fraud identification. Thereby, our results confirm the possibility of discovering click-fraud using behavioral models of an attacker.

References

1. Richter, F.: 25 percent of global ad spend goes to google or facebook (2017). https://www.statista.com/chart/12179/google-and-facebook-share-of-ad-revenue/. Accessed 10 Mar 2018
2. Fain, D.C., Pedersen, J.O.: Sponsored search: a brief history. Bull. Assoc. Inf. Sci. Technol. **32**(2), 12–13 (2006)
3. Jansen, B.J., Mullen, T.: Sponsored search: an overview of the concept, history, and technology. Int. J. Electron. Bus. **6**(2), 114–131 (2008)
4. Chertov, O., Malchykov, V., Pavlov, D.: Non-dyadic wavelets for detection of some click-fraud attacks. In: 2010 International Conference on Signals and Electronic Systems (ICSES), pp. 401–404. IEEE (2010)
5. Gerbing, D.: Time series components. Portland State University, p. 9 (2016)
6. Hamilton, J.D.: Time Series Analysis, vol. 2. Princeton University Press, Princeton (1994)
7. Gorshkov, V., Miller, N., Persiyaninova, N., Prudnikova, E.: Principal component analysis for geodinamical time series. Commun. Russ. Acad. Sci. **214**, 173–180 (2000). (in Russian)
8. Ghil, M., Allen, M., Dettinger, M., Ide, K., Kondrashov, D., Mann, M., Robertson, A.W., Saunders, A., Tian, Y., Varadi, F., et al.: Advanced spectral methods for climatic time series. Rev. Geophys. **40**(1) (2002)
9. Antoniou, I., Ivanov, V., Ivanov, V.V., Zrelov, P.: Principal component analysis of network traffic measurements: the Caterpillar-SSA approach. In: VIII International Workshop on Advanced Computing and Analysis Techniques in Physics Research, ACAT, pp. 24–28 (2002)
10. Golyandina, N., Osipov, E.: 'Caterpillar'-SSA for analysis of time series with missing values. Math. Model. Theor. Appl. **6**, 50–61 (2005). (in Russian)
11. Golandian, N., Nekrutkin, V., Stepanov, D.: Variants of 'Caterpillar'-SSA method for multidimensional time series analysis. In: II International Conference on Systems Identification and Control Processes, SICPRO , vol. 3, pp. 2139–2168 (2003). (in Russian)
12. Hassani, H.: Singular spectrum analysis: methodology and comparison. J. Data Sci. **5**, 239–257 (2007)

13. Aleksandrov, F.: Development of a software complex for automatic identification and forecasting of additive components of time series within the framework of the approach 'Caterpillar'-SSA. Ph.D. thesis (2006). (in Russian)
14. Furashev, V., Lande, D.: Practical basics of possible thread prediction via analysing interconnections of events and information space. Open information and computer integrated technologies, **42**, 194–203 (2009). (in Ukrainian)
15. Chertov, O., Tavrov, D., Pavlov, D., Alexandrova, M., Malchikov, V.: Group Methods of Data Processing. LuLu.com, Raleigh (2010)

The Model of Words Cumulative Influence in a Text

Dmytro Lande[1,2(✉)], Andrei Snarskii[1,2], and Dmytro Manko[1]

[1] Institute for Information Recording, NAS of Ukraine, Kyiv, Ukraine
dwlande@gmail.com
[2] Igor Sikorsky Kyiv Polytechnic Institute, Kyiv, Ukraine

Abstract. A new approach to evaluation of the influence of words in a text is considered. An analytical model of the influence of words is presented. The appearance of a word is revealed as a surge of influence that extends to the subsequent text as part of the approach. Effects of individual words are accumulated. Computer simulation of the spread of influence of words is carried out; a new method of visualization is proposed as well. The proposed approach is demonstrated using an example of J.R.R. Tolkien's novel "The Hobbit." The proposed and implemented visualization method generalizes already existing methods of visualizing of the unevenness of words presence in a text.

Keywords: Influence of words · Word weight · Visualization
Computer modeling · Impact model · Analytical model

1 Purpose

The influence of a unit of a text (a word in particular) on the entire text is a hot topic in the task of practicing natural language processing. This problem can be solved within the framework of the theory of information retrieval by determining "weights" of words. At the same time, influence of words on the perception of a subject reading the text with such a memory that the words having been read are forgotten, or cleared out form the memory with time, has not been ever taken into account. This work is devoted to this task. It is the solution of this problem that this work is devoted to, and the corresponding model is described below. Fragments of a text, the total weight of the influence of words, in which the highest one can be considered as the most important for subsequent analytical processing, are provided as well.

2 Approaches to Weighting Words

In the theory of information retrieval, the most common ranking of words weights is performed by the criterion of Salton TF IDF [1], where TF (Term Frequency) is the frequency of occurrence of a word within the selected document, and IDF (Inverse Document Frequency) is a function (most often, a logarithmic function) of the value, inverse to the number of documents, in which the word appeared:

© Springer Nature Switzerland AG 2019
O. Chertov et al. (Eds.): ICDSIAI 2018, AISC 836, pp. 249–256, 2019.
https://doi.org/10.1007/978-3-319-97885-7_25

$$w_i = tf_i \cdot \log \frac{N}{n_i}, \tag{1}$$

where w_i is the weight of word i, tf_i is the frequency of word t_i in a document, n_i is the number of documents in the information array, in which the word is used, N is the total number of documents in the information array. At present, there are many modifications of this algorithm, the most famous of which is Okapi BM25 [2].

Evaluation of the unevenness of the occurrence of words is also possible on the basis of purely statistical variance estimates. In work [3], such an estimate of the discriminant power of the word is given as follows:

$$\sigma_i = \frac{\sqrt{\langle d^2 \rangle - \langle d \rangle^2}}{\langle d \rangle}, \tag{2}$$

where $\langle d \rangle$ is the mean sequence value $d_1, d_2, \ldots, d_n$, n is the number of occurrences of a word t_i in the information array. If we denote the coordinates (numbers) of occurrence of a word t_i in the information array by $e_1, e_2, \ldots, e_n$, then $d_k = e_{k+1} - e_k$ $(e_o = 0)$.

Another way for determining weighted meanings of words is proposed in [4]. The idea of the work is based on the concept of graphs of horizontal visibility, in which nodes correspond not only to numerical meanings, but to words themselves. A network of words using the horizontal visibility algorithm is built in three stages. First, a number of nodes is marked on the horizontal axis, each of which corresponds to words in the order of appearance in the text, and weighted numerical estimates are plotted along the vertical axis. At the second stage, a traditional graph of horizontal visibility is constructed [5]. There is a link between the nodes, if they are in "line of sight," i.e. if they can be connected by a horizontal line that does not intersect any vertical line. Finally, the network obtained at the previous stage is compactified. All nodes with a given word are combined into one node (naturally, the index and the position number of the word disappear in this case). All links of such nodes are also combined. It is important to note that no more than one connection remains between any two nodes; multiple links are withdrawn. The result is a new network of words is a compactified graph of horizontal visibility (CHVG). The normalized degree of a node corresponding to a particular word is ascribed to the weight of the word.

3   Visualization of Words Occurrence in a Text

In order to visualize the uneven occurrence of words in the texts, in [6], the technology of spectrograms was proposed, which outwardly resembles the barcodes of goods [3]. But at the same time, this method doesn't allow for considering occurrences of words in different scales of measurements in comparison wavelet analysis.

An algorithm and several examples of visualization of word occurrences are shown in [3], depending on the width of the observation window, which we applied to the fragment of J.R.R. Tolkien's novel "The Hobbit" (Fig. 1). The spectrogram shows the numbers of occurrences of the analyzed word in the text (starting with value 1 at the very bottom) versus the width of the observation windows (occurrence of the word in this case is highlighted in light gray). If several target words occurred in the corresponding monitoring window, then it is covered darker. Expert linguist in appearance can immediately determine the degree of uniformity of occurrence in the text of the analyzed word.

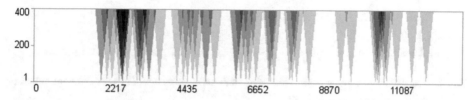

Fig. 1. Spectrogram for the lexeme "Gandalf" in the starting fragment of J.R.R. Tolkien's novel "The Hobbit"

4 The Influence Model of Words

The strength of the influence of words, V, in accordance with the proposed model increases dramatically when a word appears in the text (a point x_i), and then gradually decreases, which is set by a special parameter b. In addition, the magnitude of the influence is determined by the weights σ_i of words and determined by methods considered above. This allows one to consider this model as an extension of well-known "weight" models. The following analytical expression is proposed for determining the strength of the influence of words:

$$V(x, x_i) = \begin{cases} \sigma_i \dfrac{(x - x_i)}{b^2} \exp\left(-\dfrac{(x - x_i)}{b}\right), & \text{if } (x - x_i) > 0 \\ 0, & \text{otherwise} \end{cases}, \tag{3}$$

where x is the point at which the influence of a word is calculated, x_i is the word appearance position, σ_i is the weight of a word, b is the "memory" parameter of a word.

The code of the MATLAB program for calculating the influence of a single word is given as follows:

```
b=0.8
x=0:0.01:20
T=1
f=((x-T)/(b*b)).*exp(-(x-T)/b);
g=(sign(f)+1).*f/2;
plot(g)
```

Figure 2 shows two peaks realizing the formula (3) with different values of the parameter b.

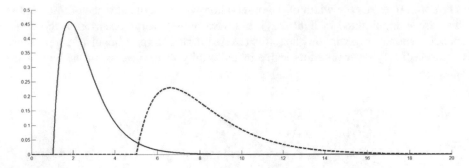

Fig. 2. Dependence of changes in word weights in accordance with (1) with different values of *b* (0.8—solid line, 1.6—dashed line)

We can assume that the power of influence of individual words does not add up, and the influence of a new word absorbs the influence of previous words. In the proposed model, it is assumed, however, that the force of influence of individual words accumulates, summing up in accordance with the following formula:

$$SV(x, X) = \sum_{x_i \in X} V(x, x_i) \tag{4}$$

The code of the MATLAB program for calculating and visualizing the cumulative effect of words in the text model for a fixed parameter *b* is as follows:

```
b=0.8
x=0:0.01:20
hold on
Slova=[1,3,8,10,14];
g=x-x;
for T=Slova
    f=((x-T)/(b*b)).*exp(-(x-T)/b);
    g=g+(sign(f)+1).*f/2;
end
plot(g)
```

Figure 3 shows the result of the "accumulation" of bursts realizing the formula (4) for a fixed value of the parameter *b*.

Calculation of the accumulative effect of words in the text model and its visualization with the spectrum of the parameter values in the form of the projection of the surface onto the plane (Fig. 4) is performed using MATLAB as follows:

```
clear
[x,b]=meshgrid(0:0.01:20,0.5:0.01:2);
Slova=[1,3,8,10,14]
f=(x)./(b.*b).*exp(-(x)./b);
g=f-f;
for T=Slova
     f=(x-T)./(b.*b).*exp(-(x-T)./b);
     g=g+(sign(f)+1).*f/2;
end
pcolor(g)
```

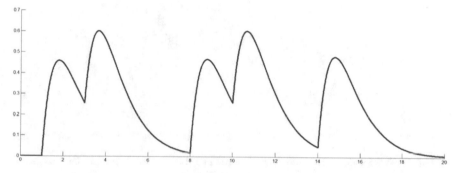

Fig. 3. Graph of accumulative influence of words in the text model

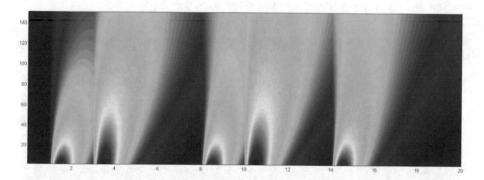

Fig. 4. Number of words in the text versus values of the parameter b

Figure 4 represents the accumulative influence of words taking into account the memory parameter in the text model. The shades of the diagram correspond to the value of the memory effect of words.

5 Example

As an example, J.R.R. Tolkien's novel "The Hobbit" was analyzed. In this case, using the CHVG method, the following reference words with weights from 0.9 to 0.6 were identified: BILBO, GANDALF, THORIN, GOBLINS, DWARVES, MOUNTAIN, DOOR, DRAGON, FOREST, ELVES, GOLLUM.

In order to visualize the power of influence of these words, a fragment of the novel was considered (the first 10,000 words). In this fragment, about 300 occurrences of these words were identified. A dependence of their accumulative influence was plotted for the case of a fixed value of $b = 0.1$ (Fig. 5).

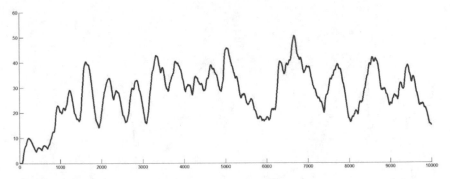

Fig. 5. The accumulative influence of words in J.R.R. Tolkien's novel "The Hobbit"

In Fig. 6, a diagram of the cumulative effect of these words is presented, taking into account the memory parameter b in the range from 0.01 to 1.

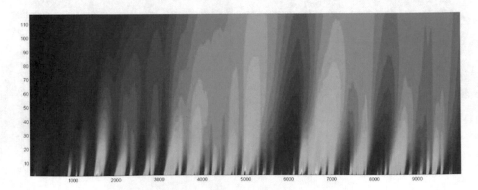

Fig. 6. The accumulative influence of words, taking into account the memory parameter in J.R. R. Tolkien's novel "The Hobbit"

The fragments of the text of the story, corresponding to the peak values in the above diagram, were analyzed.

These fragments of the highest influence of the reduced reference words can be used for the formation of short summaries of the text, as "reference" fragments in analytical and search systems in particular.

Let's reveal such fragments. Peak values: (1600–1700 words). Considered fragment is:

> "Yes, yes, my dear sir—and I do know your name, Mr. Bilbo Baggins. And you do know my name, though you don't remember that I belong to it. I am Gandalf, and Gandalf means me! To think that I should have lived to be good-morninged by Belladonna Took's son, as if I was selling buttons at the door!". "Gandalf, Gandalf! Good gracious me!…"

Peak values: (6600–6900 words). Considered fragment is:

> "A long time before that, if I know anything about the loads East," interrupted Gandalf. "We might go from there up along the River Running," went on Thorin taking no notice, "and so to the ruins of Dale-the old town in the valley there, under the shadow of the Mountain. But we none of us liked the idea of the Front Gate. The river runs right out of it through the great cliff at the South of the Mountain, and out of it comes the dragon too-far too often, unless he has changed."

6 Conclusions

In this paper, a new approach to the evaluation of the influence of words in texts is proposed, an analytical model of the influence of words is presented. The appearance of a word is viewed as a surge of influence that extends to all subsequent text in the proposed model. In this case, the effects of individual words are accumulated. Also, the visualization method presented in the paper, generalizes the existing methods of visualizing the unevenness of the occurrence of words in a text.

The proposed approach can be used to identify a fragment of a text, which is closely related to the most rated words; thereby the most accurately reflects its semantic structure. This allows generating short summaries of texts and digests based on news reports, snippets for search engines, etc.

It should also be noted that in the work, the word is considered as a unit of meaning in the text. If we consider not a single text, but the flow of news messages, and consider an event as a unit, then the proposed model can be generalized for the analysis of text streams in content monitoring systems without changing the mathematical formalism.

References

1. Salton, G., McGill, M.J.: Introduction to Modern Information Retrieval, p. 448. McGraw-Hill, New York (1983)
2. Lv, Y., Zhai, C.X.: Lower-bounding term frequency normalization. In: Proceedings of CIKM 2011, pp. 7–16 (2011)
3. Ortuño, M., Carpena, P., Bernaola, P., Muñoz, E., Somoza, A.M.: Keyword detection in natural languages and DNA. Europhys. Lett. **57**, 759–764 (2002)

4. Lande, D.V., Snarskii, A.A.: Compactified HVG for the language network. In: International Conference on Intelligent Information Systems: The Conference is Dedicated to the 50th Anniversary of the Institute of Mathematics and Computer Science, 20–23 August 2013, Chisinau, Moldova, Proceedings IIS/Institute of Mathematics and Computer Science, pp. 108–113 (2013)
5. Luque, B., Lacasa, L., Ballesteros, F., Luque, J.: Horizontal visibility graphs: exact results for random time series. Phys. Rev. E **80**, 046103-1–046103-11 (2009)
6. Yagunova, E., Lande, D.: Dynamic frequency features as the basis for the structural description of diverse linguistic objects. In: CEUR Workshop Proceedings. Proceedings of the 14th All-Russian Scientific Conference "Digital Libraries: Advanced Methods and Technologies, Digital Collections", 15–18 October, Pereslavl-Zalessky, Russia, pp. 150–159 (2012)
7. Chui, C.K.: An Introduction to Wavelets, p. 366. Academic Press, Cambridge (1992)

Mathematical and Computer Models of Message Distribution in Social Networks Based on the Space Modification of Fermi-Pasta-Ulam Approach

Andriy Bomba[1], Natalija Kunanets[2], Volodymyr Pasichnyk[2], and Yuriy Turbal[3(✉)]

[1] Rivne State University of Humanities, Rivne, Ukraine
[2] Lviv Polytechnic National University, Lviv, Ukraine
[3] National University of Water and Environmental Engineering, Rivne, Ukraine
turbaly@gmail.com

Abstract. The article proposes a new class of models for message distribution in social networks based on specific systems of differential equations, which describe the information distribution in the chain of the network graph. This class of models allows to take into account specific mechanisms for transmitting messages. Vertices in such graphs are individuals who, receiving a message, initially form their attitude towards it, and then decide on the further transmission of this message, provided that the corresponding potential of the interaction of two individuals exceeds a certain threshold level. Authors developed the original algorithm for calculating time moments of message distribution in the corresponding chain, which comes to the solution of a series of Cauchy problems for systems of ordinary nonlinear differential equations. These systems can be simplified, and part of the equations can be replaced with the Boussinesq or Korteweg-de Vries equations. The presence of soliton solutions of the above equations provides grounds for considering social and communication solitons as an effective tool for modeling processes of distributing messages in social networks and investigating diverse influences on their dissemination.

Keywords: Social network · Shock wave · Message distribution
Fermi-Pasta-Ulam problem · Graph of message flow · Excitation
Activity threshold · Cauchy problem · Differential equation

1 Introduction

In recent years, a number of new challenges appear in the humanities and in the fields of data analysis, mathematical modeling, physics, computer technology, etc. [1–5], which are related to the development of the Internet and social networks. New interdisciplinary research areas have emerged, such as Big Data, Data Mining, Cloud Computing, which are becoming increasingly widespread, particularly in solving many tasks related to the research and analysis of social networks. Functionally, the most

O. Chertov et al. (Eds.): ICDSIAI 2018, AISC 836, pp. 257–266, 2019.
https://doi.org/10.1007/978-3-319-97885-7_26

important process that is implemented in social networks is the message dissemination that significantly affects both the attitudes and behavior of individuals and the formation of public opinion of groups and communities on certain issues. This is an effective motive for creation of new models, which have appeared in recent years in a large number. We are talking about models for distributing messages [6–8] and models for forming the thought of both individuals and communities in general [9–12]. Analysis confirms the validity of the assertion that all models should be divided according to the level of refinement into the corpuscular, in which it is possible to identify an individual by certain multiple characteristics, and generalized models describing the characteristics of groups of individuals or the community as a whole, and which allow us to form general ideas of processes of message distribution or formation of public opinion (for example, the question can be about the number of individuals who spread rumors etc.).

Generalized models, for example, include so-called epidemic models [1], models of innovation diffusion [13–17], the Delay-Kendall model [7], models based on the concept of message density [12, 18]. Corpuscular models include a number of models that use cellular automata [13, 19, 20], cascading models of various types [9], models of network autocorrelation [21–23], adaptive and imitation behavior model [22], "Game Name" model [13], quantum models similar to Ising models [24–29], the message distribution model in society [30–32].

Probabilistic approaches, in particular, Markov chains [21], are widely used in simulation of social and communication processes, in particular various stochastic influences. Classes of tasks of forming and managing public opinion are important to solve problems that arise when it is necessary to change the opinion of individuals or target groups in a certain way due to the influence of certain agents [21, 33, 34].

Along with a number of advantages, each of the above-mentioned model classes has its own limits of application, within which their adequacy is maximal. Therefore, in many cases, there is a need to develop new approaches to modeling that would combine the benefits of models of message dissemination processes and models of public opinion formation processes.

The aim of this work is to develop a new approach to modeling the process of message dissemination in social networks based on the procedures for synthesizing existing patterns of message dissemination and processes of forming public opinion using the benefits of the latter.

At the heart of the approach proposed by the authors is the combination of two basic essences. The first essence is the process of making individual decisions on message distribution. This process resembles the process of transferring excitation to the nerve cell: if an input signal exceeds a certain threshold, a cell forms a certain signal at the output. The second essence is the well-known Fermi-Pasta-Ulam model [35], which is used to simulate the processes of wave transmission in chains. The application of this approach to social networks is also logical, since a set of individuals transmitting messages in a social network can also be described as a certain chain. If we take into account the excitation threshold of the message distributing individual, we can formulate certain modifications of the Fermi-Pasta-Ulam model, for example, considering the cases when the system of balls analyzed in this model is located on brittle rods.

The basis for creation of a new class of models for message dissemination in social networks is, in particular, [36], in which a model is proposed that can be attributed to discrete analogies of diffusion models of knowledge dissemination.

2   Characteristics of the Process of Message Distribution

Let us consider the graph $G = (U, V)$ that describes the social network. Let each graph vertex be an object, which can receive messages of type i from others and transmit them. This object is any vector $\bar{x}$, $\bar{x} \in U \subset R^n$, whose components may include geographical coordinates of the individual, IP addresses of computers, from which the connection in the social network is done, etc. The corresponding relationships will be described by the edges of the unoriented graph G. Every message over time causes a certain reaction (excitement) of the individual. In the case when the excitement exceeds a certain threshold (we correspond the level of threshold to the edges), the individual becomes active and generates a message of the type that is passed on to its network partners. Thus, we construct the graph of message distribution $G(i)$. To simplify the algorithm of information chains formation, we can propose a graph $\tilde{G}(i)$, in which the same vertex (that models an individual) can simultaneously be in several information chains (Fig. 1).

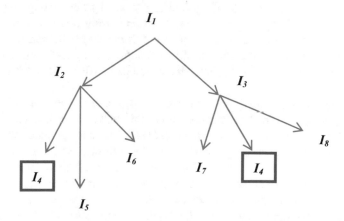

Fig. 1. Illustration of message dissemination

The process of distributing messages can be represented as paths outgoing from the root graph top. In this case, it the shortest path can be formed from the root to all other vertices of the graph. If an individual receives a message, it can generate a new message in the same context or forward the received message without changes, provided that its social and communication potential exceeds a certain threshold. The message raises the growth of the social and communication potential of an individual according to a certain law (excitation equation of the axon), depending on the mass of the message. Receiving a re-message may also increase the social and communication potential and further re-transmit messages or participate in discussions, forums, etc.

Thus, we can model an individual as a special "smart" neuron characterized by a special constant-excitation level, which changes continually with time and the mechanism for transmitting a message in a set of synapses, provided that the threshold level of excitation is exceeded (with the corresponding delay in signal transfer over the synapse).

When the message is transmitted from an individual to their partners, each one receives a message at a different time. In this context, for example, we can choose the Erlang model, when the moments of receiving or transmitting messages form the easiest flow. At the same time, when an individual sends a message at the beginning of a certain time interval, then, after some interval, having no corresponding level of excitation, messages from an individual can continue to be received. Such an effect will be called phantom message distribution in the social network.

3 Model of the Process of Information Interaction

Let us consider the chain $(x_1, x_2, \ldots, x_n)$ in graph $G(i)$. The process of information interaction can be described as follows. We introduce a certain function $u(x, i, t)$ that describes the excitation level of each individual (in the context of message transmission), its deviation from the state of equilibrium caused by the message $i' \in I$ at some point in time. We will assume that excitation extends from one individual to another through the message transmission of a type $i' \in I$. If $u(x, i, t) > 0$, then we assume that an element x, as a result of the growth of its excitation, will pass its excitement to the following objects, if $u(x, i, t) < 0$, then, the same effect will be done on the previous individual.

The excitement of an individual is determined by the excitement of their immediate partners. So, x_1, x_2 are direct partners at initial time $u(x_1, i, t) > 0$, $u(x_2, i, t) = 0$. We match an edge (x_1, x_2) with some function $\delta(x_1, x_2, i)$ that simulates the excitation transmission threshold caused by the message $i' \in I$ from x_1 to x_2. Let $f(u(x_2, i, t) - u(x_1, i, t))$ be some interaction strength between adjacent elements x_1 and x_2 depending on the level of their excitement. We will assume that at the time when $f(0 - u(x_1, i, t)) = \delta(x_1, x_2, i)$ affects the element x_1 on the element x_2. The influence force at the initial moment of time $t f(0 - u(x_1, i, t))$ will determine its excitement. At this time, the value begins to change $u(x_2, i, t)$. Consequently, taking into account the threshold, we can define the force of interaction as follows:

$$F(u(x_2, i, t) - u(x_1, i, t)) = \begin{cases} f(u(x_2, i, t) - u(x_1, i, t)), t \geq \tau_1, u(x_1, i, t) \neq 0, \\ f(-u(x_1, i, t)), t < \tau_1, u(x_1, i, t) \neq 0, u(x_2, i, t) = 0 \\ 0, u(x_1, i, t) = 0, \end{cases} \quad (1)$$

In addition, we suppose that, at this moment of time, information exchange begins the transmission of messages $i' \in I$ from an individual x_1 to x_2, which increases the excitement of an individual x_2. We introduce into the model some analogy of the concept of an individual "mass" in the context of message distribution; the larger its mass, the slower an individual responds to the change in the excitement of its neighbors. In this case, based on the analogy of the second law of Newton, we can write the equation system of the dissemination of social and communication excitation:

$$m_k u''(x_k, i, t) = F(u(x_{k+1}, i, t) - u(x_k, i, t)) - F(u(x_k, i, t) - u(x_{k-1}, i, t)),$$
$$k = \overline{1, n}, (x_k, x_{k+1}) \in G(i). \tag{2}$$

Without loss of generality, we assume that $u(x_k, i, t) = 0$, $u'(x_k, i, t) = 0$, $u(x_0, i, t)$ is a given function defining the initial perturbation. It is the time of activation $\tau_k = \min\{t : f(0 - u(x_{k-1}, i, t)) = \delta(x_{k-1}, x_k, i)\}$. Then, at the time interval $[0, \tau_1]$, the social and communicative potential of an individual x_1 is unchanged, it is defined as the solution to the following obvious Cauchy problem:

$$m_1 u''(x_1, i, t) = 0, u(x_1, i, 0) = 0, u'(x_1, i, 0) = 0.$$

From the moment of time $\tau_1 = \min\{t : f(0 - u(x_0, i, t)) = \delta(x_0, x_1, i)\}$, the social and communication potential of an individual begins to change, and its change in the interval of time $[\tau_1, \tau_2]$ will be determined as a solution of the following problem:

$$m_1 u''(x_1, i, t) = f(0 - u(x_1, i, t)) - f(u(x_1, i, t) - u(x_0, i, t)), \tau_1 \leq t \leq \tau_2 \tag{3}$$

Initial conditions are $u(x_1, i, \tau_1) = 0, u'(x_1, i, \tau_1) = 0$.

Having solved (3), we can determine the time of activation τ_2 by the following formula:

$$\tau_2 = \min\{t : f(0 - u(x_1, i, t)) = \delta(x_1, x_2, i)\}.$$

Similarly, we get the general Cauchy problem for the single chain in the graph at $\tau_k \leq t \leq \tau_{k+1}$:

$$\begin{cases} m_1 u''(x_1, i, t) = f(u(x_2, i, t) - u(x_1, i, t)) - f(u(x_1, i, t) - u(x_0, i, t)), \\ m_2 u''(x_2, i, t) = f(u(x_3, i, t) - u(x_2, i, t)) - f(u(x_2, i, t) - u(x_1, i, t)), \\ \ldots \\ m_{k-1} u''(x_{k-1}, i, t) = f(u(x_k, i, t) - u(x_{k-1}, i, t)) - f(u(x_{k-1}, i, t) - u(x_{k-2}, i, t)), \\ \qquad m_k u''(x_k, i, t) = f(0 - u(x_k, i, t)) - f(u(x_k, i, t) - u(x_{k-1}, i, t)). \end{cases} \tag{4}$$

Initial conditions $u(x_k, i, \tau_k) = 0, u'(x_k, i, \tau_k) = 0$, $u(x_{r-1}, i, \tau_r), u'(x_{r-1}, i, \tau_r)$ are known, $r = \overline{1, k}$. Then,

$$\tau_{k+1} = \min\{t : f(0 - u(x_k, i, t)) = \delta(x_k, x_{k+1}, i)\}. \tag{5}$$

Let $f(x) = \alpha x + \beta x^2$. Then, (4) will be written as follows:

$$\begin{cases} m_{r-1}u''(x_{r-1},i,t) = \alpha(u(x_r,i,t) - 2u(x_{r-1},i,t) + u(x_{r-2},i,t)) + \beta(u(x_r,i,t) \\ \quad - 2u(x_{r-1},i,t) + u(x_{r-2},i,t))(u(x_r,i,t) - u(x_{r-2},i,t)), \tau_k \leq t, r = \overline{1,k} \\ m_k u''(x_k,i,t) = -\alpha(2u(x_k,i,t) - u(x_{k-1},i,t)) - \beta u(x_{k-1},i,t)(2u(x_k,i,t) \\ \quad - u(x_{k-1},i,t)), \tau_k \leq t \leq \tau_{k+1}. \end{cases} \quad (6)$$

Thus, we received a series of Cauchy problems, the solution to which will help to find a sequence of time moments $\tau_1, \tau_2, \ldots$ that describe the beginning of the growth of the social and communication potential (times of activation) of relevant individuals in the chain or the transition to the next layer of the message flow. Obviously, for any moment of time t, the value $\tau_{\max}(t) = \max_i\{\tau_i : \tau_i \leq t\}$ determines the front of the wave of message dissemination.

Taking into account specificity of data representation $\tilde{G}(i)$, we can get the general system as follows:

$$m_k u''(x_k,i,t) = \Xi_{out}\{F(u(x_p,i,t) - u(x_k,i,t)), (x_k,x_p) \in G(i)\} \\ - \Xi_{inp}\{F(u(x_k,i,t) - u(x_p,i,t)), (x_p,x_k) \in G(i)\}, x_k \in U, \quad (7)$$

where operator Ξ_{out} described the combined influence of the object x_k for all partners x_p, $(x_k,x_p) \in G(i)$, Ξ_{inp} describes the combined influence of all partners x_p for the object x_k, $(x_p,x_k) \in G(i)$. We can consider the following operators:

$$\Xi_{out}\{F(u(x_p,i,t) - u(x_k,i,t)), (x_k,x_p) \in G(i)\} \\ = F(u(x_{p_0},i,t) - u(x_k,i,t)), \tau(x_{p_0}) = \min_p \tau(x_p),$$

$$\Xi_{inp}\{F(u(x_k,i,t) - u(x_p,i,t)), (x_p,x_k) \in G(i)\} \\ = F(u(x_k,i,t) - u(x_{p_0},i,t)), \tau(x_{p_0}) = \min_p \tau(x_p)$$

or

$$\Xi_{out}\{F(u(x_p,i,t) - u(x_k,i,t)), (x_k,x_p) \in G(i)\} \\ = \sum_{p:(x_k,x_p)\in G(i)} \beta_p(t)F(u(x_p,i,t) - u(x_k,i,t)),$$

$$\Xi_{inp}\{F(u(x_k,i,t) - u(x_p,i,t)), (x_p,x_k) \in G(i)\} \\ = \sum_{p:(x_p,x_k)\in G(i)} \alpha_p(t)F(u(x_k,i,t) - u(x_p,i,t)),$$

which describe some collective effects and interpersonal characteristics.

4 Numerical Results

Let us consider a part of any social network containing 20 persons, whose numbers are 1, 2, 3, ..., 20 respectively, $U = \{1,2,\ldots,20\}$. Then we can describe the structure of corresponding relationships as a matrix of incidence $R = (r_{ij})_{i,j=1}^{n}$,

$$r_{ij} = \begin{cases} 0, (i,j) \notin V, \\ \delta(i,j), (i,j) \in V, \end{cases}$$

where $\delta(i,j)$ are the corresponding thresholds. We can correspond to every person an object $p(j)$, which can be denoted as an array $p(j) = (p_1^j, p_2^j, \ldots, p_{N_j}^j)$, where p_1^j is the indicator of activation, p_2^j is the number of the object that activated the object j, p_3^j is the time of the object j activation, p_4^j is the number of differential equations in system (3), which is necessary for finding the force of influence for the object j, $p_5^j, p_6^j \ldots, p_{p_4^j}^j$ are initial conditions for the Cauchy problem.

Then we can propose the following algorithm:

1. Consider first a person and find the "temporary" times of activation for all corresponding partners according to the matrix of incidence R using (3).
2. Find the number of object j with minimum of nonzero "temporary" time of activation $\tau(j)$ and activate them.
3. Form all elements of vector $p(j) = (p_1^j, p_2^j, \ldots, p_{N_j}^j), p_1^j = 1$.
4. Form the Cauchy problem for all partners corresponding to j, according to the matrix of incidence R, solve them and find "temporary" times of activation according to (5).
5. Repeat steps 2–4 while there are non-activated objects.

A fragment of the graph $\tilde{G}(i)$, which is constructed according to the matrix of incidence is shown in Fig. 2. The corresponding elements of the matrix of incidence are as follows:

$$\delta_{1,2} = 0.015, \delta_{1,7} = 0.061, \delta_{1,8} = 0.049, \delta_{1,16} = 0.015, \delta_{1,17} = 0.037,$$
$$\delta_{1,18} = 0.043, \delta_{2,13} = 0.052, \delta_{4,14} = 0.083, \delta_{6,11} = 0.0028, \delta_{6,20} = 0.0309,$$
$$\delta_{7,10} = 0.084, \delta_{7,13} = 0.0905, \delta_{8,6} = 0.0349, \delta_{8,13} = 0.0928, \delta_{12,19} = 0.0767,$$
$$\delta_{16,4} = 0.0732, \delta_{16,9} = 0.0162, \delta_{17,5} = 0.0629, \delta_{17,15} = 0.0812, \delta_{18,12} = 0.0482.$$

Using the algorithm described above, we can find the moments of activation for every object. The process of activation can be illustrated in Fig. 2. Every vertex of graph in Fig. 2 is marked with two numbers. The first one is element index, and the second one is the time of activation (in brackets).

We can assign each objects index to the activation time. Then we obtain the set of activation times: 0, 0.095, 0, 0.700, 0.623, 0.536, 0.133, 0.126, 0.406, 0.973, 0.876, 0.637, 1.046, 1.940, 0.927, 0.096, 0.118, 0.123, 2.318, 1.537. Thus, we can investigate the process of information wave propagation in real-time mode.

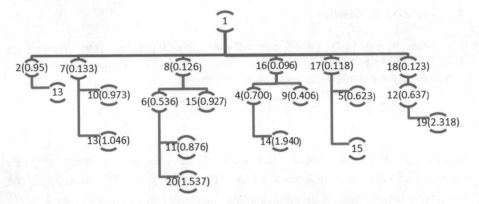

Fig. 2. Fragment of the graph $\tilde{G}(i)$

5 Conclusions

In the paper, a new class of computing models of message distribution in social networks was constructed, in which it was possible to systematically combine approaches used to model the processes of message dissemination and the processes of formation the thought of both individuals and public opinion in a particular community.

It should be noted that individuals in this model are considered as neurons that take qualified decisions regarding the further message distribution of a certain type. This approach, based on its essential grounds, is as close as possible to the real processes that occur in social networks, which allows us to conclude that the adequacy of the proposed class of models is high.

The following mechanical analogy of the proposed class of models appears to be quite constructive when the balls of a certain mass are connected by the springs and further fixed on the bare rods. In this submission, the rods model the threshold level of an individual's excitement during transmitting messages. If the message is transmitted without taking into account the threshold of excitation, we get the classic Fermi-Pasta-Ulam model.

References

1. Horkovenko, D.K.: Overview of the models of information distribution in social networks. Young scientist, no. 8, pp. 23–28 (2017). (in Russian)
2. Cha, M., Haddadi, H., Benevenuto, F., Gummadi, K.P.: Measuring user influence in twitter: the million follower fallacy. In: ICWSM 2010 (2010)
3. Goetz, M., Leskovec, J., Mcglohon, M., Faloutsos, C.: Modeling blog dynamics. In: ICWSM (2009)
4. Leskovec, J., Backstrom, L., Kleinberg, J.: Meme-tracking and the dynamics of the news cycle. In: KDD 2009 (2009)

5. Liben-Nowell, D., Kleinberg, J.: Tracing information flow on a global scale using Internet chain-letter data. PNAS **105**(12), 4633–4638 (2008)
6. Nosova, M.V., Sennikova, L.I.: Modeling the information dissemination in decentralized network systems with irregular structure. New information technologies and automated systems, no. 17 (2014). (in Russian)
7. Daley, D.J., Kendall, D.G.: Stochastic rumors. J. Inst. Math. Appl. **142**, 42–55 (1965)
8. Kempe, D., Kleinberg, J., Tardos, E.: Maximizing the spread of influence through a social network. In: Proceedings of the 9-th ACM SIGKDD International Conference on Knowledge Discovery and Data Mining, pp. 137–146 (2003)
9. Goldenberg, J., Libai, B., Muller, E.: Talk of the network: a complex systems look at the underlying process of word-of-mouth. Mark. Lett. **2**, 11–34 (2001)
10. Hethcote, H.W.: The mathematics of infectious diseases. SIAM Rev. **42**(4), 599–653 (2000)
11. Isea, R., Mayo-García, R.: Mathematical analysis of the spreading of a rumor among different subgroups of spreaders. Pure Appl. Math. Lett. **2015**, 50–54 (2015)
12. Kuizheva, S.K.: About the mathematical tools of the research of social and economic systems. Terra Economicus **12**(2, part 3), 46–51 (2014). (in Russian)
13. Lomakin, S.H., Phedotov, A.M.: Analysis of the model of information distribution in the cellular automata network. In: Bulletin of Novosibirsk State University. Information Technology, pp. 86–97 (2014). (in Russian)
14. Minaev, V.A., Ovchinskii, A.S., Skryl, S.V., Trostianskii, S.N.: How to control mass consciousness: modern models, p. 200 (2013). (in Russian)
15. Baronchelli, A., Felici, M., Caglioti, E., Loreto, V., Steels, L.: Evolution of opinions on social networks in the presence of competing committed groups. J. Stat. Mech. http://arxiv.org/abs/1112.6414
16. Dallsta, L., Baronchelli, A., Barrat, A., Loreto, V.: Non-equilibrium dynamics of language games on complex networks. http://samarcanda.phys.uniroma1.it/vittorioloreto/publications/language-dynamics/
17. Lu, Q., Korniss, G., Szymanski, B.K.: Naming games in two-dimensional and small-world-connected random geometric networks. http://journals.aps.org/pre/abstract/10.1103/PhysRevE.77.016111
18. Kuizheva, S.K.: The Korteweg-de Vries equation and mathematical models of social and economic systems. Bulletin of the Adyghe State University. Series 4: natural-mathematical and technical sciences, vol. 154, pp. 20–26 (2015). (in Russian)
19. Baronchelli, A.: Role of feedback and broadcasting in the naming game. Phys. Rev. E **83**, 046103 (2011)
20. Lobanov, A.I.: Models of cellular automata. Computer studies and modeling, no. 3, pp. 273–293 (2010). (in Russian)
21. Hubanov, H.A., Novikov, D.A., Chshartishvili, A.H.: Social networks: modeling of information influence, management and confrontation, p. 228 (2010). (in Russian)
22. Myerson, R.B.: Game Theory: Analysis of Conflict. Harvard Univ Press, London (1991)
23. Hubanov, H.A.: Review of online reputation/trust systems. In: Internet Conference on Governance Issues, p. 25 (2009). (in Russian)
24. Altaiskii, M.V., Kaputkina, N.E., Krylov, V.A.: Quantum neural networks: current state and prospects of development. Phys. Elem. Part. At. Nucl. **45**, 43 (2014). (in Russian)
25. Cabello, A.: Quantum social networks. J. Math. Phys. A **45**, 285101 (2012)
26. Beck, F.: Synaptic quantum tunnelling in brain activity. Neuroquantology **6**(2), 140–151 (2008)
27. Mikhailov, A.P., Petrov, A.P., Marevtseva, N.A., Tretiakova, I.V.: Development of the model for the information dissemination in the society of 2014. Math. Model. **26**(3), 65–74 (2014). (in Russian)

28. Mikhailov, A.P., Izmodenova, K.V.: About the optimal control of the propagation process of the formation. Math. Model. **17**(5), 67–76 (2005). (in Russian)
29. Mikhailov, A.P., Marevtseva, N.A.: Information fight models. Math. Model. **23**(10), 19–32 (2011). (in Russian)
30. Delitsyn, L.L.: Quantitative models of the spread of innovations in the field of information and telecommunication technologies, p. 106 (2009). (in Russian)
31. Friedkin, N.E.: Structural cohesion and equivalence explanations of social homogeneity. Soc. Methods Res. **12**, 235–261 (1984)
32. Prudkovskii, P.A.: The theory of nonlinear waves. W06 Waves in Chains
33. Bomba, A., Turbal, Y., Turbal, M.: Method for studying the multi-solitone solutions of the Korteveg de-Vries type equations. J. Differ. equ. **2**, 1–10 (2015)
34. Bomba, A.Ya., Turbal, Yu.V.: Data analysis method and problems of identification of trajectories of solitary waves. J. Autom. Inf. Sci. 5, 34–43 (2015)
35. Novokshenov, V.Y.: Introduction to the theory of solitons, Izhevsk, Institute for Computer Research, p. 96 (2002). (in Russian)
36. Bomba, A., Nazaruk, M., Kunanets, N., Pasichnyk, V.: Constructing the diffusion-liked model of bicomponent knowledge potential distribution. Int. J. Comput. **16**(2), 74–81 (2017)

Social Network Structure as a Predictor of Social Behavior: The Case of Protest in the 2016 US Presidential Election

Molly Renaud, Rostyslav Korolov$^{(\boxtimes)}$, David Mendonça, and William Wallace

Department of Industrial and Systems Engineering,
Rensselaer Polytechnic Institute, Troy, NY, USA
korolr@rpi.edu

Abstract. This research explores relationships between social network structure (as inferred from Twitter posts) and the occurrence of domestic protests following the 2016 US Presidential Election. A hindcasting method is presented which exploits Random Forest classification models to generate predictions about protest occurrence that are then compared to ground truth data. Results show a relationship between social network structure and the occurrence of protests that is stronger or weaker depending on the time frame of prediction.

Keywords: Social network · Protest · Collective action · Mobilization

1 Introduction

From the Arab Spring [1], to demonstrations in Latin America [2], to waves of protests in the United States [3], online social media is now recognized as a driver of protest. A prominent approach to linking social media to protest has been through Natural Language processing, in which message content, author profile, keywords, and the like are viewed as possible indicators of intent to protest [1,3,4]. However, NLP techniques require substantial model training, and concerns may be raised that conclusions from the study of any one protest may not generalize to others, particularly due to the use of specialized corpuses.

This research explores predominantly structural properties of social networks in order to predict protest occurrence. It employs keyword-defined Twitter datasets associated with the 2016 US Presidential Election and politically neutral topics such as weather for the period of October 26, 2016 through February 1, 2017. Research questions are presented (Sect. 2), followed by Data and Methods (Sect. 3), Results (Sect. 4), Discussion (Sect. 5), and Conclusions (Sect. 6).

2 Background

Previous research has demonstrated the possibility of a relationship between social media and protest occurrences [5,6]. Civil unrest is associated with information cascades or activity bursts in social media, and these phenomena may be used to predict protests, or at least peaks of protest activity [3,7].

© Springer Nature Switzerland AG 2019
O. Chertov et al. (Eds.): ICDSIAI 2018, AISC 836, pp. 267–278, 2019.
https://doi.org/10.1007/978-3-319-97885-7_27

There are many definitions of protests in the literature, but for the purposes of this research a protest is defined in the context of a social movement, which is "one of the principal social forms through which collectivities give voice to their grievances and concerns about rights, welfare, and well-being of themselves and others by engaging in various types of collective action, such as protesting in the streets, that dramatize those grievances and concerns and demand that something be done about them" [8]. Other definitions exist, but this one identifies a number of the major issues that can be found in one alternate definition or another: collective action, change-oriented rhetoric, some level of organization, and a level of temporal continuity [9–11]. To distinguish from other sorts of collective action, this definition specifies that a protest is "noninstitutionalized," in that it goes beyond what is allowed by law [8]. This aspect of protest and civil demonstration affects an individual's decision to take part.

In theory, an individual's decision to participate in a protest is preceded by a four-step process called action mobilization [12], consisting of sympathy to the cause, awareness of an upcoming protest, motivation to take part, and the ability to take part. "Trigger events" are those said to cause an emergence of spontaneous protests that may condense some of the action mobilization steps [13]. Regardless, the decision to join a protest is not only shaped by an individuals' views but also by social context within one or more social networks [14]. Protests are a form of collective action, and may involve multiple social networks. Mobilization was previously shown to be associated with protest occurrences [3].

Studies that address the role of social media that explicitly acknowledge the underlying social network research on anti-government sit-ins in Istanbul [15], protest movements in Spain [16] and online protest in Venezuela [17]. Studies involving the structure of the networks in social media and their temporal evolution leading to collective action are less common [15]. These studies are of a descriptive nature and the authors do not attempt use network structure or dynamics to identify or predict protest or other forms of collective action. The work presented here addresses this gap in the literature by examining a relationship between networks derived from social media and protest occurrences. A control study involving the linguistic analysis identifying mobilization is then performed to clarify the roles of social network and mobilization in protest development. The following research questions are used to frame and focus this research.

RQ1: What is the observed relationship between social networks constructed from social media and the actual occurrence of protests?

A positive answer to **RQ1** is not sufficient for behavior prediction. To enable prediction, the relationship between the social networks observed in the past and the current behavior must be considered.

RQ2: What is the relationship between social network structure prior to protest and the actual occurrence of protests?

Mobilization is associated with protest occurrence [3]. It is interesting to analyze the combined effect of social network and mobilization on protest occurrences and the possible inter-relation among the three phenomena.

RQ3: How does the inclusion of mobilization in the model affect observed relationships between networks and protests?

3   Data and Methods

Three social media datasets are collected from Twitter using the Apollo Social Sensing Tool [18]. The tool is based on the Twitter REST API, which returns at least 1% of the tweets matching the query (the upper bound on the number of the returned results depends on the popularity of the topic). Multiple Twitter datasets are used to address the proposed research questions. The *Election* dataset contains messages filtered using any one or more of the keywords "Trump," "Clinton," and "election." The approximately 2.5 million tweets in this dataset cover the period from October 26, 2016 to December 20, 2016. The *Neutral* dataset contains approximately 3 million messages filtered using any and all of the keywords "food," "travel," and "weather" from October 27, 2016 to February 1, 2017. This dataset provides a baseline in that structural changes within it are not likely to occur in response to political exigencies. Lastly, the *Test* dataset contains about 1.8 million messages using the same keyword filters as the *Election* set, though from the period December 21, 2016 through January 31, 2017.

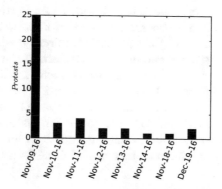

Fig. 1. Frequency of protests following the President Election ('Election' dataset).

The Twitter data are separated by geographic origin of message, using only locations where protests occurred following the election. Locations are extracted from the user profiles using a dictionary of popular naming variations (e.g. "Los Angeles, CA"; "LA"; "Los Angeles, California", etc.) In addition, information about protests in the United States that are associated with the outcome of the

2016 Presidential Election are collected for the period from October 26, 2016 to January 31, 2017. The data regarding protests in the United States are collected from online news outlets across the country, cross-checked using multiple sources when possible. A protest's occurrence is characterized by geographic location and date, where the date is the 24-h period of time between 7:00 AM EST and 6:59 AM EST next day. The frequencies of the protests in 39 cities we have considered are shown on Figs. 1 and 2.

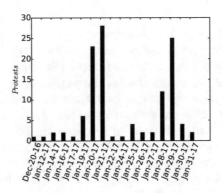

Fig. 2. Frequency of protests in December 2016 – January 2017 ('Test' dataset).

This study examines the relationship between social network structure and protest occurrence. 18 network measures are taken, relating to size, structure and geography, supplemented with two measures of mobilization for a control study (Table 1). To calculate these, each Twitter user is treated as a vertex. An edge is added between two vertices if the interaction between them (in the form of a re-tweet or direct mention) is observed. Undirected edges are used here, as some of the network measures can only be calculated for undirected graphs.

As the population size and the number of social media users in each geographic location are different, data are normalized for each location by taking the difference between the feature value and the sample mean and dividing by the sample standard deviation. This results in the features measured in standard deviations instead of absolute values. A practical limitation of this method is that it requires maintaining a history of social media activity.

3.1   Research Hypotheses

With the measures defined, research questions may now be reformulated as testable hypotheses. Denoting the set of characteristics of the social network as W, the measure of the protest occurrence as A, and the function mapping one to the other as $f : W \to A$, **RQ1** question can be answered by testing the following hypothesis:

H1: $\exists f : W \to A$, here and further the social network is assumed to be derived from the protest-related social media content.

RQ2 may be answered by testing the second hypothesis:

H2: $\exists f_k : W_{t-k} \to A_t$, where t is the observation period, and k is the lag.

Table 1. Network measures calculated from collected tweets.

Structure measurements	Geographic structure measurements	Size measurements	Mobilization measurements
Total density	Local users	Users	Mobilization tweets
Connected components	All local users	Unique links	Mobilizers
Avg. connected component size	Local tweets		
Average degree	Local density		
Communities	Local links		
Modularity	Bridges		
Communities per user	Undefined links		
Average community size			
Maximum community size			

Three hypotheses are formed to answer **RQ3**. Denote the previously considered set of social network characteristics with measurements of mobilization added to them as M and $g_k : M_{t-k} \to A_t$. Let $\epsilon(f)$ and $\epsilon(g)$ be classification errors obtained from applying the appropriate function. Additionally, let $\epsilon(b)$ and $\epsilon(h)$ represent classification errors obtained from classification models stratified by mean and high levels of mobilizations, respectively.

H3A: If $\epsilon(f) > \epsilon(g)$, then the network structure and mobilization independently contribute to the effect (protest occurrence) and may be confounding factors.

H3B: If $\epsilon(f) \approx \epsilon(g)$, this would suggest that the effect of mobilization has already been captured by network structure measurements, providing the evidence that the social network can predict mobilization, which is a known antecedent of the protest [3, 12].

H3C: If $\epsilon(b) \approx \epsilon(h)$, this would provide the evidence that the social network and mobilization are not confounding factors, additionally supporting **H3B**.

3.2   Methods

The methodological approach in addressing the research questions involves hypothesis testing via the application of Random Forest classification methods [19]. Random Forests construct a large number of decision trees using a subset of the data for each tree, which provides an internal method of cross-validation [19]. These trees are aggregated to create the most accurate classification model for input variables (network parameters in this case). Each random forest model outputs a confusion matrix indicating the number of true positives (TP), false positives (FP), true negatives (TN), and false negatives (FN). These quantities are used to calculate several measures of goodness for the model: accuracy, true positive rate (TPR), and false positive rate (FPR):

$$Accuracy = \frac{TP+TN}{TP+TN+FP+FN},$$

$$TPR = \frac{TP}{TP+FN}, \quad FPR = \frac{FP}{TP+FN}.$$

To examine the relationship between social network measurements and protest occurrences (**H1**), the Random Forest model using parameters of the social network as features and protest occurrences as the dependent variable is formed using both the *Neutral* dataset and the *Election* dataset. As the *Neutral* dataset is drawn from keywords that in theory do not relate to protests, there should be approximately equal classifications of true positives, false positives, true negatives, and false negatives. To show the applicability and robustness of the model to different protest scenarios, the model was retrained with the *Test* dataset, which should display similar relationships.

To test the effect of a time lag on the relationships between social network structures and protest occurrences (**H2**), the classification models are trained for six different time periods. Here, a lag of "0" indicates the current time period (i.e., when the analysis is run). Several random forest classification models are trained on the social media and protest occurrence data during the same twenty-four hour period. Additionally, similar random forest classification models are trained on the social network data in k 24-h periods before a protest is observed, where k varies from zero to five. A lag of $k = 0$ denotes "today" (i.e., the day being analyzed). As a robustness check, the analysis performed to address **H2** is also performed on the *Test* dataset. This is to ensure that the models are not over-fitted to one dataset. This is a variation of cross-validation methods frequently used in machine learning applications [20].

Two mobilization variables are calculated and added to the network measures dataset for the control study (**H3A, H3B**): mobilization tweets and mobilizers. The "mobilization tweet" variable is a count of the number of tweets expressing mobilization. The "mobilizers" variable denotes the number of authors of mobilization tweets, which may differ from the actual number of tweets expressing mobilization. These variables are calculated using a human-annotated NLP classifier described in some detail in [3]. Following the addition of these two variables, the random forest classification models are retrained and tested again on the *Election* dataset. These results are compared to those produced by models without mobilization variables.

To address **H3C**, the random forest classification models trained without mobilization variables are retrained using the data stratified by the number of Twitter users expressing mobilization in the network. The dataset is divided into three groups: high mobilization (more than one standard deviation above the mean), medium mobilization (within one standard deviation from the mean), and low mobilization (more than one standard deviation below the mean). These results are compared to the previously formed classification models, including and excluding mobilization variables. All data preparation and modeling are done using the R library randomForest [21].

4   Results

The results are summarized in Table 2, which displays the accuracy rates of models created to explore the research questions developed above. The "Lag (k)" column of Table 2 indicates the number of days that the network measures have been lagged (e.g. $k = 0$ indicates the day the analysis is run, whereas $k = 1$ indicates the day following the analysis, and so on). The rest of the table indicates the confusion matrix as output by the model (TN, FP, TP, FN), along with quantities derived from values in the matrix: accuracy, the true positive rate (TPR), and the false positive rate (FPR).

4.1   RQ 1 Results

As seen in Panel 1 of Table 2, the models formed using the *Neutral* dataset perform similarly to random chance. They show approximately equal classifications of each category (TP, FP, TN, FN) and the Accuracy within 0.05 of 50% for a majority of cases, which implies no relationship between social networks based on random topics of communication and protest occurrences, as expected. Classification models trained on the *Election* dataset, shown in Panel 2 of Table 2, display remarkably better relationships between network measures and protest occurrences: there are substantially more correct classifications (TP, TN) than false classifications (FP, FN). The accuracy for these models is much better than those trained on the *Neutral* dataset, particularly when the Lag (k) is low (less than four). This is also true of the TPR: when k is low, the TPR is above 50%. This suggests that **H1** is correct and that networks derived from protest-related social media posts are associated with protest occurrences.

4.2   RQ 2 Results

As seen in Panel 2 of Table 2, accuracy decreases as k increases. This suggests that the more in advance of the protest date (temporally-speaking), the weaker the relationship between network structure and protest occurrence. The TPR also appears to decrease when Lag increases, while the FPR seems to fluctuate with no obvious pattern. This confirms **H2**, that the relationship between network structure and protest occurrence changes over time. These trends hold when the classifiers are rerun on the *Test* dataset, as shown in Panel 1 of Table 3. These results indicate the possibility of predicting the protest using the proposed model when $k < 4$, or up to four days in advance of a protest. When $k = 5$, accuracy of classification models (both in Panels 2 and 3) drops substantially; moreover, FPR increases, indicating that the relationship between network structure and protest occurrences is not detectable at this point.

Table 2. Comparison of different classification models.

	Lag (k)	TN	FP	TP	FN	Total	Accuracy	TPR	FPR
RQ1: Neutral-trained models	0	12	13	14	11	50	0.520	0.560	0.520
	1	15	10	12	13	50	0.540	0.480	0.400
	2	8	17	7	18	50	0.300	0.280	0.680
	3	15	10	12	13	50	0.540	0.480	0.400
	4	10	15	14	11	50	0.480	0.560	0.600
	5	15	10	18	7	50	0.660	0.720	0.400
RQ1, RQ2: Election (no mobilization variables)	0	22	3	19	6	50	0.820	0.760	0.120
	1	16	9	17	8	50	0.660	0.680	0.360
	2	17	8	17	8	50	0.680	0.680	0.320
	3	18	7	14	11	50	0.640	0.560	0.280
	4	13	12	16	9	50	0.580	0.640	0.480
	5	11	14	11	14	50	0.440	0.440	0.560
RQ3: Election (with mobilization variables)	0	21	4	18	7	50	0.780	0.720	0.160
	1	15	10	18	7	50	0.660	0.720	0.400
	2	18	7	16	9	50	0.680	0.640	0.280
	3	16	9	15	10	50	0.620	0.600	0.360
	4	18	7	18	7	50	0.720	0.720	0.280
	5	13	12	11	14	50	0.480	0.440	0.480
RQ4: Medium mobilization	0	21	4	19	6	50	0.800	0.760	0.160
	1	9	5	3	11	28	0.429	0.214	0.357
	2	15	4	13	6	38	0.737	0.684	0.211
	3	16	5	16	5	42	0.762	0.762	0.238
	4	13	5	13	5	36	0.722	0.722	0.278
	5	15	8	12	11	46	0.587	0.522	0.348
RQ4: High mobilization	0	19	2	16	5	42	0.833	0.762	0.095
	1	18	5	17	6	46	0.761	0.739	0.217
	2	18	0	14	4	36	0.889	0.778	0.000
	3	13	4	12	5	34	0.735	0.706	0.235
	4	14	5	11	8	38	0.658	0.579	0.263
	5	9	4	9	4	26	0.692	0.692	0.308

4.3   RQ 3 Results

The classification models trained above are retrained with the two mobilization variables added to features, with results as shown in Panel 3 of Table 2. The accuracy of these models appears similar to that of classifiers without information on mobilization, supporting **H3B**. TPR and FPR also follow similar trends as noted above: TPR shows a weak inverse relationship to the Lag, while FPR fluctuates without any apparent trends. Similar results are also seen when the classification models are rerun on the *Test* dataset, as shown in Table 3.

After stratifying the data as described in the Sect. 3.2, no data points with corresponding protest occurrences are observed in the low mobilization group. Thus, only results for the high and medium level datasets are given here. The accuracy for these classification models is shown in Panels 5 and 6 of Table 2. TPR and FPR follow similar trends to previous results, with the exception of the classification model in Panel 4 with $k = 2$. The accuracy for the models trained on the "high mobilization" dataset does not appear to differ substantially from those of the medium mobilization classification models, supporting **H3C**.

5 Discussion

The results above indicate a relationship between network structure and protest occurrence. Interestingly, there does not appear to be a difference between the accuracy of classification models using only network structure and classification models using both network structure and mobilization information. This shows that the two methods can be used interchangeably. The results also suggest that the social network can be a predictor of mobilization, which, in turn, is a predictor of the protest. This follows from the fact that mobilization, known to be a predictor of protest [3], does not appear to be a confounding factor with the social network.

Classification models trained using networks relevant to protests (from the *Election* dataset) display a significant relationship between network measures and protest occurrences, especially as compared to classification models trained using networks unrelated to protests (from the *Neutral* dataset). The latter shows equal classification of true positives, false positives, true negatives, and false negatives, suggesting the importance of capturing the relevant communications.

The classification models trained using the *Election* dataset, without mobilization variables included (Panel 2 of Table 2), show a decrease in accuracy as time lag increases: 82% accurate with no lag, as compared to 44% accurate five days before the protest occurs. This supports **H2**, and suggests that the relationship between the network structure changes over time. The addition of mobilization variables does not change this relationship: accuracy continues to decrease with increased lag, as seen in Panel 3 of Table 2.

However, these models are trained on balanced samples: equal numbers of data points with and without protests. In reality, there are more dates/locations without protests occurring, which creates a heavily imbalanced dataset. To run accurate classification models on imbalanced datasets, a penalty for false negatives may be imposed: this discourages the model from classifying everything as "no protest" and still getting a reasonable accuracy. Failure to predict an unexpected protest may result in injuries or damage. Regardless of this imbalance, classification models presented here do demonstrate a relationship between social network structures and protest occurrences. Training classification models on imbalanced data is an avenue for future research, in order to clarify the

Table 3. Robustness check: training models on *'Test'* dataset for comparison.

	Lag (k)	TN	FP	TP	FN	Total	Accuracy	TPR	FPR
RQ1, RQ2: Test models (no mobilization variables)	0	80	33	73	40	226	0.677	0.646	0.292
	1	71	40	72	39	222	0.644	0.649	0.360
	2	72	40	71	41	224	0.638	0.634	0.357
	3	62	48	47	63	220	0.495	0.427	0.436
	4	55	57	64	48	224	0.531	0.571	0.509
	5	59	51	68	42	220	0.577	0.618	0.464
RQ3: Test models (with mobilization variables)	0	79	34	72	41	226	0.668	0.637	0.301
	1	68	43	68	43	222	0.613	0.613	0.387
	2	72	40	72	40	224	0.643	0.643	0.357
	3	65	45	66	44	220	0.595	0.600	0.409
	4	62	50	64	48	224	0.563	0.571	0.446
	5	52	58	66	44	220	0.536	0.600	0.527

relationship between network measures, reduce the amount of data processing needed, and enable the usage of proposed models for protest prediction.

Future work may also consider predicting attributes of protests (e.g. size, level of violence, etc.) instead of a binary variable. This would require a more accurate source of information on protests than news media. Additionally, future research may examine if the relationship between network structure and protest occurrences is observable outside of the United States.

6 Conclusion

The antecedents and consequences of protest behavior are of concern to society as a whole. The growth of social media provides another avenue of exploration to help identify behavior related to protests. By examining the structure of social networks as related in tweets related to the 2016 US Presidential Election, a relationship is identified between network structure and protest occurrence. Social media, and networks derived from social media, factor into mobilization of a group, and mobilization is theorized to precede a protest. Thus, the analysis of social media activity using network-based models shows a relationship to protest occurrences. Testing of the models developed here provides evidence of this relationship, which is observable in advance of protest occurrences. This may be of use in preparing for such an event, and help minimize injuries and property damage. The existence of the relationship between social network structure and present protests suggests future research directions aimed at improvements in the prediction of collective action based on analysis of network structure.

Acknowledgments. This research is sponsored by the Army Research Laboratory, accomplished under Cooperative Agreement Number W911NF-09-2-0053 (the ARL

Network Science CTA). The views and conclusions contained in this document are those of the authors and should not be interpreted as representing the official policies, either expressed or implied, of the Army Research Laboratory or the U.S. Government. The U.S. Government is authorized to reproduce and distribute reprints for Government purposes notwithstanding any copyright notation hereon.

References

1. Steinert-Threlkeld, Z.C., Mocanu, D., Vespignani, A., Fowler, J.: Online social networks and offline protest. EPJ Data Sci. **4**(1), 19 (2015)
2. Compton, R., Lee, C., Xu, J., Artieda-Moncada, L., Lu, T.C., De Silva, L., Macy, M.: Using publicly visible social media to build detailed forecasts of civil unrest. Secur. Inform. **3**(1), 4 (2014)
3. Korolov, R., Lu, D., Wang, J., Zhou, G., Bonial, C., Voss, C., Kaplan, L., Wallace, W., Han, J., Ji, H.: On predicting social unrest using social media. In: 2016 IEEE/ACM International Conference on Advances in Social Networks Analysis and Mining (ASONAM), pp. 89–95, August 2016
4. Kallus, N.: Predicting crowd behavior with big public data. In: Proceedings of the 23rd International Conference on World Wide Web, pp. 625–630. ACM (2014)
5. Oh, O., Eom, C., Rao, H.R.: Research note—role of social media in social change: an analysis of collective sense making during the 2011 Egypt revolution. Inf. Syst. Res. **26**(1), 210–223 (2015)
6. Tufekci, Z., Wilson, C.: Social media and the decision to participate in political protest: observations from Tahrir Square. J. Commun. **62**(2), 363–379 (2012)
7. Cadena, J., Korkmaz, G., Kuhlman, C.J., Marathe, A., Ramakrishnan, N., Vullikanti, A.: Forecasting social unrest using activity cascades. PLoS ONE **10**(6), e0128879 (2015)
8. Snow, D.A., Soule, S.A., Kriesi, H.: The Blackwell Companion to Social Movements. Wiley, London (2008)
9. Dalton, R., Van Sickle, A., Weldon, S.: The individual-institutional nexus of protest behaviour. Br. J. Polit. Sci. **40**(1), 51–73 (2010)
10. Furusawa, K.: Participation and protest in the European union and the 'outsider' states. Contemp. Polit. **12**(2), 207–223 (2006)
11. Dalton, R.J.: Citizen Politics: Public Opinion and Political Parties in Advanced Industrial Democracies. CQ Press, Thousand Oaks (2013)
12. Klandermans, B., Oegema, D.: Potentials, networks, motivations, and barriers: steps towards participation in social movements. Am. Sociol. Rev. **52**(4), 519–531 (1987)
13. Kuran, T.: Sparks and prairie fires: a theory of unanticipated political revolution. Public Choice **61**(1), 41–74 (1989)
14. Klandermans, B., Van der Toorn, J., Van Stekelenburg, J.: Embeddedness and identity: how immigrants turn grievances into action. Am. Sociol. Rev. **73**(6), 992–1012 (2008)
15. Barberá, P., Wang, N., Bonneau, R., Jost, J.T., Nagler, J., Tucker, J., González-Bailón, S.: The critical periphery in the growth of social protests. PLoS ONE **10**(11), e0143611 (2015)
16. Borge-Holthoefer, J., Rivero, A., García, I., Cauhé, E., Ferrer, A., Ferrer, D., Francos, D., Iniguez, D., Pérez, M.P., Ruiz, G., et al.: Structural and dynamical patterns on online social networks: the Spanish May 15th movement as a case study. PLoS ONE **6**(8), e23883 (2011)

17. Morales, A., Losada, J., Benito, R.: Users structure and behavior on an online social network during a political protest. Phys. A: Stat. Mech. Appl. **391**(21), 5244–5253 (2012)
18. Le, H.K., Pasternack, J., Ahmadi, H., Gupta, M., Sun, Y., Abdelzaher, T., Han, J., Roth, D., Szymanski, B., Adali, S.: Apollo: towards factfinding in participatory sensing. In: Proceedings of the 10th ACM/IEEE International Conference on Information Processing in Sensor Networks, pp. 129–130, April 2011
19. Bosch, A., Zisserman, A., Munoz, X.: Image classification using random forests and ferns. In: 2007 IEEE 11th International Conference on Computer Vision, pp. 1–8, October 2007
20. Kohavi, R.: A study of cross-validation and bootstrap for accuracy estimation and model selection. In: Proceedings of the 14th International Joint Conference on Artificial Intelligence, IJCAI 1995, San Francisco, CA, USA, vol. 2, pp. 1137–1143. Morgan Kaufmann Publishers Inc., Los Altos (1995)
21. Liaw, A., Wiener, M.: Classification and regression by randomForest. R News **2**(3), 18–22 (2002)

Internet Data Analysis for Evaluation of Optimal Location of New Facilities

Liubov Oleshchenko$^{(\boxtimes)}$, Daria Bilohub, and Vasyl Yurchyshyn

Department of Computer Systems Software, Igor Sikorsky Kyiv Polytechnic Institute, Kyiv, Ukraine
oleshchenkoliubov@gmail.com, dasha.bilogub@gmail.com, vasil.yurchishin@gmail.com

Abstract. Today, most purchases and orders for various services are made via the Internet. Network users are looking for the products they need, rest facilities and evaluate the positive and negative reviews from other users. By using web analytics tools, we can understand the search prospects of the requests, the likes and the popularity of a particular institution in a certain area of the city, as well as understand which areas of the city need such institutions most of all. Thus, we can create software to determine the best places to build new recreational facilities that will take into account the search needs, territorial peculiarities and the popularity of institutions - potential competitors. In this article, current software solutions and their disadvantages for solving the problem of determining the optimal location of the future institution of rest within the city are analyzed. The data on positive reviews, search and site visits statistics is analyzed. On the basis of the obtained data, a probabilistic model of optimal placement of the institution is constructed taking into account the distance of the institutions and their possible range of influence. Software solution, which allows you to simulate the optimal location of the future rest establishment, using only Internet data is proposed.

Keywords: Internet data analysis · Visualization · Web statistics
Mapping data · Software · Google maps · Huff model · Search query stats
Hospitality probability · Web services · Web resource analysis
R · C#

1 Introduction

The location of service and leisure establishments determines the potential number of customers and turnover. Forecasting turnover is the central and most difficult procedure during choosing a location. The demand for institution services is geographically oriented. It is necessary to determine the maximum distance that the client agrees to overcome to visit the institution, calculate the population that lives within the circle with the given radius and determine the number of competitors in that region. These values then are used as the basis for forecasting the turnover of a new institution. There are software solutions that perform the analysis by the determined factors and analyze the terrain, using 3D simulation for a more visual view. The disadvantages of these

© Springer Nature Switzerland AG 2019
O. Chertov et al. (Eds.): ICDSIAI 2018, AISC 836, pp. 279–291, 2019.
https://doi.org/10.1007/978-3-319-97885-7_28

systems are that the data used for the analysis may be outdated (or the user needs to enter the data for analysis himself) and the lack of free version of the product.

At the moment, most of the orders for rest facilities (ticket booking in cinemas, etc.) are carried out over the Internet. Users of the relevant institutions' sites leave feedback and evaluate institutions using online services. There is an opportunity to track the number of visitors to the site of the institution and its popularity on the net, thus comparing the probability of visiting these institutions. Thus, the application that is developed should be an application program interface, which enables entrepreneurs to determine the optimal location of the institution using data from the Internet. Also, the system should provide the opportunity for ordinary users to view information about institutions, identify the nearest institutions of competition. To fulfill the task, the following requirements were developed.

1. The system must update the data by downloading them from the Internet. That is, the user should not enter any data except the location that should be analyzed. System founds all data via the Internet;
2. The system must be synchronized with Google Maps;
3. The system should be available for use on any OS, regardless of the computer software;
4. The system should have a convenient interface. That is, the user can work with the program, without having special skills for this.

To design new entertainment centers and recreation facilities, system needs to analyze the following factors:

1. The area of the object.
2. The number of competitors nearby and their popularity according to on-line data;
3. Environmental factors (the presence of harmful plants nearby, etc.);
4. The opportunity for visitors to ride from different parts of the city.

2 Available Software Solutions

2.1 ArcGIS

ArcGIS Spatial Analyst offers a wide range of powerful features for spatial modeling and analysis. You can create, query, and analyze cell-based raster data; perform integrated raster/vector analysis; receive new information from existing data; request information in several layers of data; and fully integrate cellular raster data with traditional vector data sources. The latest version of the application is integrated into the Python environment. ArcGis Spatial Analyst integrates user data that allows you to interact with different types of data. Images, models of heights and surfaces can be combined with automated design (CAD), vector data, data on the Internet and many other formats to provide an integrated analysis environment. ArcGis Spatial Analyst can create bitmap data from any source of any point, line, or polygon function, such as ArcInfo, curved files, location databases, CAD files, vector file files, and ArcGis themes created using tabular data. In addition, data in the standard format can be imported, including TIFF, JPEG, BMP, USGS DEM, DTM, NIMA DTED, general ASCII, MrSID and others [1, 2].

ArcGIS for the server allows you to work directly with spatial data stored by managed commercial databases that support spatial types. ArcGIS for Server Basic Edition allows you to enable read-only features for your data. These office functions will allow you to display and query the map information database from browsers and mobile devices. All databases are read-only in ArcGIS for Server Basic Edition and are available in read-write mode in ArcGIS for Server Standard and Advanced editions (except for non-read-only Netezza). There is no mobile version of the application.

2.2   Map Info

Pitney Bowes MapInfo provides the analytical tools and services that enable clients to find developments and locations where success is almost guaranteed. These products and services help determine the correct number of units, and the best locations for those units.

The Predictive Analytics Portal is a Pitney Bowes MapInfo enterprise location intelligence platform. The portal has the functionality of a desktop solution but in a web based platform that enables users to view customer data, demographics, mapping functionality, through to predictive analytics using generic or custom models for market site selection and or market analysis. This technology allows decision makers throughout a company, regardless of their physical location to access the powerful reporting and mapping capabilities to make strategic marketing, site selection and market evaluation decisions anytime and anywhere. The information that drives the process is centralized in a single location thereby reducing the potential for discrepancies and misdirected efforts, allowing for a single source of truth [3].

2.3   The Tango Geospatial Platform

Tango's Market Planning & Site Selection enables retailers to evaluate their existing store portfolio, new market opportunities and the impact of competitive moves to optimize their real estate plans, while identifying the highest and best use of their store development capital. The Tango Geospatial Platform offers proactive customer and location analytics, specially designed GIS and location development in one package. Key features:

- Managing location data. Contains location information, including stores, goals, perspectives, sites, rental data and competitors, as well as geographic information, including shopping areas, regions, markets, and territories.
- Availability of qualitative location data. Built-in processes and data quality tools help discover new data, as well as problems and abnormalities within existing data, which ensures data accuracy across the entire system.
- Geocoding. The ability to convert address data into latitude and longitude is available.
- Available plug-in for Land-based Geographic Information Services (GIS). Mobile version is available [4].

2.4   Oracle Locator

With the Oracle Locator and spatial option, each Oracle database includes built-in location features that allow any business application to include location information and the realization of competitive advantages. Oracle Spatial includes support for all types and models of geospatial data, including vector and bitmap data, as well as topological and network models that meet the needs of the GIS system such as land management, utilities and protection/internal security [5].

Oracle Locator supports three basic geometric forms that represent geographic and location data:

- points: they can represent locations such as buildings, fire hydrants, utility poles, oil rigs, boxcars, or roaming vehicles;
- lines: they can represent things like roads, railroad lines, utility lines, or fault lines;
- polygons and complex polygons with holes: they can represent things like outlines of cities, districts, flood plains, or oil and gas fields. A polygon with a hole might geographically represent a parcel of land surrounding a patch of wetlands [6].

3   Research Methods

According to statistics, about 60% of users choose a rest institution to take into account its reviews. The more reviews about the entertainment center, the more precisely you can determine its popularity, if the institution has the majority of negative reviews, then it attracts less consumers. To get this information, consumers are turning to a mapping service (for example, Google Maps suggests giving feedback and rating about the site) [7].

Equally important is the attendance of the site. The logic of thinking is based on the above assumption: the greater attendance is - the more popular institution is. Consumers turn to the site for different information. Research shows that 29% of site visitors pay attention to the proposed prices, 15% for the range of services, 13% for the presence of shares [8].

Also, using a mapping service, it is easy to calculate the distance between entertainment centers, using the map scale. The annex proposes to consider the distance between institutions in order to analyze the need of the construction in this region. So, the first selected criterion was the assessment of visitors to the selected location, which can be obtained through a mapping service.

The next criterion is the attendance of the site for recreational facilities, because consumers use the site to learn about the pricing policy of the institution, to review the set of services, availability of shares. As the next criterion, the remoteness of the institutions was chosen, to determine the density of the location of the entertaining establishments and the expediency of considering the chosen district for opening a business.

3.1 Mapping Data Analysis

As a mapping service, it was decided to choose Google Maps. As mentioned above, the project is being developed for the Windows platform with a usage of the programming language C#, also Google Maps has an API that could be easily integrated with this language. Let's consider in more detail the chosen cartographic service. Google Maps is a suite of apps built on the basis of Google's free cartographic service and technology. The service is a map and satellite images of the whole world.

The service has an integrated business directory and road map, with route search. Google actively co-operates with hundreds of transport departments around the world to provide its users with information of timetables and public transit routes. Google Maps presents traffic schedules for more than a million halts located in over 800 major cities around the world, including 23 Ukrainian cities. Google Street View gives users the opportunity to "get around" in a three-dimensional street projection over the Internet. This functionality is possible by means of round-the-field photography of special equipment in real-time. Today it is a free service, but the ability to add ads left for the future. The Google Static Maps API allows building static maps using custom URL [9].

3.2 Analysis of Visits to Institutions' Websites

Google Analytics is a handy and feature rich service for analyzing web sites and applications. Allows webmasters to check the state of indexing and optimize the visibility of their websites. Google Webmaster Tools help:

- track the appearance of pages on your site in the index of the search engine;
- totals of site attendance;
- popularity of pages and sections of sites;
- view links to your pages from other resources;
- analyze access errors to pages that the search engine has encountered when indexing the site and much more.

Thus, with such a wide range of tools, it is possible to get not only the number of visits to the site, but also to determine how the user went to the web page and how often searched the search engine [10].

All statistics are collected on a special company server using the Google Analytics Custom Meter. The counter code (written in Javascript) is placed on all pages of the researched web resource.

Visitors can get to the site in different ways - from search engines for a given key query, from links to other web resources, by mailing list, by directions in various information products, directly by typing an address, etc. As soon as they appear on the site page with the Google Analytics code, tracking information about the user's actions and the visitor itself (for example, page URL, temporal data, visitor screen resolution, etc.) begins - creates a special set of cookies for user identification. The visitor tracking code of the Google Analytics counter is forwarding information to the corporation data collection servers in a fraction of a second. At certain intervals (typically for a small site up to 50,000 pages this time in one hour), the company processes the data and updates the custom reports Google Analytics accordingly. Reports may not immediately appear, but with some delay (from 3 to 24 h depending on the load) [11].

To use this service, the Google Analytics Reporting API Client Library for .NET has been selected in its project, which has enough documentation and examples [12].

3.3   Analysis of Positive Reviews

For feedback analysis, data will be collected from the institution's web site and from Google Maps (Fig. 1). Users will first use their site before visiting an institution. In the case of cinemas, the site provides the opportunity to buy or book tickets, find out about new items and promotions. Visitors are equally important to give feedback on the film, the service, the institution as a whole, and these reviews can be found on the site's website. Because owners are primarily interested in the fact that many users have left comments for, in order to improve the recreation center. The collection of this information will be done using Google Analytics, which is considered above. Google Maps is a well-developed mapping system with a description for any institution. With the help of location data, the system receives information about the location of any person and if, at that moment, he used the Google Maps application on his smart phone, the utility will offer feedback and rating about the place you visit. Many users leave a lot of reviews every day with the estimates of institutions. Because many people use this service to build routes (including from their location to a particular facility), users pay attention to the assessment of this recreation center. So, this information should also be taken into account.

Fig. 1. Map of cinemas of the Kyiv region. Evaluate and comment visitors using the Google Maps service

In Fig. 2 statistics of visits to cinema sites in the surveyed area of the studied city is shown.

3.4   Probabilistic Huff Model

One of the most productive borrowings made by economists from the exact sciences is the theory of gravity. Models of spatial interaction, built on its basis, give an opportunity to study a wide range of issues related to the characteristics of placements and the mutual influence of a range of socio-economic facts. Application of mathematical modeling is possible not only for the analysis of processes and phenomena in international economic relations, but also in market research, in particular for modeling the spatial organization of sales markets. Proposed by David Huff in 1963, the model for determining the location of a commercial object most optimally in terms of profit is successfully applied to this day. Being criticized and unlucky shortcomings, Huff's model attracts researchers with its simplicity and versatility [13].

Fig. 2. Statistics of visiting cinema sites per month in Kyiv.

The basic idea of the model is the definition of the attractiveness of the trading object, which is directly proportional to the size of the object and inversely proportional to the distance between the buyer and the trade object. That is, this model of spatial interaction calculates the probability based on consumer gravity at each location of origin that patronizes each institution in the data set of the institution. Of these probabilities, the sales potential can be calculated for each place of origin on the basis of available income, population, or other variables.

In the model of D. Huff to assess the retail industry, consumers have a number of alternative purchasing opportunities. Each trading point has a certain probability of a permanent visit, the usefulness of the retail point depends on its size (S) and the distance to it (D). The probability of P_{ij} is the consumer's visit to a facility is equal to its utility (U_{ij}) to the amount of utility of all facilities considered by the consumer (the model of multiplier interaction):

$$P_{ij} = \frac{U_{ij}}{\sum_{k=1}^{n} U_{ik}} = \frac{S_j^{\alpha} \cdot D_{ij}^{\beta}}{\sum_{k=1}^{n} S_k^{\alpha} \cdot D_{ik}^{\beta}}, \tag{1}$$

where P_{ij} is the probability of a customer visiting from the city i, point j; n is the set of competing points (or cities) in the region, U_{ij} is the value of j for the individual in and, S_j is the area of the institution j, D_{ij} is the distance between the consumer in and that point j, α and β are the sensitivity parameters [14].

In general, the expression for the probability of choosing point of sale has the form:

$$P_{ij} = \frac{\prod_{k=1}^{q} x_{k_{ij}}^{\beta_k}}{\sum_{k=1}^{n} \prod_{k=1}^{q} x_{k_{ij}}^{\beta_k}}, \tag{2}$$

where x_{kij} is k^{th} variable, that describes the point in the situation of choice i, β_k is the indicator of the sensitivity of the utility function relative to the k variable (or the level of elasticity for this variable), q is the number of variables in functions of utility.

Also is used the utility function of the multivariate type:

$$P_{ij} = \frac{\exp\left(\sum_k \beta_k - x_{k_{ij}}\right)}{\sum_{j=1}^{n} \exp\left(\sum_k \beta_k - x_{k_{ij}}\right)}. \tag{3}$$

The choice of functions is determined by the nature of the dependence of market share on the values of factors. To estimate the coefficients β_k, it is necessary to know the factors x_{kij} and probabilities of the choice P_{ij}. The values of these variables are tied to certain "selection situations," that is, situations in which the same conditions of choice are assumed. As a rule, factors that distinguish a huge variety of situations in artificial groups or "situation of choice" are simultaneously included in the Eqs. (2 and 3) as regressors.

Each situation of choice is associated with the corresponding probabilities P_{ij} and variables x_{kij}. Since actual values of probability are unknown, they are replaced by averaged values for each situation estimates of the distribution of consumer preferences π_{ij}. In the case when a plurality of choice situations coincides with a plurality of respondents, probabilities are estimated by distributing the number of visits made by the consumer.

After receiving an array of data prepared for analysis, the calculation of particle P estimates and consumer parameters is performed. With these data, the analyst estimates the model with the least squares or maximum likelihood methods.

3.5   Experimental Results of Research

In the previous study we used GIS to determine the locations of new gas stations. To find the optimal location, we compared the characteristics of existing retail outlets with their values for potential locations. A transport buffer for each gas station was used to sample the characteristics. Each characteristic was assigned a weighted coefficient, which was taken into account during calculation of the overall assessment of the location. Comparison of the characteristics of the locations was used in the selection of operating gas stations, similar in basic characteristics.

This method allowed to show on the map which of the neighboring areas are influenced by each of the filling stations in the form of a set of probability values. The level of these values is depicted on the contour maps, while taking into account the influence of the average markets of sales and influence. The ArcGis statistical tools and the Huff gravity model evaluated the locations of potential gas stations.

Multiple regression connects the trade turnover with independent variables characterizing the location of the gas station with the help of a linear equation. As a result of the analysis, a place is selected that provides the maximum expected value of fuel sales. Multiple regression gives a good forecast of the expected turnover of the gas station at the established limit of error. ArcGis geo-processing tools allow you to perform a regression analysis based on the method of the least squares. For analysis of the results of multiple regression, a statistical report is compiled comprising: regression coefficients, mean square error, t-statistics, probability, R^2, F-statistics, factor scatter diagrams, estimation of spatial autocorrelation of residues, residue map. ArcGis modeling includes the following steps:

1. Determining the market boundary on a line that combines two competitors.
2. Scrolling the range of two competitors on the plane.
3. Transition to ranges for the three closest competitors.
4. Simulation of the network of competitors in the region.
5. Simulation of the network of competitors on a heterogeneous road-network of the region.
6. Accounting of the distribution of fuel consumers.
7. Economic analysis based on the conducted network analysis.

We used the ARCGIS Spatial Analyst module to simulate the optimal location of new gas stations in the Chernihiv region. To do this, we digitized the map of the region, divided it into administrative units (districts), mapped the network of existing gas stations. In the estimation of the optimal location of the gas station, the analysis of the demand for fuel of the existing gas stations, the condition and traffic of the highways, the availability of road transport, the level of average monthly salary and incomes, carried out digitization of the street and road network of Chernihiv region and the network of gas stations. A cartographic model with a network of streets was displayed on the territory of the regional center, which shows the network of existing gas stations. Figure 3 depicts the received cartographic model with the zones of influence of existing gas stations with a radius of 10 km and the proposed locations of new gas stations. As can be seen from Fig. 3, none of the proposals is included in the specified zones.

4 Discussion and Conclusion

Despite the advantages, the above-mentioned applications have a number of drawbacks. In terms of software development:

- programs require powerful software from a computer that is problematic for ordinary users;
- manual input, which is usually unknown to the user.

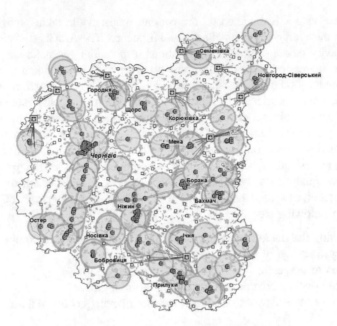

Fig. 3. Areas of the sale of existing gas stations and the proposal of the location of new gas stations according to the Huff model (Chernihiv region, Ukraine).

In terms of the user:

- has an incomprehensible interface;
- need some skills to work properly with the program.

Given these disadvantages, it was decided to create software that has a user-friendly interface, will perform all calculations quickly, without the need for certain hardware. As mentioned above, the criteria for the analysis were changed.

Consequently, we can conclude that the application is aimed at analyzing the popularity of institutions, and not population by district. This choice makes the software developed for free, since all the information is open source and does not require any extra costs.

The development consists in the creation of modules:

- to collect all the necessary information, present it in a convenient way for future use and save it in the database;
- analysis of the data obtained according to the different criteria outlined above;
- user Interface.

Analyzing the existing software market, one can conclude that there are a lot of them, but not all perform functions that are necessary for ordinary users, including private entrepreneurs. The main disadvantage of these programs is that the user needs to manually enter statistics and those that do not take into account various factors that may affect the placement of rest facilities. The main advantages are that these applications allow you to display 3D models and that the analysis is performed not only with the parameters mentioned above, but also taking into account the relief. So, the solution to this problem is the creation of software that will receive all the necessary information using the Internet. Data analysis is performed using the probabilistic Huff model, which allows determining the relative proportions in which consumers are distributed among the zones of different recreation facilities.

As a solution to all of the above-mentioned shortcomings of existing solutions, an application is proposed for calculating the optimal location of recreational facilities based on the Huff model. Given that the main disadvantage of existing analogs is that the user needs to manually enter data, the application created should itself receive all the necessary information from the Internet. From the user's side, you need to enter parameters such as the type of institution and the radius of service (Fig. 4).

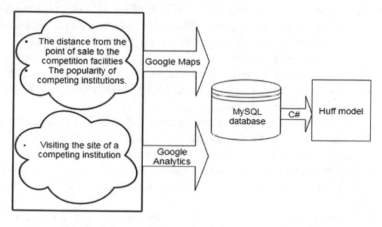

Fig. 4. Scheme of the work of developed application

The application is developed in C# because C# is a fully featured object-oriented language that supports encapsulation, imitation and polymorphism. It has component support, reliable and stable thanks to the use of a "garbage collection", exception handling, security types.

Different technologies are Windows Forms, WPF are used to create graphical interfaces using the .Net platform. However, Windows Forms or forms remain a simpler and more comfortable platform. Thanks to the use of C#, acceleration of the algorithm is achieved, since R has certain drawbacks compared to C#: most functions of R are significantly overloaded. In R, the overload is implemented using one function with many optional parameters named; many R-commands devote little attention to memory management, and therefore R can very quickly consume all available memory. This may be a limitation when performing data extraction [15–17].

MySQL is used as a database, since MySQL is characterized by high speed, stability and ease of use. The capabilities of the MySQL server are to support an unlimited number of users simultaneously working with the database; the number of rows in the tables can reach 50 million; high speed execution of commands; the availability of a simple and effective security system.

All data is stored in JSON format, which provides quick data exchange with a small amount of data between client browsers and AJAX web services. Google Maps is selected as the map used in the app. This particular version of the maps was chosen because there are ready-made C# libraries for working with maps.

The Google Analytics service is chosen to collect site statistics because it provides up-to-date user activity data on the site, how it went over to it, and how many visitors were within a certain time frame. From the standpoint of the developer, this service is convenient because C# has ready-made libraries for working with these utilities. Google Analytics is used to collect user activity data from websites, mobile apps, and real-world and digital environments. To track the site, Google Analytics uses Java-Script to collect information.

When placing this script on a page, he begins collecting various information about the user's actions on the site. JavaScript can collect information directly from a site, for example, URLs that are viewed by a user. The script can receive information about the user's browser, such as browser settings, its name, and the name of the device and operating system used to login to the site.

The JavaScript script can transmit information about sources that contain links to this site from which the user has visited the site. All these data are packaged and sent to Google services in the queue for processing.

To track the number of orders executed on the site, the item for sale is configured to send the data packet every time the purchase was made. Such a package may include information about the location of the store, the article, the date of purchase, etc.

The main disadvantage of the product being developed is that it is developed only for Windows, but the solution to this problem can be considered as further development of the product.

At the moment, the project is aimed at data analysis for recreation facilities, however, as a further development of the application, it will be possible to model accommodation of public catering establishments, pharmacies, hospitals, supermarkets, refueling stations, etc.

Also, one can consider the option of combining this analysis with the analysis offered by analogues. Thus, it will be possible to achieve even more detailed modeling according to different criteria.

References

1. Reference on summary of lectures on discipline "GIS in the tasks of computer monitoring". http://kruzhan.vk.vntu.edu.ua/file/49d34849e12b4257705624e19cce3bb0.pdf. Accessed 9 Nov 2017
2. ArcGIS Spatial Analyst: Advanced GIS Spatial Analysis, December 2001. https://www.esri.com/library/whitepapers/pdfs/arcgis_spatial_analyst.pdf. Accessed 10 Nov 2017
3. Pitney Bowes MapInfo view. Transforming Location into Success. http://www.pbinsight.co.in/files/resource-library/resource-files/Transforming_Location_into_Success__11-08.pdf. Accessed 8 May 2018
4. Geospatial Location Platform. https://tangoanalytics.com/products/geospatial-location-platform/. Accessed 25 Dec 2017

5. Oracle Locator (2010). https://docs.oracle.com/cd/E18283_01/appdev.112/e11830/sdo_locator.htm. Accessed 25 Dec 2017
6. Oracle spatial option and oracle locator. Location Features in Oracle Database 10g. http://www.oracle.com/technetwork/database/enterprise-edition/10g-spatial-locator-ds-092286.html. Accessed 8 May 2018
7. 8 Secrets That Make Customers Buy, December 2015. https://geniusmarketing.me/lab/8-sekretov-kotorye-zastavlyayut-klientov-pokupat/. Accessed 20 Mar 2018
8. What motivates Internet store customers? (2016). https://4service-group.com/ru/about/press-about-us/chto-motiviruet-klientov-internet-magazinov. Accessed 20 Mar 2018
9. How does Google Maps Work? 8 January 2017. http://www.makeuseof.com/tag/technology-explained-google-maps-work/. Accessed 20 Feb 2018
10. Basics of working with web analytics. How Google Analytics works, November 2013. https://blog.aweb.ua/osnovy-raboty-s-veb-analitikoj-kak-rabotaet-google-analytics/. Accessed 15 Dec 2017
11. How Google Analytics works, February 2013. http://www.workformation.ru/google-analytics.html#h2_1. Accessed 10 Feb 2018
12. Google API Client Libraries .Net, March 2017. https://developers.google.com/api-client-library/dotnet/guide/aaa_client_secrets. Accessed 15 Feb 2018
13. Golikov, A.: Economic and mathematical modeling of world economic processes, pp. 143–151 (2006)
14. Simulation of location choice processes of commercial, November 2016. http://elartu.tntu.edu.ua/bitstream/123456789/20641/2/GEB_2016n2_%2851%29_Beley_O-Simulation_of_location_choice_103-112.pdf. Accessed 20 Nov 2017

Novel Theoretical and Practical
Applications of Data Science

Algebra of Clusterizable Relations

Boris Kulik[1] and Alexander Fridman[2]([✉])

[1] Institute of Problems in Mechanical Engineering,
Russian Academy of Sciences (RAS), St. Petersburg, Russia
ba-kulik@yandex.ru
[2] Institute for Informatics and Mathematical Modelling,
Kola Science Centre of RAS, Apatity, Russia
fridman@iimm.ru

Abstract. Relations are usually represented in a space of attributes whose values differ only in names similar to algebra of sets. The order of the values or any other preference measures are not significant for such attributes. The paper proposes a mathematical model based on n-tuple algebra (NTA), for relations in which the values of attributes are ordered. For this case, a mathematical tool has been developed that can be used to perform not only the previously discussed methods and means of logical-semantic analysis on the basis of NTA, including analysis of defeasible reasoning and logic-probabilistic analysis, but also to analyze the order and connectivity of structures and implement clustering methods. The concept of granules is introduced, the power of connectivity between the granules is defined, and methods to calculate distances between the disconnected granules are proposed. The obtained dependencies make it possible to extend the scope of classification techniques.

Keywords: N-tuple algebra · Logical-semantic analysis · Ordered attribute
Clusterization

1 Introduction

In artificial intelligence, an interest has recently appeared in works (for example, see [1–3]), which suggest to construct systems according to certain principles of the cognitive approach, where a certain order of elements is specified (in particular, corresponding with some preference measures). The main area of scientific research and application of such systems is the problem of categorization, that is, classification and identification of objects whose characteristics are defined in *conceptual spaces* (CSs) proposed by Gärdenfors [1] as the basic apparatus for cognitive research.

In this case, heuristic measures of distances between categories of elements of the system and between elements within one category are used. The mathematical justification for such systems is very weak, so there is a great variety of different theoretical approaches, and it is difficult to choose the most appropriate one.

The given paper considers a class of relations with ordered values of all attributes. Such relations are often encountered, and they comprise not only quantitative attributes (age, distance, price, etc.), but also qualitative attributes that are specified by names. For values of such attributes, a certain order can be chosen (for example, based on

O. Chertov et al. (Eds.): ICDSIAI 2018, AISC 836, pp. 295–304, 2019.
https://doi.org/10.1007/978-3-319-97885-7_29

expert estimates). Another method of ordering the values of an attribute is the *method of interval quantization* set forth in [4]. This method allows to represent a system of arbitrary intervals on a coordinate axis as a linearly ordered set of quanta. Let us consider this method in more detail.

Let there be a closed interval Ω on a numerical axis and a finite set $E = \{E_i\}$ of closed intervals such that $E_i \subseteq \Omega$. The boundaries of the intervals are given by the sets of coordinates of the initial and final points. If all the points denoting the boundaries are considered in ascending order, then we obtain a partition of the system of intervals into quanta, i.e. neighboring points and open intervals between them (points can be considered as degenerate intervals with zero measure). An example of the quantization procedure for four intervals E_1, E_2, E_3, E_4 is shown in Fig. 1. For clarity, the intervals E_i are moved above the horizontal coordinate axis.

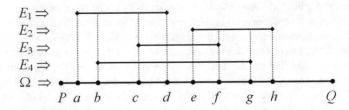

Fig. 1. Quantizing an interval system

Here the interval Ω bounded by the points P and Q contains interior points $a, b,..., h$ and open intervals between them (a, b), (b, c), (c, d), etc. For simplicity, we can ignore all points and consider only open intervals as quanta, especially since the absence of points does not affect metric properties of objects. For example, the closed interval E_3 can be represented as an ordered set $\{(c, d), (d, e), (e, f)\}$ of open disjoint intervals (quanta).

Relations with such attributes can be described by n-tuple algebra (NTA) [4]. This algebra is a mathematical system for analysis of diverse information representable in the form of relations. Relations in NTA are modeled using four types of structures called *NTA-objects*. Each NTA-object is immersed into a specific space of *attributes*. Names of NTA-objects contain an identifier followed by a sequence of attribute names enclosed in square brackets. These attributes form the *relation diagram* in which this object is specified. For instance, the name $R[XYZ]$ denotes an NTA-object R defined in the attribute space $X \times Y \times Z$, while $R[XYZ] \subseteq X \times Y \times Z$. NTA-objects are expressed by matrix-like structures in which the cells contain *components*, i.e. subsets of the domains of the corresponding attributes rather than single elements. NTA-objects with the same relation diagram are called *homotypic* ones.

Completeness of this system (namely, feasibility of operations and checks of relations for any collections of NTA-objects) is proved as well as its isomorphism to algebra of sets. Universality of NTA is due to the fact that it realizes a generalized algebra of arbitrary n-ary relations, which turned out to be interpretations of predicates and formulas of mathematical logic. Combinations of NTA operations and relations allow implementing almost all necessary operations and checks upon conventional data and knowledge structures, including logical inference, quantifications, operations of join and composition, transitive closure, etc.

Properties of NTA are based on properties of the Cartesian product of sets. Below we brief necessary information for understanding NTA. The basic structure in NTA is a tuple of components bounded by square brackets and called *C-n-tuple*. For example, the *C-n*-tuple $R[XYZ] = [A\ B\ C]$ is the relation equal to the Cartesian product $A \times B \times C$ where $A \subseteq X$; $B \subseteq Y$; $C \subseteq Z$. In NTA, elements of Cartesian products of sets are called *elementary n-tuples*. The union of homotypic *C*-n-tuples, expressed as a matrix, is called a *C-system*. For instance, $Q[XYZ] = \begin{bmatrix} A_1 & A_2 & A_3 \\ B_1 & B_2 & B_3 \end{bmatrix}$ is a *C*-system. It can be transformed into an ordinary relation (i.e., into a set of elementary *n*-tuples) as follows: $Q[XYZ] = (A_1 \times A_2 \times A_3) \cup (B_1 \times B_2 \times B_3)$ where A_1, $B_1 \subseteq X$; A_2, $B_2 \subseteq Y$; A_3, $B_3 \subseteq Z$.

Developing the theoretical grounds of NTA until now, we assumed that the domains of attributes of NTA-objects are ordinary sets with equivalent elements. Studies have shown that NTA structures acquire new properties, if values of attributes are ordered. Name this modification of NTA *algebra of clusterizable relations* (ACR).

2   Analysis of Connections in ACR

In the further text, we will use quanta to represent not only "interval" attributes, but also any ordered attributes with a finite number of values. In such attributes, quanta are named by numbers 1, 2, 3, …. If no length of quanta is specified, it is assumed to equal 1 for each quantum.

If lengths of quanta are known and different (for example, when the attribute values result from quantizing a certain set of intervals), such attributes need a *normalization of intervals* in order to provide average length of every quantum equal 1.

Let the total length L of an interval containing quanta and the lengths of quanta l_i ($i = 1, 2, …, M$) be given. Then during normalization, the total *measure of the normalized interval* is assumed equal to M, and measures of normalized quanta (n_i) are calculated by the formula $n_i = \frac{l_i \cdot M}{L}$.

Consider the structure shown in Fig. 2, where the domains of attributes X and Y are represented by a system of quanta with names 1, 2, 3, …, 9. The shaded areas display an NTA-object $R[XY]$, which can be expressed as the *C*-system

$$R[XY] = \begin{bmatrix} \{1,2,3,7\} & \{1,8,9\} \\ \{3,4,5,8,9\} & \{2,3,4,7,8\} \end{bmatrix}. \tag{1}$$

Despite attribute values have a linear order, its violations often occur in the space of several attributes. For example, two "points" with coordinates (2, 4) and (3, 2) are incomparable in order. Obviously, in a space with dimension 2 or more, the order of elements is established as dominance. The *dominance relation* is well known to be such a correlation between vectors that $(a_1, a_2, ..., a_n) \leq (b_1, b_2, ..., b_n)$ is true if and only if $a_i \leq b_i$ for all $i = 1, 2, ..., n$.

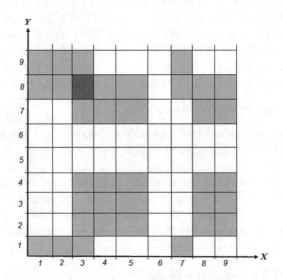

Fig. 2. Image of the NTA-object $R[XY]$

Consider the possible relations for components of the same attribute. Components can be *whole* (for instance, $\{2, 3, 4, 5\}$) or *separable* (for example, $\{2, 3, 6, 7, 8\}$). A whole component is called a *range*.

Definition 1. The *ordering relation* for a pair of components (C_i, C_j) of the same attribute is defined as $C_i < C_j$ only under the condition that $max(C_i) < min(C_j)$.

It is easy to prove that for a pair of ranges (I_r, I_s) of the same attribute, if $I_r \cap I_s = \varnothing$, then either $I_r < I_s$ or $I_s < I_r$.

Definition 2. The *distance* $D(I_r, I_s)$ between ranges I_r and I_s of the same attribute under the condition $I_r < I_s$ is defined as $D(I_r, I_s) = min(I_s) - max(I_r) - 1$.

For example, the distance between the pair of ranges $\{2, 3\}$ and $\{5, 6, 7\}$ is equal to 1, and the distance between the pair of ranges $\{1, 2, 3\}$ and $\{4, 5\}$ is 0.

For normalized attributes, the distance between ranges is calculated by a formula different from the one in Definition 2. Let the non-intersecting ranges I_r and I_s contain some sets of consecutive integers in a normalized attribute. For these ranges under the condition $I_r < I_s$, we compute two numbers: $S = min(I_s) - 1$; $R = max(I_r) + 1$.

Definition 3. The distance $D(I_r, I_s)$ between ranges I_r and I_s of a normalized attribute under the condition $I_r < I_s$ is: $D(I_r, I_S) = \sum_{i=R}^{S} n_i$. If $R > S$, then $D(I_r, I_s)$.

The previously determined ordering relation for ranges can be weakened if we define it for intersecting ranges.

Definition 4. The *quasi-ordering relation* π for intervals (I_r, I_s) under the condition $I_r \cap I_s \neq \varnothing$ is a relation such that $I_r \pi I_s$ if and only if $min(I_r) \leq min(I_s)$ and $max (I_r) \leq max(I_s)$.

In order to investigate different variants of ordering in n-dimensional space as well as to solve other problems in similar structures (for instance, computing the connectivity power of objects, clustering problems, etc.), a more convenient representation of NTA-objects is needed. In this case, the inconvenience of expressing their structure lies in the fact that some of the C-n-tuples from an NTA-object can be separable since their components (for example, $\{2, 3, 4, 7, 8\}$) contain interruptions. To convert the original C-system into an equivalent C-system, where each C-n-tuple has no interruptions (for example, the C-n-tuple $[\{1, 2\}\ \{8, 9\}]$), we act as follows. Let the C-n-tuple $Q[X_1 X_2 \ldots X_n] = [C_1\ C_2 \ldots C_n]$ be given, in which each component C_i $(i = 1, 2, \ldots, n)$ is splitted into k_i disjoint non-empty sets C_{ij} $(j = 1, 2, \ldots, k_i)$. Then the C-system that contains $k_1 \times k_2 \times \ldots \times k_n$ C-n-tuples written as elements of the Cartesian product

$$\{C_{11}, \ldots, C_{1k_1}\} \times \{C_{21}, \ldots, C_{2k_2}\} \times \ldots \times \{C_{n1}, \ldots, C_{nk_n}\} \tag{2}$$

will be equivalent to the initial C-n-tuple Q.

To transform each separable C-n-tuple of the C-system $R[XY]$ specified in (1) into a set of whole C-n-tuples, we split each separable component into a set of ranges and transform the C-n-tuples with separable components into C-systems in accordance with (2). In particular, after uniting the decomposed C-systems from (1), we obtain:

$$R[XY] = \begin{bmatrix} \{1,2,3\} & \{1\} \\ \{1,2,3\} & \{8,9\} \\ \{7\} & \{1\} \\ \{7\} & \{8,9\} \\ \{3,4,5\} & \{2,3,4\} \\ \{3,4,5\} & \{7,8\} \\ \{8,9\} & \{2,3,4\} \\ \{8,9\} & \{7,8\} \end{bmatrix}.$$

Another possible way of transforming a C-system into a union of whole C-n-tuples is to represent it as a set of elementary tuples, but in some cases this method is much more complex. For the example with the C-system $R[XY]$, this decomposition requires 36 elementary tuples.

C-n-tuples, in which all components are ranges, represent a completely filled hyperrectangle in n-dimensional space. Let us call this rectangle a *bar*. Consider different variants of connections between bars. In our two-dimensional example (Fig. 2), there are three variants of relations between pairs of bars:

(1) two bars have a non-empty intersection, such as the pair $[\{1, 2, 3\}\ \{8, 9\}]$ and $[\{3, 4, 5\}\ \{7, 8\}]$; their intersection is the elementary bar $[\{3\}\ \{8\}]$;

(2) the bars touch each other by facets (in our case, the facet is one-dimensional), such bars in our example include pairs
($[\{7\}\ \{8, 9\}]$; $[\{8, 9\}\ \{7, 8\}]$) and ($[\{1, 2, 3\}\ \{1\}]$; $[\{3, 4, 5\}\ \{2, 3, 4\}]$);

(3) the bars touch each other at one point (pair [{7} {1}] and [{8, 9} {2, 3, 4}]).

Obviously, a pair of bars is not connected, if the distance between components for at least one pair in the compared bars is greater than 0. For instance, the pair [{3, 4, 5} {7, 8}] and [{7} {8, 9}] is not connected, because the distance between the first components is 1.

Let us generalize the obtained results for bars of arbitrary dimension. Let the bars be distributed in an n-dimensional space. Consider variants of connections between bars from a separately taken pair of them. To do so, it is necessary to calculate relations between pairs of components belonging to the same attribute in all attributes. These pairs either have a non-empty intersection or are at a certain distance and one of the elements of the pair precedes the other element.

Definition 5. A pair of bars is called a *connected* one, if each pair of its components has either a non-empty intersection or a zero distance.

Definition 6. For a pair of connected bars in an n-dimensional space, let k be the number of pairs of components with a non-empty intersection. For such bars, the *power of connectedness* equals k.

Definition 6 yields that bars with power of connectedness 0 touch at one point, and the intersection is nonempty for bars with the power of connectedness n. The number k determines the dimension of adjoining facets for connected bars.

Definition 7. Bars or aggregates of connected bars are called *granules*.

In clusterizable relations, an arbitrary set of granules is an algebraic system in which operations coincide with operations defined for NTA-objects, while ordering and quasi-ordering relations are added to NTA relations (equality and inclusion), as well as the connectedness and proximity relations considered below.

Consider a technique to construct a connectivity graph for bars. Let the C-system Q in an n-dimensional space be decomposed in accordance with (2) into m bars B_i ($i = 1, 2, \ldots, m$). For all possible pairs of these bars, we construct a *connectivity graph* G with weighted edges where a pair of bars (B_i, B_j) is connected by an edge of weight k if and only if these bars are connected and their connectivity power equals k. Clearly, the number of checks required to form a graph G is $m(m - 1)/2$.

If there is at least one edge of weight k in a graph G, then the *graph of the kth connectivity power* ($0 \leq k \leq n - 1$) is an unweighted graph G_k, which contains all edges of the graph G with weights $w \geq k$, while all edges with weights smaller than k are removed. If there are no edges of weight z in a graph G, the connectivity graph of the zth power G_z is assumed to not exist for the given NTA-object.

Definition 8. If a graph G_k can be constructed for an NTA-object, the set of *granules of the kth connectivity power* ($0 \leq k \leq n - 1$) comprises all isolated vertices and connected components of the graph G_k.

Definition 8 yields that the number of granules of the kth connectivity power does not increase, and in many cases decreases with decreasing k. This is due to the fact that new edges are added to the graph G_k in the transition from it to the graph G_{k-i} ($i \geq 1$). For example with the NTA-object $R[XY]$ (Fig. 2), the number of granules of the 1st connectivity power is 5, and the number of granules of zero power is 4.

If a clustering problem presupposes each granule to form a cluster, connected components of a certain graph G_k will have the power equal to the number of clusters. If the criterion for the clustering quality is a given range of possible clusters, the desired solution can be obtained by varying the connectivity power of the granules.

3 Calculation of Proximity Measures Between Granules

Consider the case of sparse structures, when the connections of all orders are taken into account, but the number of granules is large and it is required to unite unconnected close granules into one cluster. To do this, we can determine the distance between pairs of granules and clusterize the system according to known methods of clustering by distance [1, 5].

For each attribute in a granule, we define the minimum and maximum elements. For example, the structure in Fig. 2 contains two granules expressed by C-systems

$$R_1[XY] = \begin{bmatrix} \{1,2,3\} & \{8,9\} \\ \{3,4,5\} & \{7,8\} \end{bmatrix}; \; R_2[XY] = \begin{bmatrix} \{7\} & \{8,9\} \\ \{8,9\} & \{7,8\} \end{bmatrix}.$$

Comparing the corresponding components, we get for $R_1[XY]$ $\min(X) = 1$, $\max(X) = 5$, $\min(Y) = 7$, $\max(Y) = 9$, and for $R_2[XY]$ $\min(X) = 7$, $\max(X) = 9$, $\min(Y) = 7$, $\max(Y) = 9$. Using this data, we can construct a minimal bar for each granule, which includes this granule. Let us call such bar a *granule universe* (GU). For example, GU for the granule $R_1[XY]$ is expressed by the C-n-tuple $U(R_1)[XY] = [\{1, 2, 3, 4, 5\} \; \{7, 8, 9\}]$, and for the $R_2[XY]$ it is the C-n-tuple $U(R_2)[XY] = [\{7, 8, 9\} \; \{7, 8, 9\}]$.

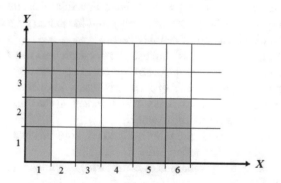

Fig. 3. Granules with intersecting universes

It should be considered that universes of unbound granules can have a non-empty intersection. An example structure is shown in Fig. 3 and expressed by the C-system

$$Q[XY] = \begin{bmatrix} \{1,5,6\} & (1,2) \\ \{3,4\} & \{1\} \\ \{1,2,3\} & \{3,4\} \end{bmatrix}.$$

Granules with intersected or connected universes can be considered connected and interpreted as one granule. Proximity measures between unconnected universes of granules can be defined in many ways. Consider two of them. In the first method, the measure of proximity is the size of the gap between unbound bars; in the second method, coordinates of centers of gravity are calculated for unbound bars, these coordinates determine distances between these centers.

The *value of the gap* is calculated using the distance between the corresponding pairs of components. In this case, only distances greater than 0 are taken into account. Depending on the meaning of the problem, several variants of the distance calculation are possible: the mean square of the distances, the Manhattan distance [6], etc. For our spatial model, the root-mean-square distance is more acceptable since it corresponds to the diagonal of the bar located between the nearest points of the compared bars.

Let us consider an example. For the structure $R[XY]$ (Fig. 2), we choose granules

$$R_1[XY] = \begin{bmatrix} \{1,2,3\} & \{8,9\} \\ \{3,4,5\} & \{7,8\} \end{bmatrix}; R_4[XY] = \begin{bmatrix} \{7\} & \{1\} \\ \{8,9\} & \{2,3,4\} \end{bmatrix}$$

and calculate their universes:

$$U(R_1[XY]) = [\{1, 2, 3, 4, 5\} \{7, 8, 9\}]; U(R_4[XY]) = [\{7, 8, 9\} \{1, 2, 3, 4\}].$$

For attributes X and Y, distances between their components are $D_X = 1$; $D_Y = 2$. Then the rms distance $\sqrt{D_X^2 + D_Y^2} = 2.24$. Note that this measure of proximity can violate axioms of distance (the axiom of identity and the axiom of triangle inequality), with the exception of the axiom of reflexivity.

The density within quanta is assumed to be constant and measures of quanta be known. Then for elementary bars that form granules or their universes, it is easy to calculate *coordinates of their centers of gravity* by using known methods of analytical geometry or theoretical mechanics. Distances between the centers of gravity of the granules comply with all axioms of distance.

If the distances between the granules are known, conventional methods of classification can be used [5–7]. Let us consider an example.

Based on the connectivity relations, the following granules were obtained for the NTA-object $R[XY]$ (1):

$$R_1[XY] = \begin{bmatrix} \{1,2,3\} & \{8,9\} \\ \{3,4,5\} & \{7,8\} \end{bmatrix}; R_2[XY] = \begin{bmatrix} \{7\} & \{8,9\} \\ \{8,9\} & \{7,8\} \end{bmatrix}; R_3[XY] = \begin{bmatrix} \{1,2,3\} & \{1\} \\ \{3,4,5\} & \{2,3,4\} \end{bmatrix};$$
$$R_4[XY] = \begin{bmatrix} \{7\} & \{1\} \\ \{8,9\} & \{2,3,4\} \end{bmatrix}.$$

For these granules, we compute their universes $U(R_i[XY])$ (abbreviated as U_i further on) and construct the table of distances between all pairs of universes assuming that lengths of all quanta are the same and equal to 1 (see Table 1).

Table 1. Distances between universes of granules

U_1–U_2	U_1–U_3	U_1–U_4	U_2–U_3	U_2–U_4	U_3–U_4
1	2	2.24	2.24	2	1

To construct clusters, it is necessary to successively select maximal admissible distances (D_{max}) between granules in Table 1. Then the number of clusters will depend on the chosen distance. For example, if $D_{max} < 1$, the number of clusters will be 4, and if $D_{max} \leq 1$, the number of clusters will be 2.

Similar technologies are used to solve problems of cognitive classification [1–3] and induction within Carnap's inductive logics (for example, [8, 9]), with the only difference that instead of ordinary mathematical spaces the solution is sought in so-called conceptual spaces [1–3].

The main distinction between conceptual spaces and ordinary mathematical ones is that CSs often contains groups of semantically interrelated attributes, each such group defines one "quality dimension" [1–3] of the analyzed objects. An obvious example of a CS is the geometric dimensions and/or volume of an object in three-dimensional Euclidean space that are considered independent on other characteristics of this object (weight, density, etc.).

Attributes that form one CS are reasonable to place next to each other in the general list of attributes of the object. The peculiarity of CSs is that their elements may have no ordering relation. For classification purposes, CSs will require heuristic methods of establishing an order, as in conventional classification systems, or consider each attribute of a CS as independent one and use the distance measures proposed above. Besides, any projection of an NTA-object can include either all the attributes that make up a CS or none of them.

4 Conclusion

Above, we described and analyzed possibilities of applying our earlier developed *n*-tuple algebra to problems where it is possible to establish an ordering relation on the attributes domains of NTA-objects. The concept of granules was introduced; the connectivity power between the granules was determined, and methods to calculate distances between the disconnected granules were proposed. The obtained dependencies make it possible to extend the application scope of classification techniques. Moreover, the developed concept of connectivity power allows to change the number of granules included in one cluster. This way a user can adapt the number of resulting clusters according to her/his notions regarding the task under study.

We plan to conduct our future research in this field towards applying different metrics as well as investigating conceptual spaces of non-determined nature, namely fuzzy, probabilistic, etc.

Acknowledgments. The authors would like to thank the Russian Foundation for Basic Researches (grants 16-29-04424, 16-29-12901, 18-07-00132, 18-01-00076, and 18-29-03022) for partial funding of this work.

References

1. Gärdenfors, P.: Conceptual Spaces: The Geometry of Thought. A Bradford Book. MIT Press, Cambridge (2000)
2. Lakoff, G.: Women, Fire, and Dangerous Things. The University of Chicago Press, Chicago (1987)
3. Rosch, E.: Cognitive representations of semantic categories. J. Exp. Psychol. Gen. **104**, 192–232 (1975)
4. Kulik, B., Fridman, A.: N-ary Relations for Logical Analysis of Data and Knowledge. IGI Global, Hershey (2017)
5. Everitt, B.: Cluster Analysis. Wiley, Chichester (2011)
6. Krause, E.F.: Taxicab Geometry. Dover, NewYork (1987)
7. Hastie, T., Tibshirani, R., Friedman, J.: The Elements of Statistical Learning: Data Mining, Inference, and Prediction, 2nd edn. Springer, Heidelberg (2009)
8. Carnap, R.: The Continuum of Inductive Methods. University of Chicago Press, Chicago (1952)
9. Carnap, R., Jeffrey, R.C. (eds.): Studies in Inductive Logic and Probability, vol. I. University of California Press, Berkeley (1971)

Constructive Proofs of Heterogeneous Equalities in Cubical Type Theory

Maksym Sokhatskyi[✉] and Pavlo Maslianko

Igor Sikorsky Kyiv Polytechnical Institute, Kyiv, Ukraine
5ht@outlook.com, mppdom@i.ua

Abstract. This paper represents the very small part of the developed base library for homotopical prover based on Cubical Type Theory (CTT) announced in 2017. We demonstrate the usage of this library by showing how to build a constructive proof of heterogeneous equality, the simple and elegant formulation of the equality problem, that was impossible to achieve in pure Martin-Löf Type Theory (MLTT). The machinery used in this article unveils the internal aspect of path equalities and isomorphism, used e.g. for proving univalence axiom, that became possible only in CTT. As an example of complex proof that was impossible to construct in earlier theories we took isomorphism between Nat and Fix Maybe datatypes and built a constructive proof of equality between elements of these datatypes. This approach could be extended to any complex isomorphic data types.

Keywords: Formal methods · Type theory · Computer languages
Theoretical computer science · Applied mathematics · Isomorphism
Heterogeneous equality · Cubical Type Theory
Martin-Löf Type Theory

1 Introduction

After formulating Type Theory to model quantifiers using *Pi* and *Sigma* types in 1972 [1] Per Martin-Löf added *Equ* equality types in 1984 [2]. Later *Equ* types were extended to non-trivial structural higher equalities (∞-groupoid model) as was shown by Martin Hofmann and Thomas Streicher in 1996 [3]. However formal constructing of Equ type eliminators was made possible only after introducing Cubical Type Theory in 2017 [4]. CTT extends MLTT with interval $I = [0, 1]$ and its de Morgan algebra: $0, 1, r, -r, min(r, s), max(r, s)$ allowing constructive proofs of earlier models based on groupoid interpretation.

The Problem and Tools Used. In this paper, we want to present the constructive formulation of proof that two values of different types are equal using constructive heterogeneous equality in Cubical Type Checker [4][1]. In the end, we will use path isomorphism for that purposes [5].

---

[1] https://github.com/mortberg/cubicaltt

© Springer Nature Switzerland AG 2019
O. Chertov et al. (Eds.): ICDSIAI 2018, AISC 836, pp. 305–318, 2019.
https://doi.org/10.1007/978-3-319-97885-7_30

Author's Contribution. During the story of comparing two zeros, we will show the minimal set of primitives needed for performing this task in the cubical type checker. Most of them were impossible to derive in pure MLTT. We show these primitives in dependency order while constructing our proof. This article covers different topics in type theory, which is the main motivation to show how powerful the notion of equality is: (1) Complete Formal Specification of MLTT; (2) Contractability and n-Groupoids; (3) Constructive J; (4) Functional Extensionality; (5) Fibers and Equivalence; (6) Isomorphism; (7) Heterogeneous equality.

1.1 Research Formal Description

As a formal description of the research includes all cubical programs as research object, type theory in general and MLTT and CTT in particular as research subject, direct proof construction as logical method and encoded cubical base library and examples as research results.

Research Object. The homotopy type theory base libraries in Agda, Cubical, Lean and Coq. While modern Agda has the cubical mode, Coq lacks of the computational semantics of path primitives while has HoTT library. The real programming language is not enough to develop the software and theorems, the language should be shipped with base library. In this article we unveil the practical implementation of base library for cubical type checkers.

Research Subject. We will analyze the base library through the needs of particular features, like basic theorems, heterogeneous path equalities, univalence, basic HITs like truncations, run-time versions of the list, nat, and stream datatypes. We use Martin-Löf Type Theory as a subject and its extension CTT. The main motivation is to have small and concise base library for cubicaltt type checker than can be used in more complex proofs. As an example of library usage we will show the general sketch of the constructive proofs of heterogeneous equality in CTT, concentrating only on homotopical features of CTT.

Research Results. Research result is presented as source code repository that can be used by cubicaltt[2] language and contains the minimal base library used in this article. These primitives form a valuable part of base library, so this article could be considered as an brief introduction to several modules: **proto_path**, **proto_equiv**, **pi**, **sigma**, **mltt**, **path**, **iso**. But the library has even more modules, that exceed the scope of this article so you may refer to source code repository[3]. Brief list of base library modules is given in Conclusion.

Research Methods. The formal definition of MLTT theory and constructive implementation of its instance that supplied with minimal but comprehensive base library that can be used for verifying homotopical and run-time models. The type theory itself is a modeling language, logical framework and research

---

[2] http://github.com/mortberg/cubicaltt.
[3] http://github.com/groupoid/infinity.

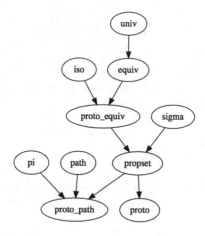

Fig. 1. Base library and its dependencies used in article

method. The MLTT is a particular type theory with *Universes*, Π, Σ and *Equ* types. This is common denominator for a series of provers based on MLTT such as Coq, Agda, Cubicaltt, Idris. In the article we will use Cubical language (Fig. 1).

2 MLTT Type Theory

MLTT is considered as contemporary common math modeling language for different parts of mathematics. Thanks to Vladimir Voevodsky this was extended to Homotopy Theory using MLTT-based Coq proof assistant[4]. Also he formulated the univalence principle $Iso(A, B) = (A = B)$ [5], however constructive proof that isomorphism equals to equality and equivalences is possible only in Cubical Type Theory [4] (Agda and Cubical type checkers).

In this section we will briefly describe the basic notion of MLTT, then will give a formal description of MLTT and the informal primitives of CTT. Only after that will start the proof of heterogeneous equality ended up with proof term. MLTT consist of Π, Σ and *Equ* types living in a Hierarchy of universes $U_i : U_{i+1}$. We will also give an extended equality *HeteroEqu* which is needed for our proof.

2.1 Syntax Notes

Types are the analogues of sets in ZFC, or objects in topos theory, or spaces in analysis. Types contains elements, or points, or inhabitants and it's denoted $a : A$ and there is definitional equality which is usually built into type checker and compare normal forms.

---

[4] http://github.com/UniMath

$$a : A \qquad \text{(terms and types)}$$

$$x = [y : A] \qquad \text{(definitional equality)}$$

MLTT type theory with Pi and $Sigma$ types was formulated using natural deduction inference rules as a language. The inference rules in that language will be translated to cubicaltt in our article.

$$\frac{(A : U_i)\ (B : A \to U_j)}{(x : A) \to B(x) : U_{max(i,j)}} \qquad \text{(natural deduction)}$$

Equivalent definition in cubicaltt (which is inconstend $U : U$ but this won't affect correctness of our proof). Here we consider Π and Pi synonimically identical.

$$Pi\,(A : U)\,(B : A \to U) : U = (x : A) \to B(x) \qquad \text{(cubicaltt)}$$

In article we will use the latter notation, the cubical syntax. The function name in cubical syntax is an inference rule name, everything from name to semicolon is context conditions, and after semicolon is a new construction derived from context conditions. From semicolon to equality sign we have type and after equ sign we have the term of that type. If the types are treated as spaces then terms are points in these spaces.

According to MLTT each type has 4 sorts of inference rules: Formation, Introduction, Eliminators and Computational rules. Formation rules are formal definition of spaces while introduction rules are methods how to create points in these spaces. Introduction rules increase term size, while eliminators reduce term size. Computational rules always formulated as equations that represents reduction rules, or operational semantics.

2.2 Pi Types

Pi types represent spaces of dependent functions. With Pi type we have one lambda constructor and one application eliminator. When B is not dependent on $x : A$ the Pi is just a non-dependent total function $A \to B$. Pi has one lambda function constructor, and its eliminator, the application [1–6].

$$Pi(A, B) = \prod_{x:A} B(x) : U, \quad \lambda x.b : \prod_{x:A} B(x)$$

$$\prod_{f:\prod_{x:A} B(x)} \prod_{a:A} fa : B(a)$$

Here we formulate the math formula of Pi and its eliminators in cubical syntax as Pi. Note that backslash '\' in cubical syntax means λ function from math notation and has compatible lambda signature.

```
Pi (A:U) (B:A->U) : U = (x:A)->B(x)
lambda (A:U) (B:A->U) (a:A) (b:B(a)): A->B(a) = \ (x:A)->b
app (A:U) (B:A->U) (a:A) (f:A->B(a)): B(a) = f(a)
```

2.3 Sigma Types

Sigma types represents a dependent Cartesian products. With sigma type we have pair constructor and two eliminators, its first and second projections. When *B* is not dependent on $x : A$ the *Sigma* is just a non-dependent product $A \times B$. *Sigma* has one pair constructor and two eliminators, its projections [1–6].

$$Sigma(A, B) = \sum_{x:A} B(x) : U, \quad (a, b) : \sum_{x:A} B(x)$$

$$\pi_1 : \prod_{f:\sum_{x:A} B(x)} A, \quad \pi_2 : \prod_{f:\sum_{x:A} B(x)} B(\pi_1(f))$$

As *Pi* and *Sigma* are dual the *Sigma* type could be formulated in terms of *Pi* type using Church encoding, thus *Sigma* is optional. The type systems which contains only *Pi* types called Pure or PTS. Here we rewrite the math formula of *Sigma* and its eliminators in cubical syntax as Sigma:

```
Sigma (A:U) (B:A->U): U = (x:A) * B(x)
pair  (A:U) (B:A->U) (a: A) (b: B(a)): Sigma A B = (a,b)
pr1   (A:U) (B:A->U) (x: Sigma A B): A = x.1
pr2   (A:U) (B:A->U) (x: Sigma A B): B (pr1 A B x) = x.2
```

2.4 Equ Types

For modeling propositional equality later in 1984 was introduced *Equ* type. [2] However unlike *Pi* and *Sigma* the eliminator J of *Equ* type is not derivable in MLTT [3–5].

$$Equ(x, y) = \prod_{x,y:A} x =_A y : U, \quad reflect : \prod_{a:A} u =_A a$$

$$D : \prod_{x,y:A}^{A:U_i} x =_A y \to U_{i+1}, \quad J : \prod_{C:D} \prod_{x:A} C(x, x, reflect(x)) \to \prod_{y:A} \prod_{p:x=_A y} C(x, y, p)$$

Eliminator of Equality has complex form and underivable in MLTT. Here we can see the formulation of *Equ* in cubical syntax as Equ:

```
Equ       (A: U) (x y: A): U = undefined
reflect   (A: U) (a: A): Equ A a a = undefined
D         (A: U)  : U = (x y: A) -> Equ A x y -> U
J         (A: U) (x y: A) (C: D A) (d: C x x (reflect A x))
                 (p: Equ A x y): C x y p = undefined
```

Starting from MLTT until cubicaltt there was no computational semantics for J rules and in Agda and Coq it was formulated using inductive data types wrapper around built-in primitives (J) in the core:

```
data Equality (A:U) (x y:A) = refl_ (_: Equ A x z)
reflection (A:U) (a:A): Equality A a a = refl_ (reflect A a)
```

Heterogeneous equality is needed for computational rule of *Equ* type. And also this is crucial to our main task, constructive comparison of two values of different types. We leave the definition blank until introduce cubical primitives, here is just MLTT signature of HeteroEqu which is underivable in MLTT.

HeteroEqu (A B:U)(a:A)(b:B)(P:Equ U A B):U = undefined

E.g. we can define Setoid specification [7] as not-MLTT basis for equality types. These signatures are also underivable in MLTT.

$$symm: \prod_{a,b:A} \prod_{p:a=_A b} b =_A a, \quad transitivity: \prod_{a,b,c:A} \prod_{p:a=_A b} \prod_{q:b=_A c} a =_A c$$

sym (A:U)(a b:A)(p:Equ A a b): Equ A b a = undefined
transitivity (A:U)(a b c:A)(p: Equ A a b)(q: Equ A b c):
             Equ A a c = undefined

2.5 Complete Formal Specification of MLTT

MLTT needn't and hasn't the underlying logic, the Logical Framework could be constructed directly in MLTT. According to Brouwer-Heyting-Kolmogorov interpretation the propositions are types, Pi is an universal quantifier, Sigma is existential quantifier. Implication is given by Pi over types, conjunction is Cartesian product of tpes and disjunction is disjoint sum of types.

So we can build LF for MLTT inside MLTT. Specification could be formulated as a single Sigma chain holding the computation system and its theorems in one package. Carrying object along with its properties called type refinement, so this type represents a refined MLTT:

```
MLTT (A:U): U
  = ( Pi_Former :    (A->U)->U)
  * ( Pi_Intro :     (B:A->U) (a:A)->B a->(A->B a))
  * ( Pi_Elim :      (B:A->U) (a:A)->(A->B a)->B a)
  * ( Pi_Comp1 :     (B:A->U) (a:A) (f:A->B a) -> Equ (B a)
                     (Pi_Elim B a (Pi_Intro B a (f a)))(f a))
  * ( Pi_Comp2 :     (B: A->U) (a:A) (f:A->B a) ->
                     Equ (A->B a) f (\(x:A)->Pi_Elim B a f))
  * ( Sig_Former :   (A->U)->U)
  * ( Sig_Intro :    (B:A->U) (a:A)->(b:B a)->Sigma A B)
  * ( Sig_Elim1 :    (B:A->U)->(_: Sigma A B)->A)
  * ( Sig_Elim2 :    (B:A->U)->(x: Sigma A B)->B (pr1 A B x))
  * ( Sig_Comp1 :    (B:A->U) (a:A) (b: B a)->Equ A a
                     (Sigma_Elim1 B (Sigma_Intro B a b)))
  * ( Sig_Comp2 :    (B:A->U) (a:A) (b:B a)->Equ (B a) b
                     (Sigma_Elim2 B (a,b)))
  * ( Id_Former :    A->A->U)
  * ( Id_Intro :     (a:A) -> Equ A a a)
  * ( Id_Elim :      (a x: A) (C: predicate A a)
```

```
                    (d:C a(Id_Intro a))(p:Equ A a x)->C x p)
*  (Id_Comp:        (x y:A)(C: D A)(p: Equ A x y)
                            (b: C x x (reflect A x))
                    (X: Equ U (C x x (reflect A x))
                            (C x y p)) ->
                HeteroEqu X b (J A x C b y p))  * Unit
```

Even more complex challenges on Equ type was introduced such as heterogeneous equality *HeteroEqu* needed to formulation of computational rule *Id_Comp* of *Equ* type. Presheaf model of Type Theory, specifically Cubical Sets with interval $[0,1]$ and its algebra was introduced to solve derivability issues. So the instance of MLTT is packed with all the type inference rules along with operational semantics:

```
instance (A: U): MLTT A
    = (Pi A, lambda A, app A, comp1 A, comp2 A,
        Sigma A, pair A, pr1 A, pr2 A, comp3 A, comp4 A,
        Equ A, reflect A, J A, comp5 A, tt)
```

3   Preliminaries

3.1   Cubical Type Theory Primitives and Syntax

The path equality is modeled as an interval $[0,1]$ with its de Morgan algebra 0, 1, r, $\min(r,s)$, $\max(r,s)$. According to underlying theory it has lambdas, application, composition and glueing of $[0,1]$ interval and Min and Max functions over interval arguments. This is enough to formulate and prove path isomorphism and heterogeneous equality.

Heterogeneous Path. The HeteroPath formation rule defines a heterogeneous path between elements of two types A and B for which Path exists $A = B$.

Abstraction Over $[0,1]$**.** Path lambda abstraction is a function which is defined on $[0,1]$: $f : [0,1] \rightarrow A$. Path lambda abstraction is an element of Path type.

Min, Max and Invert. In the body of lambda abstraction besides path application de Morgan operation also could be used: $i \wedge j$, $i \vee j$, i, $-i$ and constants 0 and 1.

Application of Path to Element of $[0,1]$**.** When path lambda is defined, the path term in the body of the function could be reduced using lambda parameter applied to path term.

Path Composition. The composition operation states that being extensible is preserved along paths: if a path is extensible at 0, then it is extensible at 1.

Path Glueing. The path glueing is an operation that allows to build path from equivalences. CTT distinguishes types glueing, value glueing and unglueing.

Here we give LALR specification of BNF notation of Cubicat Syntax as implemented in our Github repository[5]. It has only 5 keywords: **data**, **split**, **where**, **module**, and **import**.

```
def := data id tele = sum + id tele : exp = exp +
       id tele : exp where def
exp := cotele*exp + cotele→exp + exp→exp + (exp) + app + id +
       (exp,exp) + \ cotele→exp + split cobrs + exp.1 + exp.2
  0 := #empty         imp     := [ import id ]
brs := 0 + cobrs      tele    := 0 + cotele
app := exp exp        cotele  := ( exp : exp ) tele
 id := [ #nat ]       sum     := 0 + id tele + id tele | sum
ids := [ id ]         br      := ids → exp
cod := def dec        mod     := module id where imp def
dec := 0 + codec      cobrs   := | br brs
```

3.2   Contractability and Higher Equalities

A type A is contractible, or a singleton, if there is $a : A$, called the center of contraction, such that $a = x$ for all $x : A$: A type A is proposition if any $x, y : A$ are equals. A type is a Set if all equalities in A form a prop. It is defined as recursive definition.

$$isContr = \sum_{a:A}\prod_{x:A} a =_A x, \quad isProp(A) = \prod_{x,y:A} x =_A y, \quad isSet = \prod_{x,y:A} isProp\,(x =_A y),$$

$$isGroupoid = \prod_{x,y:A} isSet\,(x =_A y), \quad PROP = \sum_{X:U} isProp(X), \quad SET = \sum_{X:U} isSet(X), \dots$$

The following types are inhabited: isSet PROP, isGroupoid SET. All these functions are defined in **path** module. As you can see from definition there is a recurrent pattern which we encode in cubical syntax as follows:

```
data N = Z | S (n: N)
n_grpd (A: U) (n: N): U = (a b: A) -> ((rec A a b) n) where
   rec (A: U) (a b: A): (k: N) -> U = split
     Z -> Path A a b
     S n -> n_grpd (Path A a b) n

isContr (A: U): U = (x:A) * ((y: A) -> Equ A x y)
isProp      (A: U): U = n_grpd A Z
isSet       (A: U): U = n_grpd A (S Z)
isGroupoid  (A: U): U = n_grpd A (S (S Z))
PROP       : U = (X:U) * isProp X
SET        : U = (X:U) * isSet  X
GRPOUPOID  : U = (X:U) * isGroupoid X
```

---

[5] http://github.com/groupoid/infinity/.

3.3   Constructive J

The very basic ground of type checker is heterogeneous equality *PathP* and constructive implementation of reflection rule as lambda over interval $[0, 1]$ that return constant value a on all domain.

```
Path (A:U)(a b:A):U = PathP (<i>A) a b
HeteroEqu (A B:U)(a:A)(b:B)(P:Equ U A B):U = Path P a b
refl (A:U)(a:A):Path A a a = <i> a
sym (A:U)(a b:A) (p: Path A a b): Path A b a = <i> p @ -i
transitivity (A: U)(a b c:A)(p: Path A a b) (q: Path A b c):
    Path A a c = comp (<i> Path A a (q @ i)) p []
```

$$trans: \prod_{p:A =_U B} \prod_{a:A} B, \quad singleton: \prod_{x:A} \sum_{y:A} x =_A y$$

$$subst: \prod_{a,b:A} \prod_{p:a=_A b} \prod_{e:B(a)} B(b), \quad congruence: \prod_{f:A \to B} \prod_{a,b:A} \prod_{p:a=_A b} f(a) =_B f(b)$$

Transport transfers the element of type to another by given path equality of the types. Substitution is like transport but for dependent functions values: by given dependent function and path equality of points in the function domain we can replace the value of dependent function in one point with value in the second point. Congruence states that for a given function and for any two points in the function domain, that are connected, we can state that function values in that points are equal.

```
singl (A:U) (a:A): U = (x: A) * Path A a x
trans (A B:U) (p: Path U A B) (a: A): B = comp p a []
congruence (A B: U) (f:A->B) (a b: A)
           (p: Path A a b): Path B (f a) (f b)
         = <i> f (p @ i)

subst (A:U) (P:A->U) (a b: A)
      (p: Path A a b) (e: P a): P b
    = trans (P a) (P b) (congruence A U P a b p) e

contrSingl (A : U) (a b : A) (p : Path A a b):
           Path (singl A a) (a,refl A a) (b,p)
         = <i> (p @ i, <j> p @ i /\ j )
```

Then we can derive J using *contrSingl* and *subst* as defined in HoTT [5]:

```
J (A:U)(x y:A)(C: D A)(d:C x x (refl A x))
  (p:Path A x y): C x y p =
  subst (singl A x) T (x,refl A x)
        (y,p) (contrSingl A x y p) d where
    T (z:singl A x):U = C (z.1) (z.2)
```

These function are defined in **proto_path** module, and all of them except singleton definition are underivable in MLTT.

3.4   Functional Extensionality

Function extensionality is another example of underivable theorems in MLTT, it states if two functions with the same type and they always equals for any point from domain, we can prove that these function are equal. $funExt$ as functional property is placed in **pi** module.

$$funExt: \prod_{[f,g:(x:A)\to B(x)]}^{A:U,B:A\to U} \prod_{[x:A,p:A\to f(x)=_{B(x)}g(x)]} f =_{A\to B(x)} g$$

```
funExt (A: U) (B: A->U)
       (f g: (x:A)->B(x))
       (p: (x:A)->Path (B x) (f x) (g x)):
       Path ((y:A)->B y) f g=<i>\(a:A)->(p a)@i
```

3.5   Fibers and Equivalence

The fiber of a map $f : A \to B$ over a point $y : B$ is family over x of Sigma pair containing the point x and proof that $f(x) =_B y$.

$$fiber: \prod_{f:A->B}^{A,B:U} \prod_{x:A,y:B} \sum f(x) =_B y, \quad isEquiv: \prod_{f:A->B}^{A,B:U} \prod_{y:B} isContr(fiber(f,y))$$

$$equiv: \sum_{f:A->B}^{A,B:U} isEquiv(f) \quad pathToEquiv: \prod_{p:X=_U Y}^{X,Y:U} equiv_U(X,Y)$$

.

Contractability of fibers called is Equiv predicate. The Sigma pair of a function and that predicate called equivalence, or equiv. Now we can prove that singletons are contractible and write a conversion function $X =_U Y \to equiv(X,Y)$.

```
fiber    (A B:U)(f:A->B)(y:B):U = (x:A) * Path B y (f x)
isEquiv  (A B:U)(f:A->B):U = (y:B) -> isContr (fiber A B f y)
equiv    (A B:U):U = (f:A->B) * isEquiv A B f

singletonIsContractible (A:U) (a:A): isContr (singl A a)
   = ((a,refl A a), \ (z:(x:A) * Path A a x) ->
   contrSingl A a z.1 z.2)

pathToEquiv (A X: U) (p: Path U X A): equiv X A
   = subst U (equiv A) A X p (idEquiv A)
```

equiv type is compatible with cubicaltt type checker and it instance can be passed as parameter for Glue operation. So all *equiv* functions and properties is placed in separate **equiv** module.

3.6 Isomorphism

The general idea to build path between values of different type is first to build isomorphism between types, defined as decode and encode functions (f and g), such that $f \circ g = id_A, g \circ f = id_B$.

$$Iso(A, B) = \sum_{[f:A \to B]} \sum_{[g:B \to A]} \left(\prod_{x:A} [g(f(x)) =_A x] \times \prod_{y:B} [f(g(y) =_B y] \right)$$

$$isoToEquiv(A, B) : Iso(A, B) \to Equiv(A, B)$$

$$isoToPath(A, B) : Iso(A, B) \to A =_U B$$

lemIso proof is a bit longread, you may refer to Github repository[6]. The by proof of *isoToEquiv* using *lemIso* we define *isoToPath* as Glue of A and B types, providing $equiv(A, B)$. Glue operation first appear in proving transport values of different type across their path equalities which are being constructed using encode and decode functions that represent isomorphism. Also Glue operation appears in constructive implementation of Univalence axiom [4].

```
lemIso   (A B:U)  (f:  A->B)  (g:B->A)
         (s:  (y:B)  ->  Path B  (f(g(y)))y)
         (t:  (x:A)  ->  Path A  (g(f(x)))x)  (y:B)  (x0  x1:A)
         (p0:  Path B  y  (f(x0)))  (p1:  Path B  y  (f(x1))):
         Path  (fiber  A B  f  y)  (x0,p0)  (x1,p1)  =  undefined

isoToEquiv  (A B:  U)  (f:A->B)  (g:B->A)
         (s:  (y:B)  ->  Path B  (f(g(y)))  y)
         (t:  (x:A)  ->  Path A  (g(f(x)))  x):  isEquiv A B f =
\(y:B)  ->  ((g  y,<i>s  y@-i),\  (z:fiber  A B  f  y)  ->
   lemIso  A B  f  g  s  t  y  (g  y)  z.1  (<i>s  y@-i)  z.2)

isoToPath  (A B:U)  (f:A->B)(g:B->A)
         (s:  (y:B)  ->  Path B  (f(g(y)))y)
         (t:  (x:A)  ->  Path A  (g(f(x)))x):  Path U A B =
         <i>  Glue  B  [(i=0)->(A,f,isoToEquiv A B f g s t),
                    (i=1)->(B,idfun B,idIsEquiv B)  ]
```

Isomorphism definitions are placed in three different modules due to dependency optimization: **iso, iso_pi, iso_sigma**. Latter two contains main theorems about paths in Pi and Sigma spaces.

4 The Formal Specification of the Problem

4.1 Class of Theorems. Constructive Proofs of Heterogeneous Equalities

Speaking of core features of CTT that were unavailable in MLTT is a notion of heterogeneous equality that was several attempts to construct heterogeneous

---

[6] http://github.com/groupoid/infinity/tree/master/priv/iso.ctt.

equalities: such as John-Major Equality[7] by Connor McBride (which is included in Coq base library). As an example of library usage we will show the general sketch of the constructive proofs of heterogeneous equality in CTT, concentrating only on homotopical features of CTT.

Let us have two types A and B. And we have some theorems proved for A and functions $f : A \to B$ and $g : B \to A$ such that $f \circ g = id_A$ and $g \circ f = id_B$. Then we can prove $Iso(A, B) \to Equ(A, B)$. The result values would be proof that elements of A and B are equal—$HeteroEqu$. We will go from the primitives to the final proof. As an example we took Nat and Fix Maybe datatype and will prove $Iso(Nat, Fix(Maybe))$. And then we prove the $HeteroEqu(Nat, Fix(Maybe))$.

4.2   Problem Example. Nat = Fix Maybe

Now we can prove $Iso(Nat, Fix(Maybe))$ and $Nat =_U Fix(Maybe)$. First we need to introduce datatypes $Nat, Fix, Maybe$ and then write encode and decode functions to build up an isomorphism. Then by using conversion from Iso to $Path$ we get the heterogeneous equality of values in Nat and $Fix(Maybe)$. We can build transport between any isomorphic data types by providing ecode and decode functions.

```
data fix (F:U->U) = Fix (point: F (fix F))
data nat          = zero   | suc (n: nat)
data maybe (A:U)  = nothing | just (a: A)

natToMaybe: nat -> fix maybe = split
     zero -> Fix nothing
     suc n -> Fix (just (natToMaybe n))

maybeToNat: fix maybe -> nat = split
     Fix m -> split nothing -> zero
                    just f -> suc (maybeToNat f)

natMaybeIso: (a: nat) ->
     Path nat (maybeToNat (natToMaybe a)) a = split
          zero -> <i> zero
          suc n -> <i> suc (natMaybeIso n @ i)

maybeNatIso: (a : fix maybe) ->
     Path (fix maybe) (natToMaybe (maybeToNat a)) a = split
          Fix m -> split nothing -> <i> Fix nothing
                         just f -> <i> Fix (just (maybeNatIso f @ i))

maybenat: Path U (fix maybe) nat
     = isoToPath (fix maybe) nat
               maybeToNat natToMaybe
               natMaybeIso maybeNatIso
```

The result term of equality between two zeros of Nat and Fix Maybe is given by isomorphism.

---

[7] https://homotopytypetheory.org/2012/11/21/on-heterogeneous-equality/.

> HeteroEqu (fix maybe) nat (Fix nothing) zero maybenat

EVAL: PathP (<!0> Glue nat [(!0 = 0) -> (fix (\(A : U) ->
maybe), (maybeToNat,(\(y : B) -> ((g y,<i> (s y) @ -i),
\(z : fiber A B f y) -> lemIso A B f g s t y (g y) z.1
(<i> (s y) @ -i) z.2)) (A = (fix (\(A : U) -> maybe)),
B = nat, f = maybeToNat, g = natToMaybe, s = natMaybeIso,
t = maybeNatIso))), (!0 = 1) -> (nat,(((\(a : A) -> a)
(A = nat),(\(a : A) -> ((a,refl A a),\(z : fiber A A
(idfun A) a) -> contrSingl A a z.1 z.2)) (A = nat)))])
(Fix nothing) zero

We admit that normalized (expanded) term has the size of one printed page.
Inside it contains the encode and decode functions and identity proofs about their
composition. So we can reconstruct everything up to homotopical primitives or
replace the isomorphic encoding with arbitrary code.

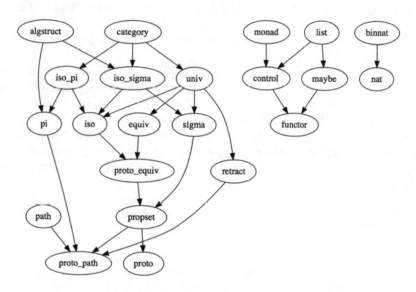

Fig. 2. Full base library dependencies

5   Conclusion

At the moment only two provers that support CTT exists, this is Agda [6] and
Cubical [4]. We developed a base library for cubical type checkers and described
the approach of how to deal with heterogeneous equalities by the example of
proving $Nat =_U Fix(Maybe)$.

Homotopical core in the prover is essential for proving math theorems in
geometry, topology, category theory, homotopy theory. But it also useful for
proving with fewer efforts even simple theorems like commutativity of Nat. By

pattern matching on the edges to can specify continuous (homotopical) transformations of types and values across paths.

We propose a general-purpose base library for modeling math systems using univalent foundations and cubical type checker.

MLTT Foundation: the set of modules with axioms and theorems for Pi, Sigma and Path types, the basis of MLTT theory. Among them: pi, sigma, proto, proto_equiv, proto_path, path, function, mltt. **Univalence Foundation**: the basic theorems about isomorphisms of MLTT types. Useful for proving transport between types, include following modules: iso, iso_pi, iso_sigma, trunc, equiv, univ. **Category Theory**: the model of Category Theory following [5] definitions. It includes: cat, pushout, fun, grothendieck. **Runtime Types**: the models for run-time systems, properties of which could be proved using univalend foundations: binnat, bool, control, either, list, maybe, stream, nat, recursion, cwf, lambek. **Set Theory**: The basic theorems about set theory and higher groupoids: hedberg, girard, prop, set. **Geometry**: Higher Inductive Types: circle, helix, quotient, retract, etc. **Algebra**: Abstract algebra, such as Monoid, Gropop, Semogroupo, Monad, etc: algebra, control.

The amount of code needed for $Nat =_U Fix(Maybe)$ proof is around 400 LOC in modules (Fig. 2).

The further development of base library implies: (1) extending run-time facilities; (2) making it useful for building long-running systems and processes; (3) implement the inductive-recursive model for inductive types (development of lambek module). The main aim is to bring homotopical univalent foundations for run-time systems and models. Our base library could be used as a first-class mathematical modeling tool or as a sandbox for developing run-time systems and proving its properties, followed with code extraction to pure type systems and/or run-time interpreters.

References

1. Martin-Löf, P., Sambin, G.: The theory of types. Studies in proof theory (1972)
2. Martin-Löf, P., Sambin, G.: Intuitionistic type theory. Studies in proof theory (1984)
3. Hofmann, M., Streicher, T.: The groupoid interpretation of type theory. In: In Venice Festschrift, pp. 83–111. Oxford University Press (1996)
4. Cohen, C., Coquand, T., Huber, S., Mörtberg, A.: Cubical type theory: a constructive interpretation of the univalence axiom (2017)
5. The Univalent Foundations Program: Homotopy Type Theory: Univalent Foundations of Mathematics. Institute for Advanced Study (2013)
6. Bove, A., Dybjer, P., Norell, U.: A brief overview of agda—a functional language with dependent types. In: Proceedings of the 22-nd International Conference on Theorem Proving in Higher Order Logics, pp. 73–78. Springer-Verlag (2009)
7. Bishop, E.: Foundations of Constructive Analysis. McGraw-Hill Series in Higher Mathematics. McGraw-Hill, New York City (1967)

Improved Method of Determining the Alternative Set of Numbers in Residue Number System

Victor Krasnobayev[1]([⊠]), Alexandr Kuznetsov[1], Sergey Koshman[1], and Sergey Moroz[2]

[1] V. N. Karazin Kharkiv National University, Kharkiv, Ukraine
v.a.krasnobaev@gmail.com, kuznetsov@karazin.ua,
s_koshman@ukr.net
[2] Kharkiv Petro Vasylenko National Technical University of Agriculture,
Kharkiv, Ukraine
frost9i@ukr.net

Abstract. The article analyzes the most well-known practical methods of determining the alternative set (AS) of numbers in a residue numeral system (RNS). The AS determining is most frequently required to perform error verification, diagnosing and correction of data in RNS, that was introduced to a minimal information redundancy in the computational process dynamics. This suggests the occurrence of only a single error in a number. The main downside of the reviewed methods is a significant time needed to determine the AS. In order to reduce time for AS determining in RNS, one of the known methods has been improved in the article. The idea of method improvement supposes preliminary correspondence table compilation (first stage tables) for each correct number out of informational numeric range of a possible set of incorrect numbers, that are not included into the range. Based on the analysis of tables content, the second stage table is being compiled, which contains the correspondence of each incorrect number out of numeric range to a possible values of correct numbers. By applying introduced method, efficiency of data verification, diagnosing and correction is increased due to time reduction of the AS numbers determining in RNS.

Keywords: Machine arithmetic · Residue numeral system · Alternative set
Data error diagnosing and correction

1 Introduction

The modern life stage of science and technology development involves perplexing tasks, which require prompt resolving. However, complexity of these tasks outstrips the pace of computational power ascension of universal computers. Therefore, principal directions in computational system enhancement of real-time data processing are its performance maximization. It's known that one of the possible directions of high-performance computational systems development is to parallelize tasks and algorithms on the level of arithmetical microoperations. One of the ways of parallel task solving is

© Springer Nature Switzerland AG 2019
O. Chertov et al. (Eds.): ICDSIAI 2018, AISC 836, pp. 319–328, 2019.
https://doi.org/10.1007/978-3-319-97885-7_31

to make a transition to the nonconventional machine arithmetic with alternative operands representation. If to choose out of numerous nonconventional arithmetics, the most practical use in computational systems belongs to the non-positional numeral system in residue classes called residue number system (RNS) [1–3].

The set of RNS beneficial properties defines the following classes of tasks, where it's significantly more efficient than a positional arithmetic: cryptographic and modular transformations (implementation of cryptographic transformations in a group of elliptic curve dots, as well as algorithm realization for hashing and random numbers generations), signal processing, image processing (compression), real-time large number of bits (hundreds of bits) integer data processing, vector and matrix informational array processing in bulk, neural computer information processing, FFT and DFT algorithm realization and optoelectronic table information processing [5–14].

In fact, its required to determine the alternative set (AS) in RNS $W(\tilde{A}) = \left\{ m_{l_1}, m_{l_2}, \ldots, m_{l_p} \right\}$ of the incorrect $\tilde{A} = (a_1, a_2, \ldots, a_{i-1}, a_i, a_{i+1}, \ldots, a_n)$ numbers. By the AS $W(\tilde{A}) = \left\{ m_{l_1}, m_{l_2}, \ldots, m_{l_p} \right\}$ of the incorrect (corrupted) number $\tilde{A} = (a_1, a_2, \ldots, a_{i-1}, a_i, a_{i+1}, \ldots, a_n)$, we understand a set $\left\{ m_{l_k} \right\}$ $(k = \overline{1, \rho})$ of ρ RNS bases, upon which the correct (not corrupted) number (codeword) $A = (a_1, a_2, \ldots, a_{i-1}, a_i, a_{i+1}, \ldots, a_n)$ may differ from the given set $\left\{ \tilde{A} \right\}$ of possible derived incorrect numbers. This also suggests that only a single (by one of the residues $m_i (i = \overline{1, n+1})$ of the number $A = (a_1, a_2, \ldots, a_{i-1}, a_i, a_{i+1}, \ldots, a_n)$ error (corruption of one of $(n + 1)$ residues) may occur in the correct number A.

Noting that AS is considered when minimal information redundancy is introduced to RNS, we apply a single $(k = 1)$ additional (check) RNS base m_{n+1} to n informational bases, given that $m_i < m_{n+1}$ $(i = \overline{1, n})$. In this case, the total amount N_{TA} of RNS codewords is $N_{TA} = \prod_{i=1}^{n+1} m_i$. The number of N_{CC} correct codewords is $N_{CC} = \prod_{i=1}^{n} m_i$, whereas the number of N_{IC} incorrect (corrupted) codewords is equal to $N_{IC} = N_{TC} - N_{CC} = N_{CC} \cdot (m_{n+1} - 1)$.

Importance of the AS determining can occur in the following principal cases. Firstly, if the error verification, diagnosing and correction of data in RNS is required. Secondly, when performing error verification, diagnose and correction of data in RNS in a process of task solving in computational process dynamics (CPD) (in real-time, i.e. without pausing computations) given minimum information redundancy [1]. One of the main requirements of the AS determining procedure in RNS is the requirement to decrease time for a given bases set determination. This requirement is especially crucial for the second case, in a process of task solving in CPD.

As can be seen, an important task is to develop new and improve existing methods of prompt AS numbers $W(\tilde{A}) = \left\{ m_{l_1}, m_{l_2}, \ldots, m_{l_p} \right\}$ determining in RNS.

2   The Main Part

All existing methods of AS numbers determining are based on the sequential determining procedure of the sought bases of AS numbers in RNS [1].

Let's review the first possible method of determining the AS $W(\tilde{A}) = \{m_{l_1}, m_{l_2}, \ldots, m_{l_p}\}$ of the incorrect number $\tilde{A} = (a_1, a_2, \ldots, a_{i-1}, a_i, a_{i+1}, \ldots, a_n)$ in RNS. This method involves AS determining by sequential checking of each RNS base m_i ($i = \overline{1, n}$). I.e. the value of the number $\tilde{A}$ is being determined by the sequential substitution of possible residues alternately by each base. Either this set contains no correct numbers at all, or a single correct number might be present. In the latter case, the obtained number is included into the AS of the controlled incorrect number $\tilde{A}$. The considered method suggests sequential verifications for each of the informational RNS bases (check bases are always contained in the AS bases). The result of such sequential verifications fully determines the AS $W(\tilde{A}) = \{m_{l_1}, m_{l_2}, \ldots, m_{l_p}\}$. The downsides of this method of the AS determining are high computational complexity and significant time consumption.

The second method for the AS determining is based on the subtraction of each of the possible projections $\tilde{A} = (a_1, a_2, \ldots, a_{i-1}, a_{i+1}, \ldots, a_n)$ of the incorrect number $\tilde{A}$, and their subsequent comparison to the value of the RNS defined informational range. As proved in [1], the necessary and sufficient condition for RNS base appearance in the AS $W(\tilde{A}) = \{m_{l_1}, m_{l_2}, \ldots, m_{l_p}\}$ of the number $\tilde{A} = (a_1, a_2, \ldots, a_{i-1}, a_l, a_{i+1}, \ldots, a_n)$ is that its projections $\tilde{A}_i$ are correct. Let's take a look at the AS numbers determining method in RNS based on this second method.

Assume that there's a need to determine the AS of the number $A_{RNS} = (0 \| 0 \| 0 \| 0 \| 5)$, that is defined in RNS by the informational bases, $m_1 = 3, m_2 = 4, m_3 - 5, m_4 = 7$ and control bases $m_k = m_5 = 11$ (k is a number of the check base, the check base is singular and is equal to $m_k = m_5 = 11$). In this case, $M = \prod_{i=1}^{n} m_i = \prod_{i=1}^{4} m_i = 420$ and $M_0 = M \cdot m_{n+1} = 420 \cdot 11 = 4620$. The orthogonal bases $B_i (i = \overline{1, n+1})$ for a given RNS are shown in Table 1.

Table 1. Orthogonal RNS bases

$B_1 = (1 \| 0 \| 0 \| 0 \| 0) = 1540, \bar{m}_1 = 1$
$B_2 = (0 \| 1 \| 0 \| 0 \| 0) = 3465, \bar{m}_2 = 3$
$B_3 = (0 \| 0 \| 1 \| 0 \| 0) = 3696, \bar{m}_3 = 4$
$B_4 = (0 \| 0 \| 0 \| 1 \| 0) = 2640, \bar{m}_4 = 4$
$B_5 = (0 \| 0 \| 0 \| 0 \| 1) = 2520, \bar{m}_5 = 6$

Data verification of $A_{RNS} = (0, 0, 0, 0, 5)$ is done preliminarily. According to the verification procedure [1, 4], the value of the original number in the decimal positional numeral system (PNS) will be determined:

$$A_{PNS} = \left(\sum_{i=1}^{n+1} a_i \cdot B_i \right) \bmod M_0 = 3360 > 420.$$

In this case, during verification, it was determined that $A_{PNS} = 3360 > M = 420$. Thus, only a single error occurrence is possible, and it's concluded that the considered number $\tilde{A}_{3360} = (0\,\|\,0\,\|\,0\,\|\,0\,\|\,5)$ is incorrect. Then the AS determining procedure of the number $\tilde{A}_{3360} = (0\,\|\,0\,\|\,0\,\|\,0\,\|\,5)$ is done. According to the second method of the AS determining, possible projections $\tilde{A}_j$ of the number $\tilde{A}_{3360} = (0\,\|\,0\,\|\,0\,\|\,0\,\|\,5)$ are: $\tilde{A}_1 = (0\,\|\,0\,\|\,0\,\|\,5)$, $\tilde{A}_2 = (0\,\|\,0\,\|\,0\,\|\,5)$, $\tilde{A}_3 = (0\,\|\,0\,\|\,0\,\|\,5)$, $\tilde{A}_4 = (0\,\|\,0\,\|\,0\,\|\,5)$ and $\tilde{A}_5 = (0\,\|\,0\,\|\,0\,\|\,0)$.

Values $\tilde{A}_{j\,PNS}$ of PNS projections are calculated using the following formula [1]:

$$\tilde{A}_{j\,PNS} = \left(\sum_{\substack{i=1; \\ j=\overline{1,\,n+1.}}}^{n} a_i \cdot B_{ij}\right) \bmod M_j = (a_1 \cdot B_{1j} + a_2 \cdot B_{2j} + \ldots + a_n \cdot B_{nj}) \bmod M_j. \quad (1)$$

According to the (1), let's calculate each of the values of $A_{j\,PNS}$ by comparing $(n+1)$ number $\tilde{A}_{j\,PNS}$ and the number $M = M_0/m_{n+1}$. If there are any numbers not included in the informational numeric range $[0,\,M)$ among the projections $\tilde{A}_{i\,PNS}$ (i.e. $\tilde{A}_{k\,PNS} \geq M$) that contains k correct numbers, then it's concluded that these k residues of the number $\tilde{A}_{RNS}$ are not corrupted. Incorrect residues can be found only among the rest $[(n+1) - k]$ of the number $\tilde{A}_{RNS}$ residues. The set of calculated quotient active bases and quotient B_{ij} orthogonal bases for a given RNS are shown in Tables 2 and 3, respectively. In this case,

$$\tilde{A}_{1PNS} = \left(\sum_{i=1}^{4} a_i \cdot B_{i1}\right) \bmod M_1 = 280 < 420.$$

It can be concluded that $\bar{a}_1$ is possibly a corrupted residue.

$$\tilde{A}_{2PNS} = \left(\sum_{i=1}^{4} a_i \cdot B_{i2}\right) \bmod M_2 = 1050 > 420.$$

In this case we get that a_2 is certainly a corrupted residue.

$$\tilde{A}_{3PNS} = \left(\sum_{i=1}^{4} a_i \cdot B_{i3}\right) \bmod M_3 = 588 > 420.$$

We get that a_3 is certainly a non-corrupted residue.

$$\tilde{A}_{4PNS} = \left(\sum_{i=1}^{4} a_i \cdot B_{i4}\right) \bmod M_4 = 60 < 420.$$

Table 2. The set of bases for a quotient orthogonal bases determination

i j	m_1	m_2	m_3	m_4	M_j
1	4	5	7	11	1540
2	3	5	7	11	1155
3	3	4	7	11	924
4	3	4	5	11	660
5	3	4	5	7	420

Table 3. Quotient orthogonal RNS bases

B_{ij} i j	1	2	3	4
1	385	616	1100	980
2	385	231	330	210
3	616	693	792	672
4	220	165	396	540
5	280	105	336	120

Conclusion: $\bar{a}_4$ is possibly a corrupted residue; $\tilde{A}_{5PNS} = \left(\sum_{i=1}^{4} a_i \cdot B_{i5} \right) \bmod M_5$. As far as $M_5 = M = 420$, the residue $\bar{a}_5$ modulo $m_k = m_5$ will be always present in the set of possibly corrupted residues of the RNS number.

In this case, for the number $\tilde{A}_{RNS} = (0, 0, 0, 0, 5)$, a certainly non-corrupted residues were determined. They are $a_2 = 0$ and $a_3 = 0$. The residues based m_1, m_4, and m_5 might be corrupted, i.e. the residues $a_1 = 0, a_4 = 0$, and $a_5 = 5$. In this case, for the number $\tilde{A}_{RNS} = (0, 0, 0, 0, 5)$, its AS is equal to the set of the RNS residues $W(\tilde{A}) = \{1, 4, 5\}$. The second method application slightly improves the AS determining process $W(\tilde{A}) = \{m_{l_1}, m_{l_2}, \ldots, m_{l_p}\}$ of the number $A = (a_1, a_2, \ldots, a_{i-1}, a_i, a_{i+1}, \ldots, a_n)$, because of the ability of simultaneous determining of the possible projections $\tilde{A}_j$ of the incorrect number. This circumstance decreases the time factor difficultness of the AS determining. However, the AS number determining procedure contains the following basic operations: number $A = (a_1, a_2, \ldots, a_{i-1}, a_i, a_{i+1}, \ldots, a_n)$ conversion from RNS to PNS; conversion of projections $\tilde{A}_i$ of the incorrect number $A = (a_1, a_2, \ldots, a_{i-1}, a_i, u_{i+1}, \ldots, a_n)$ from RNS to PNS and comparison procedure of numbers. Listed operations in RNS are referred to as non-positional operations, which require significant time and hardware investments for implementation. The downsides of the mentioned method of the AS determining are the same as in the first method, namely, significant computational complexity and significant time consumption. Therefore, there remains the task to improve the second method in terms of reducing time of the AS determining. The mission of the introduced method in this article of the AS number determining in RNS is to compile preliminary corresponding M tables (first stage tables) $A = \Phi_1(\tilde{A})$ for each correct $A = (a_1, a_2, \ldots, a_{i-1}, a_i, a_{i+1}, \ldots, a_n)$ number (contained in the numeric range $0 \div M - 1$), of the possible set $\{\tilde{A}\}$ of incorrect numbers (contained in the numeric range $M \div M_0 - 1$) with a single error (in one of the residue) occurring in the number A. Based on the analysis of the first stage tables, the second stage table is compiled, which contains the correspondence $\tilde{A} = \Phi_2(A)$ of each incorrect $\tilde{A}$ number out of the numeric range $M \div M_0 - 1$ to possible values of the corrected (non-corrupted) $A = (a_1, a_2, \ldots, a_{i-1}, a_i, a_{i+1}, \ldots, a_n)$ numbers. The number of correct A numbers corresponds to the number of the RNS bases contained in the AS $W(\tilde{A}) = \{m_{l_1}, m_{l_2}, \ldots, m_{l_p}\}$ of the A number.

Table 4. Possible PNS and RNS number variants for a given set of bases

A in PNS	m_1	m_2	m_3	A in PNS	m_1	m_2	m_3
0	0	0	0	15	1	0	0
1	1	1	1	16	0	1	1
2	0	2	2	17	1	2	2
3	1	0	3	18	0	0	3
4	0	1	4	19	1	1	4
5	1	2	0	20	0	2	0
6	0	0	1	21	1	0	1
7	1	1	2	22	0	1	2
8	0	2	3	23	1	2	3
9	1	0	4	24	0	0	4
10	0	1	0	25	1	1	0
11	1	2	1	26	0	2	1
12	0	0	2	27	1	0	2
13	1	1	3	28	0	1	3
14	0	2	4	29	1	2	4

Table 5. Tables for mapping the correct number to a possible totality of incorrect numbers

0	0	0	0
15	1	0	0
10	0	1	0
20	0	2	0
6	0	0	1
12	0	0	2
18	0	0	3
24	0	0	4

Table 7. Tables for mapping the correct number to a possible totality of incorrect numbers

2	0	2	2
17	1	2	2
12	0	0	2
22	0	1	2
20	0	2	0
26	0	2	1
8	0	2	3
14	0	2	4

Table 9. Tables for mapping the correct number to a possible totality of incorrect numbers

4	0	1	4
19	1	1	4
24	0	0	4
14	0	2	4
10	0	1	0
16	0	1	1
22	0	1	2
28	0	1	3

Table 6. Tables for mapping the correct number to a possible totality of incorrect numbers

1	1	1	1
16	0	1	1
21	1	0	1
11	1	2	1
25	1	1	0
7	1	1	2
13	1	1	3
19	1	1	4

Table 8. Tables for mapping the correct number to a possible totality of incorrect numbers

3	1	0	3
18	0	0	3
13	1	1	3
23	1	2	3
15	1	0	0
21	1	0	1
27	1	0	2
9	1	0	4

Table 10. Tables for mapping the correct number to a possible totality of incorrect numbers

5	1	2	0
20	0	2	0
15	1	0	0
25	1	1	0
11	1	2	1
17	1	2	2
23	1	2	3
29	1	2	4

It's appropriate to consider the application of the introduced method of the AS determining for a particular RNS that is defined by the informational bases $m_1 = 2, m_2 = 3$, and check bases $m_k = m_3 = m_{n+1} = 5$ ($M = 2 \cdot 3 = 6; M_0 = 30$). The set of the codewords in a positional (decimal) numeral system, as well as in RNS, is shown in Table 4. Assuming Table 4 content, by each of the correct codewords 0–5, the first stage $A = \Phi_1(\tilde{A})$ corresponding tables are compiled (Tables 5, 6, 7, 8, 9 and 10). According to these tables, the second stage $\tilde{A} = \Phi_2(A)$ (Table 11) is compiled. Table 8 shows the correspondence $\tilde{A} = \Phi_2(A)$ of each incorrect $\tilde{A}$ number out of numeric range 6–29 to possible values of corrected (non-corrupted) A numbers. Table 11 shows the algorithm for the AS $W(\tilde{A}) = \{m_{l_1}, m_{l_2}, \ldots, m_{l_p}\}$ numbers determining in RNS.

Let's review the example of the AS number determining in RNS using the introduced table method. Let's be given the incorrect number $\tilde{A}_{15} = (1 \parallel 0 \parallel 0)$ (Table 4), and the task is to determine this number's AS. Initially, six correspondence tables are compiled (Tables 5, 6, 7, 8, 9 and 10) of the first stage for each correct $A = (a_1, a_2, \ldots, a_{i-1}, a_i, a_{i+1}, \ldots, a_n)$ number (out of numeric range 0–5) for the possible set of incorrect numbers (out of numeric range 6–29) when single errors (in one of the residues) occur in the number A (Table 4). Based on the analysis of the first stage table content, the second stage table is compiled, which contains the correspondence for each incorrect number out of numeric range 6–29 to the possible corrected (non-corrupted) values $A = (a_1, a_2, \ldots, a_{i-1}, a_i, a_{i+1}, \ldots, a_n)$ of numbers. The number of correct A numbers corresponds to the number of RNS bases contained in the AS $W(\tilde{A}) = \{m_{l_1}, m_{l_2}, \ldots, m_{l_p}\}$ of the A number. Influenced by single errors occurrence, the incorrect number $\tilde{A}_{15} = (1 \parallel 0 \parallel 0)$ can be formed by the following correct A numbers.

Firstly, the correct number $A_0 = (0 \parallel 0 \parallel 0)$ (Table 5) might be corrupted in its first residue $a_1 = 0$ ($\tilde{a}_1 = 1$). Secondly, the correct number $A_3 = (1 \parallel 0 \parallel 3)$ (Table 8) might be corrupted in its third residue $a_3 - 3$ ($\tilde{a}_3 = 0$). And, lastly, the incorrect number $A_5 = (1 \parallel 2 \parallel 0)$ (Table 10) might be corrupted in its second residue $a_2 = 2$ ($\tilde{a}_2 = 0$). In this case, the AS $W(\tilde{A}) = \{m_{l_1}, m_{l_2}, m_{l_p}\}$ of the incorrect number $\tilde{A}_{15} = (1 \parallel 0 \parallel 0)$ is equal to the value $W(\tilde{A}_{15}) = \{m_1, m_2, m_3\}$ (Table 11).

Table 11. Table to determine the alternative totality

Incorrect $\tilde{A}$ number	Correct A number	AS value $W(\tilde{A})$
$\tilde{A}_6 = (0 \parallel 0 \parallel 1)$	$A_0 = (0 \parallel 0 \parallel 0)$	$\{m_3\}$
$\tilde{A}_7 = (1 \parallel 1 \parallel 2)$	$A_1 = (1 \parallel 1 \parallel 1)$	$\{m_3\}$
$\tilde{A}_8 = (0 \parallel 2 \parallel 3)$	$A_2 = (0 \parallel 2 \parallel 3)$	$\{m_3\}$
$\tilde{A}_9 = (1 \parallel 0 \parallel 4)$	$A_3 = (1 \parallel 0 \parallel 3)$	$\{m_3\}$
$\tilde{A}_{10} = (0 \parallel 1 \parallel 0)$	$A_0 = (0 \parallel 0 \parallel 0)$	$\{m_2, m_3\}$
	$A_4 = (0 \parallel 1 \parallel 4)$	

(*continued*)

(continued)

Incorrect $\tilde{A}$ number	Correct A number	AS value $W(\tilde{A})$
$\tilde{A}_{11} = (1 \parallel 2 \parallel 1)$	$A_1 = (1 \parallel 1 \parallel 1)$	$\{m_2, m_3\}$
	$A_5 = (1 \parallel 2 \parallel 0)$	
$\tilde{A}_{12} = (0 \parallel 0 \parallel 2)$	$A_0 = (0 \parallel 0 \parallel 0)$	$\{m_2, m_3\}$
	$A_2 = (0 \parallel 2 \parallel 2)$	
$\tilde{A}_{13} = (1 \parallel 1 \parallel 3)$	$A_1 = (1 \parallel 1 \parallel 1)$	$\{m_2, m_3\}$
	$A_3 = (1 \parallel 0 \parallel 3)$	
$\tilde{A}_{14} = (0 \parallel 2 \parallel 4)$	$A_2 = (0 \parallel 2 \parallel 2)$	$\{m_2, m_3\}$
	$A_4 = (0 \parallel 1 \parallel 4)$	
$\tilde{A}_{15} = (1 \parallel 0 \parallel 0)$	$A_0 = (0 \parallel 0 \parallel 0)$	$\{m_1, m_2, m_3\}$
	$A_3 = (1 \parallel 0 \parallel 3)$	
	$A_5 = (1 \parallel 2 \parallel 0)$	
$\tilde{A}_{16} = (0 \parallel 1 \parallel 1)$	$A_1 = (1 \parallel 1 \parallel 1)$	$\{m_1, m_3\}$
	$A_4 = (0 \parallel 1 \parallel 4)$	
$\tilde{A}_{17} = (1 \parallel 2 \parallel 2)$	$A_2 = (0 \parallel 2 \parallel 2)$	$\{m_1, m_3\}$
	$A_5 = (1 \parallel 2 \parallel 0)$	
$\tilde{A}_{18} = (0 \parallel 0 \parallel 3)$	$A_0 = (0 \parallel 0 \parallel 0)$	$\{m_1, m_3\}$
	$A_3 = (1 \parallel 0 \parallel 3)$	
$\tilde{A}_{19} = (1 \parallel 1 \parallel 4)$	$A_1 = (1 \parallel 1 \parallel 1)$	$\{m_1, m_3\}$
	$A_4 = (0 \parallel 1 \parallel 4)$	
$\tilde{A}_{20} = (0 \parallel 2 \parallel 0)$	$A_0 = (0 \parallel 0 \parallel 0)$	$\{m_1, m_2, m_3\}$
	$A_2 = (0 \parallel 2 \parallel 2)$	
	$A_5 = (1 \parallel 2 \parallel 0)$	
$\tilde{A}_{21} = (1 \parallel 0 \parallel 1)$	$A_1 = (1 \parallel 1 \parallel 1)$	$\{m_2, m_3\}$
	$A_3 = (1 \parallel 0 \parallel 3)$	
$\tilde{A}_{22} = (0 \parallel 1 \parallel 2)$	$A_2 = (0 \parallel 2 \parallel 2)$	$\{m_2, m_3\}$
	$A_4 = (0 \parallel 1 \parallel 4)$	
$\tilde{A}_{23} = (1 \parallel 2 \parallel 3)$	$A_3 = (1 \parallel 0 \parallel 3)$	$\{m_2, m_3\}$
	$A_5 = (1 \parallel 2 \parallel 0)$	
$\tilde{A}_{24} = (0 \parallel 0 \parallel 4)$	$A_0 = (0 \parallel 0 \parallel 0)$	$\{m_2, m_3\}$
	$A_4 = (0 \parallel 1 \parallel 4)$	
$\tilde{A}_{25} = (1 \parallel 1 \parallel 0)$	$A_1 = (1 \parallel 1 \parallel 1)$	$\{m_2, m_3\}$
	$A_5 = (1 \parallel 2 \parallel 0)$	
$\tilde{A}_{26} = (0 \parallel 2 \parallel 1)$	$A_2 = (0 \parallel 2 \parallel 2)$	$\{m_3\}$
$\tilde{A}_{27} = (1 \parallel 0 \parallel 2)$	$A_3 = (1 \parallel 0 \parallel 3)$	$\{m_3\}$
$\tilde{A}_{28} = (0 \parallel 1 \parallel 3)$	$A_4 = (0 \parallel 1 \parallel 4)$	$\{m_3\}$
$\tilde{A}_{29} = (1 \parallel 2 \parallel 4)$	$A_5 = (1 \parallel 2 \parallel 0)$	$\{m_3\}$

3   Conclusions

This article introduces the improved method for determining AS numbers in RNS. This method improvement is based on AS determining time reduction. The idea of the discussed method of AS numbers determining in RNS is to preliminary compile correspondence M tables (first stage tables) $A = \Phi_1(\tilde{A})$ for each correct $A = (a_1, a_2, \ldots, a_{i-1}, a_i, a_{i+1}, \ldots, a_n)$ number (out of numeric range $0 \div M - 1$), to the possible set of incorrect numbers (out of numeral range $M \div M_0 - 1$) when single errors occur in the number A. Based on the data analysis of the first stage table content, the second stage table is compiled, which contains the correspondence $\tilde{A} = \Phi_2(A)$ of each incorrect $\tilde{A}$ number out of numeric range $M \div M_0 - 1$ to the possible values of the corrected (non-corrupted) $A = (a_1, a_2, \ldots, a_{i-1}, a_i, a_{i+1}, \ldots, a_n)$ numbers.

Application of the proposed method, as opposed to other existing methods, allows to significantly reduce time needed for the AS number determining. Firstly, this is achieved by decreasing the amount of RNS bases being checked, that indicate the possibility of the correct number's $A = (a_1, a_2, \ldots, a_{i-1}, a_i, a_{i+1}, \ldots, a_n)$ residues to be corrupted. And, secondly, this is due to introduction of the fast (table-based) selection of preliminary calculated AS $W(\tilde{A}) = \{m_{l_1}, m_{l_2}, \ldots, m_{l_p}\}$ values. Time reduction for AS number's $W(\tilde{A}) = \{m_{l_1}, m_{l_2}, \ldots, m_{l_p}\}$ determining may subsequently increase performance of error verification, diagnosing, and correction of data in RNS.

References

1. Akushskii, I., Yuditskii, D.: Machine Arithmetic in Residual Classes: Sov. Radio, Moscow (1968, in Russian)
2. Krasnobayev, V., Koshman, S., Mavrina, M.: A method for increasing the reliability of verification of data represented in a residue number system. Cybern. Syst. Anal. **50**(6), 969–976 (2014)
3. Krasnobayev, V., Yanko, A., Koshman, S.: A Method for arithmetic comparison of data represented in a residue number system. Cybern. Syst. Anal. **52**(1), 145–150 (2016)
4. Moroz, S., Krasnobayev, V.: A data verification method in a non-positional residue number system. Control Navig. Commun. Syst. **2**(18), 134–138 (2011)
5. James, J., Pe, A., Vasu, S., Venu, V.: Application of residue number system in the generation of PN-sequences for CDMA systems. In: 2015 International Conference on Computing and Network Communications (CoCoNet), Trivandrum, pp. 670–674 (2015)
6. Manabe, T., Shibata, Y., Oguri, K.: FPGA implementation of a real-time super-resolution system with a CNN based on a residue number system. In: 2017 International Conference on Field Programmable Technology (ICFPT), Melbourne, VIC, pp. 299–300 (2017)
7. Wei, S.: Fast signed-digit arithmetic circuits for residue number systems. In: 2015 IEEE International Conference on Electronics, Circuits, and Systems (ICECS), Cairo, pp. 344–347 (2015)
8. Kuznetsov, A., Kolovanova, I., Kuznetsova, T.: Periodic characteristics of output feedback encryption mode. In: 2017 4th International Scientific-Practical Conference Problems of Infocommunications. Science and Technology (PIC S&T), Kharkov, 2017, pp. 193–198 (2017)

9. Wang, J., Ma, S., Yang, Z., Hu, J.: A systemic performance evaluation method for Residue Number System. In: 2016 2nd IEEE International Conference on Computer and Communications (ICCC), Chengdu, pp. 321–325 (2016)
10. Xiao, H., Ye, Y., Xiao, G., Kang, Q.: Algorithms for comparison in residue number systems. In: 2016 Asia-Pacific Signal and Information Processing Association Annual Summit and Conference (APSIPA), Jeju, pp. 1–6 (2016)
11. Yanko, A., Koshman, S., Krasnobayev, V.: Algorithms of data processing in the residual classes system. In: 2017 4th International Scientific-Practical Conference Problems of Infocommunications. Science and Technology (PIC S&T), Kharkov, 2017, pp. 117–121 (2017)
12. Thakur, U., et al.: FPGA based efficient architecture for conversion of binary to residue number system. In: 2017 8th IEEE Annual Information Technology, Electronics and Mobile Communication Conference (IEMCON), Vancouver, BC, pp. 700–704 (2017)
13. Asif, S., Hossain, M., Kong, Y.: High-throughput multi-key elliptic curve cryptosystem based on residue number system. In: IET Computers and Digital Techniques, vol. 11, no. 5, pp. 165–172, September 2017
14. Yatskiv, V., Tsavolyk, T., Yatskiv, N.: The correcting codes formation method based on the residue number system. In: 2017 14th International Conference the Experience of Designing and Application of CAD Systems in Microelectronics, Lviv, pp. 237–240 (2017)

Method of an Optimal Nonlinear Extrapolation of a Noisy Random Sequence on the Basis of the Apparatus of Canonical Expansions

Igor Atamanyuk[1], Vyacheslav Shebanin[1], Yuriy Kondratenko[2(✉)],
Valerii Havrysh[1], and Yuriy Volosyuk[1]

[1] Mykolaiv National Agrarian University, Mykolaiv, Ukraine
atamanyuk@mnau.edu.ua
[2] Petro Mohyla Black Sea National University, Mykolaiv, Ukraine
yuriy.kondratenko@chmnu.edu.ua

Abstract. Method of optimal nonlinear extrapolation of a random sequence provided that the measurements are carried out with an error is developed using the apparatus of canonical expansions. Filter-extrapolator does not impose any essential limitations on the class of predictable random sequences (linearity, Markovian behavior, stationarity, monotony etc.) that allows to achieve maximum accuracy of the solution of a prediction problem. The results of a numerical experiment on a computer confirmed high effectiveness of the introduced method of the prediction of the realizations of random sequences. Expression for a mean-square error of extrapolation allows to estimate the quality of a prediction problem solving using a developed method. The method can be used in different spheres of science and technics for the prediction of the parameters of stochastic objects.

Keywords: Random sequence · Canonical expansion · Extrapolation

1 Introduction

Solving a wide circle of control problems is connected with the necessity of the prediction of a future state of a controlled object by its known present and past state. Considerable number of applied problems of the prediction has to be solved under the conditions when guaranteed accuracy of the result and minimum of the volume of calculations play determinant role but at that statistical data are already available or can be accumulated if necessary. Problems of plane-to-plane navigation where timeliness and accuracy of the prediction allow to prevent dangerous development of the situation should be attributed to such problems first of all. The problem of the organization of the reliability tests [1, 2] of complicated technical objects where the use of accumulated information for the prediction allows to shorten the expensive process of tests and raise essentially the reliability of its results (similar problems arise in medical diagnostics [3], economics [4, 5] etc.) is not less important.

All stated above determines the importance of the statement and study of the problem in the context of deductive direction as a problem of extrapolation of some

© Springer Nature Switzerland AG 2019
O. Chertov et al. (Eds.): ICDSIAI 2018, AISC 836, pp. 329–337, 2019.
https://doi.org/10.1007/978-3-319-97885-7_32

concrete random sequence realization beyond the borders of the interval of its observation. For the first time, the problem of extrapolation in such formulation is solved in the works of Kolmogorov [6] and Wiener [7] for stationary random sequences. For the following years, the obtained solution turned into coherent theory of linear extrapolation [8]. However, its inherent drawbacks such as restriction of the class of studied sequences by stationary ones and inconvenience from the calculation point of view (as it is obtained in spectral terms) limited the area of its practical application. Kalman-Bucy filter [9], which allowed to refuse from the requirements of stationarity and organize the process of calculations in the most convenient recurrent form, represents a significant breakthrough in the area of linear extrapolation theory. Stated advantages provided Kalman filters with the widest selection of practical applications [8, 10], however it appeared that even this powerful apparatus has certain limitations—it is based on the assumption that a studied sequence is generated by a linear dynamic system excited by a white noise, i.e., it is Markovian.

As real sequences typically have significant aftereffect, the necessity arose to get free from these limitations as well. One of the solutions satisfying this requirement is obtained in [11] under assumption that a studied random sequence is specified by its canonical expansion [12]. Its universality is determined by that a canonical expansion exists and describes accurately any random sequence with finite variance in a studied set of points. But given solution is optimal only within the limits of linear (correlated) relations. The most general formula for the solution of the problems of nonlinear prediction is Kolmogorov-Gabor polynomial. This polynomial allows to take into consideration an arbitrary number of measurements of a random sequence and an order of nonlinearity. However, its practical application is limited with considerable difficulties of determining the parameters of an extrapolator (for example, for 12 measurements and the order of nonlinearity 4, it is necessary to obtain 1819 expressions of partial derivatives of a mean-square error of extrapolation). Thus, in spite of the stated variety of solutions, the necessity of fast, robust, and maximally accurate algorithms and forecast devices continues to be urgent at present and in the long term.

2   Problem Statement

Let's assume that, as a result of preliminary experimental studies, for the moments of time $t_i, i = \overline{1, I}$ (not necessarily with a constant step of discreteness), the variety of realizations of a certain random sequence $\{X\}$ is obtained. On the basis of the given information, sampled moment functions $M[X^{\xi_l}(i - p_{l-1})X^{\xi_{l-1}}(i - p_{l-2})\ldots X^{\xi_2}(i - p_1)$ $X^{\xi_1}(i)], \sum_{j=1}^{l} \xi_j \leq N, p_j = \overline{1, i-1}, i = \overline{1, I}$ are determined by known formulae of mathematical statistics. Random sequence describes, for example, the change of the parameter of a certain technical object in the studied set of points $t_i, i = \overline{1, I}$. During the process of exploitation, the controlled parameter is measured with a certain error and as a result a random sequence $\{Z\}$ is observed:

$$Z(\omega, t_i) = X(\omega, t_i) + Y(\omega, t_i), \ i = \overline{1, I},$$

where ω is an elementary event belonging to a certain area Ω, $Y(\omega, t_i)$ is a random sequence of measurement errors:

$$M\left[Y^{\xi_l}(i - p_{l-1})Y^{\xi_{l-1}}(i - p_{l-2}) \ldots Y^{\xi_2}(i - p_1)Y^{\xi_1}(i)\right], \sum_{j=1}^{l} \xi_j \leq N, p_j = \overline{1, i-1}, i = \overline{1, I}.$$

Let's assume that as a result of a number of consecutive measurements $k < I$ of the first values $Z(v) = z(v)$, $v = \overline{1, k}$ a concrete realization of the sequence $\{Z\}$ is obtained. On the basis of this information and mentioned above a priori data an optimal in mean-square sense nonlinear estimation $\hat{X}(i)$, $i = \overline{k+1, I}$ of future values of the corresponding realization of a studied random sequence $\{X\}$ is necessary to obtain.

Only the limitation of finite variance is imposed on the class of forecast sequences.

3 Forecasting Method

Let's introduce into consideration a mixed random sequence $\{X'\} = \{Z(1), Z(2), \ldots, Z(k), X(k+1), \ldots, X(I)\}$ combining the results of the measurements till $i = k$ as well as data about the sequence $\{X\}$ for $i = \overline{k+1, I}$.

Generalized nonlinear canonical expansion for such a sequence takes the form [13]:

$$X'(i) = M[X'(i)] + \sum_{v=1}^{i} \sum_{\xi_1^{(1)}=1}^{N-1} Q_{\xi_1^{(1)}}(v) \psi_{\xi_1^{(1)}}^{(1)}(v, i)$$

$$+ \sum_{v=2}^{i} \sum_{l=2}^{M(v)} \sum_{p_1^{(l)}=1}^{p_1^{\prime(l)}} \cdots \sum_{p_{l-1}^{(l)}=p_{l-2}^{(l)}+1}^{p_{l-1}^{\prime(l)}} \sum_{\xi_1^{(l)}=1}^{\xi_1^{\prime(l)}} \cdots \sum_{\xi_l^{(l)}=1}^{\xi_l^{\prime(l)}} Q_{p_1^{(l)} \ldots p_{l-1}^{(l)}; \xi_1^{(l)} \ldots \xi_l^{(l)}}(v) \psi_{p_1^{(l)} \ldots p_{l-1}^{(l)}; \xi_1^{(l)} \ldots \xi_l^{(l)}}^{(1)}(v, i), \ i = \overline{1, I}.$$

$$(1)$$

Parameters of model (1) are determined as follows:

$$D_{\beta_1, \ldots \beta_{n-1}; \alpha_1, \ldots \alpha_n}(v) = M[Z^{2\alpha_n}(v - \beta_{n-1}) \ldots Z^{2\alpha_1}(v)] - M^2[Z^{\alpha_n}(v - \beta_{n-1}) \ldots Z^{\alpha_1}(v)]$$

$$- \sum_{\lambda=1}^{v-1} \sum_{\xi_1^{(1)}=1}^{N-1} D_{\xi_1^{(1)}}(\lambda) \left\{ \psi_{\xi_1^{(1)}}^{(\beta_1, \ldots \beta_{n-1}; \alpha_1, \ldots \alpha_n)}(\lambda, v) \right\}^2$$

$$- \sum_{\lambda=2}^{v-1} \sum_{l=2}^{M(\lambda)} \sum_{p_1^{(l)}=1}^{p_1^{\prime(l)}} \cdots \sum_{p_{l-1}^{(l)}=p_{l-2}^{(l)}+1}^{p_{l-1}^{\prime(l)}} \sum_{\xi_1^{(l)}=1}^{\xi_1^{\prime(l)}} \cdots \sum_{\xi_l^{(l)}=1}^{\xi_l^{\prime(l)}} D_{p_1^{(l)} \ldots p_{l-1}^{(l)}; \xi_1^{(l)} \ldots \xi_l^{(l)}}(\lambda) \left\{ \psi_{p_1^{(l)} \ldots p_{l-1}^{(l)}; \xi_1^{(l)} \ldots \xi_l^{(l)}}^{(\beta_1, \ldots \beta_{n-1}; \alpha_1, \ldots \alpha_n)}(\lambda, v) \right\}^2$$

$$- \sum_{l=2}^{n-1} \sum_{p_1^{(l)}=1}^{p_1^{\prime(l)}} \cdots \sum_{p_{l-1}^{(l)}=p_{l-2}^{(l)}+1}^{p_{l-1}^{\prime(l)}} \sum_{\xi_1^{(l)}=1}^{\xi_1^{\prime(l)}} \cdots \sum_{\xi_l^{(l)}=1}^{\xi_l^{\prime(l)}} D_{p_1^{(l)} \ldots p_{l-1}^{(l)}; \xi_1^{(l)} \ldots \xi_l^{(l)}}(v) \left\{ \psi_{p_1^{(l)} \ldots p_{l-1}^{(l)}; \xi_1^{(l)} \ldots \xi_l^{(l)}}^{(\beta_1, \ldots \beta_{n-1}; \alpha_1, \ldots \alpha_n)}(v, v) \right\}^2$$

$$- \sum_{p_1^{(n)}=1}^{p_1^{*(n)}} \cdots \sum_{p_{l-1}^{(n)}=p_{l-2}^{(n)}+1}^{p_{n-1}^{*(n)}} \sum_{\xi_1^{(n)}=1}^{\xi_1^{*(n)}} \cdots \sum_{\xi_n^{(n)}=1}^{\xi_n^{*(n)}} D_{p_1^{(n)} \ldots p_{n-1}^{(n)}; \xi_1^{(n)} \ldots \xi_n^{(n)}}(v) \left\{ \psi_{p_1^{(n)} \ldots p_{n-1}^{(n)}; \xi_1^{(n)} \ldots \xi_n^{(n)}}^{(\beta_1, \ldots \beta_{n-1}; \alpha_1, \ldots \alpha_n)}(v, v) \right\}^2, \ v = \overline{1, k}.$$

$$(2)$$

$$\psi_{\beta_1,\ldots\beta_{n-1};\alpha_1,\ldots,\alpha_n}^{(b_1,\ldots b_{m-1};a_1,\ldots a_m)}(v,i) = \frac{1}{D_{\beta_1,\ldots\beta_{n-1};\alpha_1,\ldots,\alpha_n}}\{M[Z^{\alpha_n}(v-\beta_{n-1})\ldots Z^{\alpha_1}(v)Z^{a_m}(i-b_{m-1})\ldots$$
$$\ldots Z^{a_1}(i)] - M[Z^{\alpha_n}(v-\beta_{n-1})\ldots Z^{\alpha_1}(v)]M[Z^{a_m}(i-b_{m-1})\ldots Z^{a_1}(i)]$$

$$-\sum_{\lambda=1}^{v-1}\sum_{\xi_1^{(1)}=1}^{N-1} D_{\xi^{(1)}}(\lambda)\psi_{\xi_1^{(1)}}^{(\alpha_1)}(\lambda,v)\psi_{\xi_1^{(1)}}^{(b_1,\ldots b_{m-1};a_1\ldots a_m)}(\lambda,i)$$

$$-\sum_{\lambda=2}^{v-1}\sum_{l=2}^{M(\lambda)}\sum_{p_1^{(l)}=1}^{p_1^{'(l)}}\cdots\sum_{p_{l-1}^{(l)}=p_{l-2}^{(l)}+1}^{p_{l-1}^{'(l)}}\sum_{\xi_1^{(l)}=1}^{\xi_1^{'(l)}}\cdots\sum_{\xi_l^{(l)}=1}^{\xi_l^{'(l)}} D_{p_1^{(l)}\ldots p_{l-1}^{(l)};\xi_1^{(l)}\ldots\xi_l^{(l)}}(\lambda)$$
$$\times\psi_{p_1^{(l)}\ldots p_{l-1}^{(l)};\xi_1^{(l)}\ldots\xi_l^{(l)}}^{(\beta_1,\ldots\beta_{n-1};\alpha_1,\ldots\alpha_n)}(\lambda,v)\psi_{p_1^{(l)}\ldots p_{l-1}^{(l)};\xi_1^{(l)}\ldots\xi_l^{(l)}}^{(b_1,\ldots b_{m-1};a_1,\ldots a_m)}(\lambda,i)$$

$$-\sum_{l=2}^{n-1}\sum_{p_1^{(l)}=1}^{p_1^{'(l)}}\cdots\sum_{p_{l-1}^{(l)}=p_{l-2}^{(l)}+1}^{p_{l-1}^{'(l)}}\sum_{\xi_1^{(l)}=1}^{\xi_1^{'(l)}}\cdots\sum_{\xi_l^{(l)}=1}^{\xi_l^{'(l)}} D_{p_1^{(l)}\ldots p_{l-1}^{(l)};\xi_1^{(l)}\ldots\xi_l^{(l)}}(v)\psi_{p_1^{(l)}\ldots p_{l-1}^{(l)};\xi_1^{(l)}\ldots\xi_l^{(l)}}^{(\beta_1,\ldots\beta_{n-1};\alpha_1,\ldots\alpha_n)}(v,v)$$
$$\times\psi_{p_1^{(l)}\ldots p_{l-1}^{(l)};\xi_1^{(l)}\ldots\xi_l^{(l)}}^{(b_1,\ldots b_{m-1};a_1,\ldots a_m)}(v,i) - \sum_{p_1^{(n)}=1}^{p_1^{*(n)}}\cdots\sum_{p_{l-1}^{(n)}=p_{l-2}^{(n)}+1}^{p_{n-1}^{*(n)}}\sum_{\xi_1^{(n)}=1}^{\xi_1^{*(n)}}\cdots\sum_{\xi_n^{(n)}=1}^{\xi_n^{*(n)}} D_{p_1^{(n)}\ldots p_{n-1}^{(n)};\xi_1^{(n)}\ldots\zeta_n^{(n)}}(v)$$
$$\times\psi_{p_1^{(n)}\ldots p_{n-1}^{(n)};\xi_1^{(n)}\ldots\zeta_n^{(n)}}^{(\beta_1,\ldots\beta_{n-1};\alpha_1,\ldots\alpha_n)}(v,v)\psi_{p_1^{(n)}\ldots p_{n-1}^{(n)};\xi_1^{(n)}\ldots\zeta_n^{(n)}}^{(b_1,\ldots b_{m-1};a_1,\ldots a_m)}(v,i)\}, \quad v,i\leq k.$$

$$(3)$$

where
$$p_j^{'(l)} = \begin{cases} 0, & \text{at } (j\neq\overline{1,l-1})\vee(l=1); \\ i-l+j, & \text{at } j=\overline{1,l-1}, l>1; \end{cases}$$
$$\xi_\mu^{'(l)} = N-1-l+\mu-\sum_{j=1}^{\mu-1}\xi_j^{'(l)}, \mu=\overline{1,l}.$$

In (2) and (3), parameters $p_1^{*(n)},\ldots,p_{n-1}^{*(n)}$ and $\xi_1^{*(n)},\ldots,\xi_n^{*(n)}$ are calculated with the help of the following expressions:

$$p_\mu^{*(n)} = \begin{cases} p_\mu'', & \text{if } (\mu=1)\vee\left(p_{\mu-1}^{(n)}=p_{\mu-1}'', \mu=\overline{2,n}\right); \\ v-n+\mu, & \text{if } p_{\mu-1}^{(n)}\neq p_{\mu-1}'', \mu=\overline{2,n}; \end{cases} \quad (4)$$

$$\xi_i^{*(n)} = \begin{cases} \xi_i'', & \text{if } (i=1)\vee\left(\xi_{i-1}^{(n)}=\xi_{i-1}'', i=\overline{2,n}\right); \\ N-1-n+i-\sum_{j=1}^{i-1}\xi_j^{(n)}, & \text{if } \xi_{i-1}^{(n)}\neq\xi_{i-1}'', i=\overline{2,n}. \end{cases} \quad (5)$$

Values $p_\mu'', \mu=\overline{1,n-1}; \xi_i'', i=\overline{1,n}$ are the indices of a random coefficient $V_{p_1''\ldots p_{n-1}'';\xi_1''\ldots\xi_n''}(v)$, which precedes, $V_{\beta_1\ldots\beta_{n-1};\alpha_1\ldots\alpha_n}(v)$ in a canonical expansion (1) for the section t_v and the number n of the indices α_i:

1. $p_\mu'' = \beta_\mu, \mu=\overline{1,n-1}; \xi_i'' = \alpha_i, i=\overline{1,k-1}; \xi_k'' = \alpha_k-1;$

   $\xi_j'' = N-1-n+j-\sum_{m=1}^{j-1}\xi_m'', j=\overline{k+1,n};$ if $\alpha_k>1,$

   $\alpha_j = 1, j=\overline{k+1,n};$

2. $p''_\mu = \beta_\mu, \mu = \overline{1, k-1}; p''_k = \beta_k - 1; p''_j = v - n + j, j = \overline{k+1, n-1};$

$\xi''_i = N - 1 - n + i - \sum_{m=1}^{j-1} \xi''_m, i = \overline{1, n},$ if $\alpha_i = 1, i = \overline{1, n}; \beta_k > \beta_{k-1} + 1;$

$\beta_j = \beta_{j-1} + 1; j = \overline{k+1, n-1};$

3. $p''_\mu = 0; \xi''_i = 0; V_{p''_1 \dots p''_{n-1}; \xi''_1 \dots \xi''_n} = 0,$ if $\beta_\mu = \mu, \mu = \overline{1, n-1};$

$\alpha_i = 1, i = \overline{1, n}.$

Canonical expansion (1) describes accurately the values of the sequence $\{X'\}$ in the studied number of points t_i, $i = \overline{1, I}$, and provides the minimum of a mean square of the approximation in the intervals between them. Mathematical model (1) doesn't impose any essential limitations on the class of predictable random sequences (linearity, Markovian behavior, stationarity, monotonicity etc.).

Let's assume that as a result of the measurement, the value $z(1)$ of the sequence $\{X'\}$ in the point t_1 is known and consequently the values of the coefficients $Q_{\xi_1^{(1)}}(1)$, $\xi_1^{(1)} = \overline{1, N}$ are known:

$$Q_{\xi_1^{(1)}}(1) = z^{\xi_1^{(1)}}(1) - M\left[Z^{\xi_1^{(1)}}(1)\right] - \sum_{j=1}^{\xi_1^{(1)}-1} \psi_j^{(\xi_1^{(1)})}(1,1) q_j(1).$$

Substitution of $Q_1(1)$ into (1) allows to obtain a canonical expansion of an a posteriori random sequence $X'(i/z(1))$ in the points t_i, $i = \overline{k+1, I}$:

$$X'(i/z(1)) = M[X(i)] + (z(1) - M[Z(1)])\psi_1^{(1)}(1, i)$$

$$+ \sum_{\xi_1^{(1)}=2}^{N-1} Q_{\xi_1^{(1)}}(1)\psi_{\xi_1^{(1)}}^{(1)}(1, i) + \sum_{v=2}^{i} \sum_{\xi_1^{(1)}=1}^{N-1} Q_{\xi_1^{(1)}}(v)\psi_{\xi_1^{(1)}}^{(1)}(v, i)$$

$$+ \sum_{v=2}^{i} \sum_{l=2}^{M(v)} \sum_{p_1^{(l)}=1}^{p_1'^{(l)}} \cdots \sum_{p_{l-1}^{(l)}=p_{l-2}^{(l)}+1}^{p_{l-1}'^{(l)}} \sum_{\xi_1^{(l)}=1}^{\xi_1'^{(l)}} \cdots \sum_{\xi_l^{(l)}=1}^{\xi_l'^{(l)}} Q_{p_1^{(l)} \dots p_{l-1}^{(l)}; \xi_1^{(l)} \dots \xi_l^{(l)}}(v)\psi_{p_1^{(l)} \dots p_{l-1}^{(l)}; \xi_1^{(l)} \dots \xi_l^{(l)}}^{(1)}(v, i).$$

$$(6)$$

Expected value of (6) is an optimal (by the criterion of the minimum of mean-square error of extrapolation) estimation of future values of sequence $\{X'\}$ provided that one value $z(1)$ is used for the determination of the given estimation:

$$m_z^{(1;1)}(1, i) = M[X'(i/z(1))] = M[X(i)] + (z(1) - M[Z(1)])\psi_1^{(1)}(1, i). \qquad (7)$$

where $m_{x/z}^{(1;1)}(1, i)$ is optimal in a mean-square sense estimation of the future value $x(i)$ provided that for the solution of the problem of prediction the value $z(1)$ is used.

Taking into account that the coordinate functions $\psi_{\beta_1 \dots \beta_{n-1}; \alpha_1 \dots \alpha_n}^{(b_1, \dots b_{m-1}; a_1, \dots a_m)}(v, i)$ are determined from the minimum of a mean-square error of approximation in the intervals between arbitrary values $Z^{\alpha_n}(v - \beta_{n-1}) \ldots Z^{\alpha_1}(v)$ and $X^{a_m}(i - b_{m-1}) \ldots X^{a_1}(i)$, (7) can be generalized in case of prediction $x^{a_m}(i - b_{m-1}) \ldots x^{a_1}(i)$:

$$
\begin{aligned}
m_{x/z}^{(1;1)}(b_1 \ldots b_{m-1}; a_1 \ldots a_m, i) = M[X(i)] + (z(1) - M[Z(1)]) \\
\times \psi_1^{(b_1, \ldots b_{m-1}; a_1, \ldots a_m)}(1, i).
\end{aligned}
\tag{8}
$$

Measurement $z(1)$ also allows to concretize the value $Q_2(1) = z^2(1) - (z(1) - M[Z(1)]) \psi_1^{(2)}(1, 1)$ in (6):

$$
\begin{aligned}
X'(i/z(1), z^2(1)) = {} & M[X(i)] + (z(1) - M[Z(1)]) \psi_1^{(1)}(1, i) \\
& + \left(z^2(1) - (z(1) - M[Z(1)]) \psi_1^{(2)}(1, 1) \right) \psi_2^{(1)}(1, i) \\
& + \sum_{\xi_1^{(1)}=3}^{N} Q_{\xi_1^{(1)}}(1) \psi_{\xi_1^{(1)}}^{(1)}(1, i) + \sum_{v=2}^{i} \sum_{\xi_1^{(1)}=1}^{N-1} Q_{\xi_1^{(1)}}(v) \psi_{\xi_1^{(1)}}^{(1)}(v, i) \\
& + \sum_{v=2}^{i} \sum_{l=2}^{M(v)} \sum_{p_1^{(l)}=1}^{p_1'^{(l)}} \cdots \sum_{p_{l-1}^{(l)}=p_{l-2}^{(l)}+1}^{p_{l-1}'^{(l)}} \sum_{\xi_1^{(l)}=1}^{\xi_1'^{(l)}} \cdots \sum_{\xi_l^{(l)}=1}^{\xi_l'^{(l)}} Q_{p_1^{(l)} \ldots p_{l-1}^{(l)}; \xi_1^{(l)} \ldots \xi_l^{(l)}}(v) \psi_{p_1^{(l)} \ldots p_{l-1}^{(l)}; \xi_1^{(l)} \ldots \xi_l^{(l)}}^{(1)}(v, i).
\end{aligned}
\tag{9}
$$

Application of expected value operator to a posteriori sequence $X'(i/z(1), z^2(1))$ gives the algorithm of extrapolation by two values $z(1), z^2(1)$ taking into account (8):

$$
\begin{aligned}
m_{x/z}^{(2;1)}(b_1 \ldots b_{m-1}; a_1 \ldots a_m, i) = m_z^{(1;1)}(b_1 \ldots b_{m-1}; a_1 \ldots a_m, i) + \left(z^2(1) - m_z^{(1;1)}(2, 1) \right) \\
\times \psi_2^{(b_1, \ldots b_{m-1}; a_1, \ldots a_m)}(1, i).
\end{aligned}
\tag{10}
$$

Generalization of the obtained regularity allows to form the method of prediction for an arbitrary number of known measurements:

$$
\begin{aligned}
& m_{x/z}^{(\beta_1 \ldots \beta_{n-1}; \alpha_1 \ldots \ldots \alpha_n, v)}(b_1 \ldots b_{m-1}; a_1 \ldots a_m, i) = m_{x/z}^{(\beta_1^* \ldots \beta_{n-1}^*; \alpha_1^* \ldots \ldots \alpha_n^*, v)}(b_1 \ldots b_{m-1}; a_1 \ldots a_m, i) + \\
& \left[z^{\alpha_n}(v - \beta_{n-1}) \ldots z^{\alpha_1}(v) - m_{x/z}^{(\beta_1^* \ldots \beta_{n-1}^*; \alpha_1^* \ldots \ldots \alpha_n^*, v)}(\beta_1 \ldots \beta_{n-1}; \alpha_1 \ldots \alpha_n, v) \right] \\
& \times \psi_{\beta_1 \ldots \beta_{n-1}; \alpha_1 \ldots \ldots \alpha_n}^{(b_1 \ldots b_{m-1}; a_1 \ldots a_m)}(v, i), \text{ if } \beta_1^* \ldots \beta_{n-1}^*; \alpha_1^* \ldots \alpha_n^* \neq 0; \\
& m_{x/z}^{(\beta_1 \ldots \beta_{n-1}; \alpha_1 \ldots \ldots \alpha_n, v)}(b_1 \ldots b_{m-1}; a_1 \ldots a_m, i) \\
& = m_{x/z}^{(p_1'^{(n-1)} \ldots p_{n-2}'^{(n-1)}; \xi_1'^{(n-1)} \ldots \xi_{n-1}'^{(n-1)}, v)}(b_1 \ldots b_{m-1}; a_1 \ldots a_m, i) \\
& + \left[z^{\alpha_n}(v - \beta_{n-1}) \ldots z^{\alpha_1}(v) - m_{x/z}^{(p_1'^{(n-1)} \ldots p_{n-2}'^{(n-1)}; \xi_1'^{(n-1)} \ldots \xi_{n-1}'^{(n-1)}, v)}(\beta_1 \ldots \beta_{n-1}; \alpha_1 \ldots \alpha_n, v) \right] \\
& \times \psi_{\beta_1 \ldots \beta_{n-1}; \alpha_1 \ldots \ldots \alpha_n}^{(b_1 \ldots b_{m-1}; a_1 \ldots a_m)}(v, i), \text{ if } \beta_1^* \ldots \beta_{n-1}^*; \alpha_1^* \ldots \alpha_n^* = 0.
\end{aligned}
\tag{11}
$$

Expression $m_x^{(\beta_1\ldots\beta_{n-1};\alpha_1\ldots\ldots\alpha_n,v)}(b_1\ldots b_{m-1};a_1\ldots a_m,i)$ for $\beta_1 = p_1^{'(N-1)},\ldots,\beta_{N-2} = p_{N-2}^{'(N-1)}; \alpha_1 = \xi_1^{'(N-1)},\ldots,\alpha_{N-1} = \xi_{N-1}^{'(N-1)}$ and $m=1,\, a_1 = 1$ is a conditional expectation of a random sequence $X'(i) = X(i),\, i = \overline{k+1,I}$.

Then, the mean-square error of extrapolation with the help of (11) in the moments of time $t_i,\, i = \overline{k+1,I}$ for k known measurements $z(j),\, j = \overline{1,k}$ and N order of nonlinearity will be written down as

$$E^{(k,N)}(i) = M\left[(X(i) - M[X(i)])^2\right] - \sum_{v=1}^{k}\sum_{\xi_1^{(1)}=1}^{N-1} D_{\xi_1^{(1)}}(v)\left\{\psi_{\xi_1^{(1)}}^{(1)}(v,i)\right\}^2$$

$$-\sum_{v=2}^{k}\sum_{l=2}^{M(v)}\sum_{p_1^{(l)}=1}^{p_1^{'(l)}}\ldots\sum_{p_{l-1}^{(l)}=p_{l-2}^{(l)}+1}^{p_{l-1}^{'(l)}}\sum_{\xi_1^{(l)}=1}^{\xi_1^{'(l)}}\ldots\sum_{\xi_l^{(l)}=1}^{\xi_l^{'(l)}} D_{p_1^{(l)}\ldots p_{l-1}^{(l)};\xi_1^{(l)}\ldots\xi_l^{(l)}}(v)\left\{\psi_{p_1^{(l)}\ldots p_{l-1}^{(l)};\xi_1^{(l)}\ldots\xi_l^{(l)}}^{(1)}(v,i)\right\}^2.$$

$$(12)$$

When choosing the values k and N in addition to (12), the value relative to the gain in the accuracy of extrapolation can be analyzed by consecutively increasing the values of the given parameters:

$$\delta^{(N+1,k)}(i) = \left\{E^{(N,k)}(i) - E^{(N+1,k)}(i)\right\}\Big/ M\left[\{X(i) - M[X(i)]\}^2\right]$$
and
$$\delta^{(N,k+1)}(i) = \left\{E^{(N,k)}(i) - E^{(N,k+1)}(i)\right\}\Big/ M\left[\{X(i) - M[X(i)]\}^2\right]$$

If measurements are carried out without an error, the predictive model (1) coincides with nonlinear algorithm of extrapolation obtained in [3]. In a case of absence of multiple nonlinear stochastic relations, the method is simplified to the polynomial exponential Wiener-Hopf filter-extrapolator [14, 15].

The essence of the method of the prediction on the basis of (11) is in the consequent implementation of the following stages:

Stage 1. Accumulation of the realizations of a studied random sequence in ideal conditions (without hindrances, by the measurement instrumentation of increased accuracy) and in the conditions, in which the method is supposed to be applied (with hindrances in the communication channels, errors of measurements).
Stage 2. On the basis of collected statistical information, estimation of the multitude of discretized moment functions $M\left[X^l(v)X^p(\mu)\ldots X^s(i)\right], M\left[Z^l(v)Z^p(\mu)\ldots Z^s(i)\right]$ and $M\left[Z^l(v)Z^p(\mu)\ldots X^s(i)\right]$.
Stage 3. Forming of a canonical expansion (1).
Stage 4. Calculating parameters of the algorithm of prediction (11).
Stage 5. Estimation of future values of extrapolated realization on the basis of the predictive model (11).

4   Results of a Numerical Experiment

Method of extrapolation synthesized in the work was tested on the following model of a random sequence:

$$X(i+1) = \cos(X(i) + X(i-1) + X(i-2)).$$

In the first three sections, a random sequence is distributed regularly on the segment $[-1;\ 1]$. The error of the measurements $Y(i)$, a normally distributed random value on $[-0,2;0,2]$, was imposed on sequence $\{X\}$.

As a result, for the multiple solutions of the prediction problem, the mean-square extrapolation errors were obtained (Fig. 1 for sections i = $\overline{9,12}$) using the Wiener method, the third-order Kalman method, and the generalized method (11) (for the highest order $N = 4$ of nonlinear relationship).

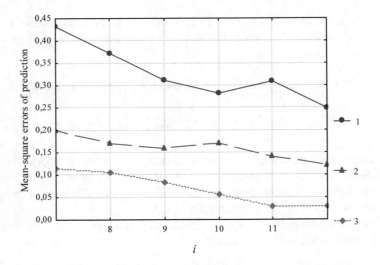

Fig. 1. Mean-square errors of prediction (1: Wiener method; 2: Kalman method; 3: method (11))

Application of nonlinear filter-extrapolator (11) for the forecast allowed to obtain a significant gain in the accuracy of extrapolation in comparison with the rest of the methods at the expense of taking into full consideration the stochastic relations and characteristics of the errors of measurement.

5   Conclusion

In the paper, a discrete optimal in mean-square sense method of nonlinear extrapolation of a noisy random sequence is obtained. Universality of the obtained solution is determined with that a canonical expansion exists and describes any random sequence with finite variance accurately in the points of discreteness. The algorithm allows application of stochastic relations of random order of nonlinearity and arbitrary number of measurement results.

Taking into consideration the recurrent character of the measurements of extrapolator parameters, its implementation on a computer is quite simple.

References

1. Atamanyuk, I., Kondratenko, Y.: Computer's analysis method and reliability assessment of fault-tolerance operation of information systems. In: Batsakis, S., et al. (eds.) ICT in Education, Research and Industrial Applications: Integration, Harmonization and Knowledge Transfer. Proceedings of the 11th International Conference ICTERI-2015, CEUR-WS, Lviv, Ukraine, 14–16 May, vol. 1356, pp. 507–522 (2015)
2. Kondratenko, Y., Korobko, V., Korobko, O., Kondratenko, G., Kozlov, O.: Green-IT approach to design and optimization of thermoacoustic waste heat utilization plant based on soft computing. In: Kharchenko, V., Kondratenko, Y., Kacprzyk, J. (eds.) Green IT Engineering: Components, Networks and Systems Implementation. Studies in Systems, Decision and Control, vol. 105, pp. 287–311. Springer, Cham (2017). https://doi.org/10.1007/978-3-319-55595-9_14
3. Atamanyuk, I., Kondratenko, Y.: Calculation method for a computer's diagnostics of cardiovascular diseases based on canonical decompositions of random sequences. In: Batsakis, S., et al. (eds.) ICT in Education, Research and Industrial Applications: Integration, Harmonization and Knowledge Transfer. Proceedings of the 11th International Conference ICTERI-2015, CEUR-WS, Lviv, Ukraine, 14–16 May, vol. 1356, pp. 108–120 (2015)
4. Poltorak, A., Volosyuk, Y.: Tax risks estimation in the system of enterprises economic security. J. Econ. Ann.-XXI **158**(3–4(2)), 35–38 (2016)
5. Atamanyuk, I., Kondratenko, Y., Sirenko, N.: Forecasting economic indices of agricultural enterprises based on vector polynomial canonical expansion of random sequences. In: 5th International Workshop on Information Technologies in Economic Research (ITER) in ICTERI 2016, Kyiv, 21st–24th June, pp. 458 468 (2016)
6. Kolmogorov, A.N.: Interpolation and extrapolation of stationary random sequences. J. Proc. Acad. Sci. USSR. Math. Ser. **5**, 3–14 (1941)
7. Wiener, N.: Extrapolation, Interpolation, and Smoothing of Stationary Time Series: With Engineering Applications. MIT Press, New York (1964)
8. Box, G.E.P., Jenkins, G.M.: Time–Series Analysis, Forecasting and Control. Holden-Day, San Francisco (1970)
9. Kalman, R.E.: A new approach to linear filtering and prediction problems. Trans. ASME Ser. D J. Basic Eng. **82**(Series D), 35–45 (1960)
10. Simon, D.: Training fuzzy systems with the extended Kalman filter. Fuzzy Sets Syst. **132**, 189–199 (2002)
11. Kudritsky, V.D.: Filtering, extrapolation and recognition realizations of random functions. FADA Ltd., Kyiv (2001)
12. Pugachev, V.S.: The Theory of Random Functions and its Application. Fizmatgiz, Moscow (1962)
13. Atamanyuk, I.P.: Algorithm of extrapolation of a nonlinear random process on the basis of its canonical decomposition. J. Cybern. Syst. Anal. **2**, 131–138 (2005)
14. Atamanyuk, I.P.: Optimal polynomial extrapolation of realization of a random process with a filtration of measurement errors. J. Autom. Inf. Sci. **41**(8), 38–48 (2009). Begell House, USA
15. Atamanyuk, I.P.: Polynomial algorithm of optimal extrapolation of stochastic system parameters. J. Upr. Sist. Mashiny (1), 16–19 (2002). (in Russian)

Regularization of Hidden Markov Models Embedded into Reproducing Kernel Hilbert Space

Galyna Kriukova$^{(\boxtimes)}$ and Mykola Glybovets

National University of Kyiv-Mohyla Academy, Kyiv, Ukraine
kriukovagv@ukma.edu.ua, glib@ukma.kiev.ua
http://ukma.edu.ua/

Abstract. Hidden Markov models (HMMs) are well-known probabilistic graphical models for time series of discrete, partially observable stochastic processes. In this paper, we discuss an approach to extend the application of HMMs to non-Gaussian continuous distributions by embedding the belief about the state into a reproducing kernel Hilbert space (RKHS), and reduce tendency to overfitting and computational complexity of algorithm by means of various regularization techniques, specifically, Nyström subsampling. We investigate, theoretically and empirically, regularization and approximation bounds, the effectiveness of kernel samples as landmarks in the Nyström method for low-rank approximations of kernel matrices. Furthermore, we discuss applications of the method to real-world problems, comparing the approach to several state-of-the-art algorithms.

Keywords: Hidden Markov model · Regularization
Reproducing kernel Hilbert space · Online algorithm

1 Introduction

Development of proper models for time series of stochastic semi-observable processes is crucial for solving a wide variety of problems in learning theory. Observed data from system don't depict the true states, but rather noisy variates of them. Moreover, the observed state space is generally only a subset of the actual state space, as the sensory facilities of most systems are limited.

Hidden Markov models (HMM) are applied to various learning problems, including prediction, classification, clustering, identification, segmentation, reinforcement learning, pattern recognition, time series change point detection, and as an online algorithms they are widely used for dynamic data stream mining [1,8]. Basic assumption for HMM is that to obtain the current hidden state we need only a fixed number of preceding hidden states (Markovian property for transition model), and an observation depends conditionally on its corresponding hidden state (observation model). Accordingly, HMM has a bunch of

O. Chertov et al. (Eds.): ICDSIAI 2018, AISC 836, pp. 338–347, 2019.
https://doi.org/10.1007/978-3-319-97885-7_33

disadvantages, among which large number of unstructured parameters, limitations caused by Markov property for first order HMMs, and the most critical is that only a small portion of distributions may be represented by HMM due to the assumption of discrete number of hidden states.

In our study we consider a non-parametric HMM that extends traditional HMMs to structured and non-Gaussian continuous distributions by means of embedding HMM into Reproducing Kernel Hilbert Space (RKHS). Much recent progress has been made for Hilbert space embedding in probabilistic distributions and their application to HMM [5,14,17,19,20]. Due to interference and ill-posedness of the inverse problem arising at learning of embedded HMM into RKHS, regularization is required. Proposed training algorithms [5,14,19] use L_1, L_2 and truncated spectral regularization to invert the corresponding kernel matrix. In our research, we consider more general regularization techniques [2], specifically Nyström-type subsampling [11].

This paper is organized as follows. In Sect. 2 we develop basic structure and theoretical background of the method. Section 3 describes the experimental framework used to evaluate the performance, pros and cons of the method.

2   Embedding HMM into RKHS

In the standard type of HMM, there is a hidden random variable $x(t)$, $x(t) \in \{x_1^t, x_2^t, \ldots, x_n^t\}$ and random variable $y(t)$ of the corresponding observation at time t. HMM is defined by emission probabilities $P(y(t)|x(t))$, transition probabilities $P(x(t)|x(t-1))$ and initial state probabilities. Viterbi training, expectation-maximization (EM) and spectral algorithms are widely used to train HMM. A priori assumptions on distribution model (i.e. Gaussian mixture) lead to narrowing of the suite of considered probability densities. Employment of RKHS embedding for probability distributions allows to generalize machine learning methods to arbitrary probability densities, not only Gaussian ones, by providing a uniform representation of functions and, consequently, probability densities as elements of a RKHS.

Here we briefly remind method for RKHS embedding for distributions and HMM described in [17,19,20].

Definition 1. *RKHS is a Hilbert space of functions* $f : \Omega \to \mathbb{R}$ *with a scalar product* $\langle \cdot, \cdot \rangle$ *that is implicitly defined by Mercer kernel function* $k : \Omega \times \Omega \to \mathbb{R}$ *as* $\langle \varphi(x), \varphi(y) \rangle = k(x, y)$, *where* $\varphi(x)$ *is a feature mapping into space corresponding to the kernel function. According to reproducing property* $\forall x \in \mathcal{X}, \forall f \in \mathcal{H} \ \langle f, k(\cdot, x) \rangle_{\mathcal{H}} = f(x)$ *we have for any element* f *from RKHS* $f(y) = \sum_{i \in I} \alpha_i k(x_i, y)$, $\alpha_i \in \mathbb{R}$.

Kernel functions have been thoroughly explored since initiative paper [7], and they have been defined on various structured objects, such as strings and graphs, although standard Gaussian radial basis function kernel is widely used as well.

Joint and conditional distributions may be embedded into a RKHS and manipulate the probability densities, by means of the chain, sum and Bayes' rule, entirely in Hilbert space.

Remark 1. Given a set of feature mappings $\Phi = [\varphi(x_1), \ldots, \varphi(x_m)]$ any distribution $q(x)$ may be embedded as a linear combination $\hat{\mu}_q = \Phi\beta$, with weight vector $\beta \in \mathbb{R}^m$. The mean embedding of a distribution can be used to evaluate expectation of any function f in the RKHS, e.g. if $f = \Phi\alpha$, then

$$\mathbb{E}_q[f(x)] = \langle \hat{\mu}_q, f \rangle = \langle \Phi\beta, \Phi\alpha \rangle = \beta^\top \Phi^\top \Phi\alpha = \beta^\top \mathsf{K}\alpha, \tag{1}$$

where $\mathsf{K} = \Phi^\top \Phi$ is Gramian matrix, $\mathsf{K}_{ij} = k(x_i, x_j)$.

Theorem 1 ([17]). *Assume $k(x, x')$ is bounded. With probability $1 - \delta$*

$$\|\hat{\mu}_q - \mu_q\|_{\mathcal{H}} = O\left(m^{-1/2}\sqrt{-\log \delta}\right).$$

Now we are ready to consider RKHS embedding for HMM.

Definition 2. *Assuming RKHS $\mathcal{F}$ with kernel $k(x, x') = \langle \varphi(x), \varphi(x') \rangle_{\mathcal{F}}$ defined on the observations, and RKHS $\mathcal{G}$ with kernel $l(h, h') = \langle \phi(h), \phi(h') \rangle_{\mathcal{G}}$ defined on the hidden states, observable operator $\mathcal{A}_x : \mathcal{G} \to \mathcal{G}$ is defined as*

$$\mathcal{A}_x\phi(h_t) = p(X_t = x|h_t)\mathbb{E}_{H_{t+1}|h_t}[\phi(H_{t+1})].$$

The observation operator is defined as a conditional operator $\mathcal{C}_{X_{t+1}|H_{t+1}} = \mathcal{C}_{X_t|H_t}$ that maps distribution function over hidden states embedded into $\mathcal{G}$ to a distribution function over emissions embedded in $\mathcal{F}$.

Straightforward from Theorem 1 we have

Corollary 1. *Assume $k(x, x')$ and $l(x, x')$ are bounded. Then with probability $1 - \delta$*

$$\|\hat{\mathcal{C}}_{XY} - \mathcal{C}_{XY}\|_{\mathcal{F} \otimes \mathcal{G}} = O\left(m^{-1/2}\sqrt{-\log \delta}\right).$$

For conditional embedding operator use of regularization is needed. Thus, for Tikhonov regularization and given regularization parameter λ we have

Corollary 2. *Assume $k(x, x')$ and $l(x, x')$ are bounded. Then with probability $1 - \delta$*

$$\|\hat{\mu}_{Y|x} - \mu_{Y|x}\|_{\mathcal{G}} = O\left(\sqrt{\lambda} + \sqrt{\frac{-\log \delta}{\lambda m}}\right).$$

Appropriate value of regularization parameter λ may be selected by means of classical approaches, such as Morozov's discrepancy principle, or using Linear Functional Strategy considered in [10]. Moreover, other regularization techniques may be successfully applied for regularization, such as Nyström subsampling [11] or regularization family $\{g_\lambda\}$ [13].

Definition 3 ([13], Definition 2.2). *A family $\{g_\lambda\}$ is called a regularization on $[0, a]$, if there are constants $\gamma_{-1}, \gamma_{-1/2}, \gamma_0$ for which*

$$\sup_{0<t\leq a} |1 - tg_\lambda(t)| \leq \gamma_0, \quad \sup_{0<t\leq a} |g_\lambda(t)| \leq \frac{\gamma_{-1}}{\lambda}, \quad \sup_{0<t\leq a} \sqrt{t}|g_\lambda(t)| \leq \frac{\gamma_{-1/2}}{\sqrt{\lambda}}.$$

It is clear that by taking $g_\lambda(t) = (t + \lambda)^{-1}$ we get widely used regularized matrix inversion. Note, that $g_\lambda(t) = \frac{1}{t}$ for $t \geq \lambda$, and 0 otherwise, corresponds to the regularization by means of spectral cut-off scheme. For details on regularization families and corresponding approximation bounds we refer to [13].

Nyström subsampling is a learning scheme applied in RKHS setting for matrix inversion, where the kernel matrix is replaced with a smaller matrix obtained by column subsampling [18,22]. Note, that arbitrary regularization family may be applied after Nyström subsampling.

This approximation-preserving reduction allows us to extend Corollary 2 to general regularization scheme and gives us an estimation of the emission probability distribution with an approximation error of order $O(\lambda^{1/2} + (\lambda m)^{-1/2})$.

In order to evaluate transition probability distribution, we use Algorithm 1 and Theorem 1 from [17] extended for general regularization scheme, that gives us bound for

$$\|\mu_{X(t+1)|\{X(1),X(2),...,X(t)\}} - \hat{\mu}_{X(t+1)|\{X(1),X(2),...,X(t)\}}\|_{\mathcal{F}}$$

of order

$$O(t(\lambda^{1/2} + (\lambda m)^{-1/2})).$$

In order to reduce the computational complexity, for each of kernel matrices used in Algorithm 1 [17] we apply Nyström subsampling, considering instead of $m \times m$ matrix $m' \times m$ for $m' << m$ (as a rule $m' \approx \sqrt{m}$). Note also, that embedding into RKHS reduces HMM training to linear operations of kernel and feature matrices for fixed sampling basis, which consequently reduces computational complexity. Further, we apply Linear Functional Strategy [10] to eliminate role of subsample selection and regularization parameter choice.

3   Applications and Numerical Experiments

Need for online denoising and data stream segmentation occurs in various real-life problems. For various health-care problems it becomes vital, i.e. for nocturnal hypoglycemia prevention for diabetis patients either wearing continuous glucose monitoring devices, or self-monitoring glucometers [3,12]. Applications of HMM embedded in RKHS to robot vision, slot car inertial measurement and audio event classification were shown [19] as exceeding previous state-of-the-art solutions, including ordinary HMM as well.

3.1 Map Matching

Reliable online localization algorithm is required by real-time applications, such as traffic sensing, routing time prediction and recommendations. State-of-the-art online map-matching algorithms employ HMMs [4], where hidden state and observation are connected with various errors of sensors, imprecise measurements and imperfect maps. Most existing approaches use Viterbi algorithm, using various sliding windows to improve performance. In HMM-based map-matching algorithms, candidate paths are sequentially generated and evaluated on the basis of their likelihoods. When a new trajectory point is acquired, past hypotheses of the map-matched route are extended to account for the new observation. One of the advantages of the approach is corresponding likelihood estimation, which can be considered as a way for uncertainty quantification. In common setting, to meet the requirement of HMM on limited state space, for each observation point only a fixed number of position candidates is considered. It leads to considering only one candidate on each map-graph edge (as a rule, the closest one to the observation point). Moreover, emission and transition probabilities for HMM are predetermined by the authors, and tuning of corresponding parameters is required. It implies distinctive drawbacks, such as cutting-off the angle at crossroads, extra U-turns, back-and-forth "jumps" on road segment, etc.

For our experiments we need an accurate dataset with noised and ground-truth positions. To generate it we used traffic simulator SUMO — Simulation of Urban MObility [9] and OpenStreetMap data [16]. It enables us to generate accurate map ground-truth positions (taking into account road network information from OpenStreetMap), corresponding exact GPS positions and model other observations. Then we added noise to these accurate observations. Noise was modelled according to our assumptions and evaluated on known dataset [15]. Online algorithm was set as a sequence of trajectory reconstructions for sliding window of up to 50 previous observation points (to limit number of layers in corresponding HMM). Performance was measured in terms of accuracy (hitting the correct road-segment) and RMSE for point-to-point correspondence of map-matched and ground-truth positions. As a baseline we used [15] algorithm (with heuristics for probability distribution assumptions). For our algorithm we used 50 trajectories for training HMM, and then applied it for the following trajectories. For training we applied RKHS with Gaussian kernel with various values of $\sigma \in \{0.5, 1, 5, 10\}$, and applied Linear Functional Strategy [10] for inverted kernel matrices. Corresponding learning algorithm we denote by EHMM. Then to the same Gaussian kernel with LFS we apply Nyström-type regularization with subsample size equal to $\sqrt{n}$ for n items in training-set, resulting algorithm we denote by EHMMN. For baseline and proposed algorithms' outputs we compared accuracy (hit rate, value from 0 to 1) and root mean square error (RMSE) for point-to-point correspondences in meters. Results are presented in Fig. 1.

We observe dramatic improvement of performance for proposed algorithms comparing to baseline, moreover, EHMMN outperformed EHMM in terms of RMSE, showing the same accuracy, and with reduced complexity. Here we have to notice, that baseline's performance suffers because of fixed heuristics applied

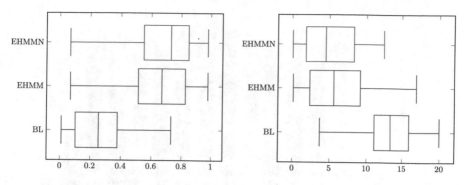

Fig. 1. Performance of baseline algorithm (BL), embedded HMM with Tikhonov regularization (EHMM), and embedded HMM with Nyström-type subsampling regularization (EHMMN) in terms of accuracy (left) and RMSE (right)

for determining emission and transition probability distribution of HMM. Retrained HMM by means of EM algorithm (Gaussian mixture) shows better performance, although implemented algorithms (EHMM and EHMMN) outperform it as well, apparently it is because of implicit ensemble of Gaussian kernels in the implemented solutions.

3.2 Seizure Prediction on Electroencephalography Signal

The electroencephalography (EEG) is crucial tool for monitoring brain activity in various clinical applications. The typical EEG data contains a set of signals measured with electrodes placed on the human scalp. Brain state recognition from EEG signals requires specific signal processing and pre-processing, features extraction, and classification tools. We evaluate our approach on seizure prediction on Electroencephalography (EEG) signal. For more details on experiment setting we refer to [21]. As hidden variable we consider seizure risk, and observation is given by EEG signal and processed cumulative features for given sliding time-interval. In [21] we applied ranking algorithm for seizure risk prediction, and achieved successive prediction rate approximately 83% for time horizon up to 1 min. That algorithm required a lot of pre-processing and calibration, therefore could not be considered as real-time application. In our current investigation we applied embedding of HMM in RKHS for corresponding hidden state and observation model. We've achieved the same accuracy for the same time horizon (see Fig. 2), whereas performance increased dramatically.

3.3 Temporal Audio Segmentation

Embedding of HMM into RKHS with Tikhonov regularization was studied in [5], although without implicit naming. In [5] experimental results were presented both on segmentation of whole audio track from TV show and on speaker diarization withing the interview segments. Namely, two soundtracks of the

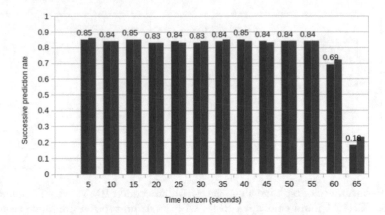

Fig. 2. Accuracy of seizure prediction depending on prediction of time-horizon for baseline algorithm (blue) and proposed method (red)

French 1980s entertainment TV-shows ("Le Grand Echiquier") of approximately three hours each, labelled with characteristic, i.e. "applause", "movie", "music", "speech", "speaker turns". After data preprocessing, every 10 ms of audio where matched to 13-dimensional vector. The experimental results from [5] are presented in Table 1, where the following methods were compared: *regularized kernel Fisher discriminant ratio* (KFDR), which is basically embedded HMM with truncated spectral regularization, *Maximum Mean Discrepancy* (MMD), *Kernel Change Detection* (KCD) algorithms and standard supervised HMM.

Table 1. Best Precision and Recall for benchmarked methods, for both semantic segmentation and speaker segmentation tasks [5]

	Semantic segmentation		Speaker segmentation	
	Precision	Recall	Precision	Recall
KFDR	0.72	0.63	0.89	0.90
MMD	0.71	0.58	0.76	0.73
KCD	0.65	0.63	0.78	0.74
HMM	0.73	0.65	0.93	0.96

Authors mention in [5], that HMM outperformes all the algorithms, but it is explained by rather unrealistic training procedure, as all speakers and possible labels involved are explicitly modelled beforehand in the speech sections, whereas proposed method demonstrated competitive performance with a completely unsupervised approach.

Unfortunately, we didn't manage to find the mentioned dataset to reproduce the results. Therefore, we applied the same preprocessing technique to dataset [6]. The Free Spoken Digit Dataset consists of 1500 audio records in wav files at 8 kHz of English pronunciations of digits by 3 speakers (50 of each digit per speaker). Corresponding digit are easily labelled, as each file is named in the following format: {digitLabel}_{speakerName}_{index}.wav.

We split preprocessed dataset into training and testing set, as it was proposed by its owners, namely 10% (records with indices 0-4 inclusive) of recordings for training set, and 90% for testing (indices 5-49).

In this setting, we compare HMM trained with EM algorithm, embedded HMM, and embedded HMM with regularization by means of Linear Functional Strategy over regularized with Nyström subsampling solutions. Corresponding results are presented in Table 2.

Table 2. Precision and Recall for audio segmentation task

	Speaker segmentation	
	Precision	Recall
HMM	0.82	0.71
EHMM	0.89	0.78
EHMMN	0.88	0.81

4   Conclusion

We consider a Hibert space embedding of HMMs as an extention of traditional HMMs to continuous observation distributions. In this setting we apply more advanced regularization techniques comparing to Tikhonov regularization. Simultaneous regularization by means of Nyström-type subsampling and improved optimization technique enable us to use this approach for online data stream mining. Combining Nyström-type subsampling and Linear Functional Strategy apparently reduce error and its variation, presumably, due to boosting effect, although detailed investigation is needed. As further steps of research we consider multi-penalty regularization for multi-dimensional observation models. Note, that combining modern kernel methods, regularization techniques and graphical models significantly improves well-known algorithms, preserving the advantages of each one.

Acknowledgment. Galyna Kriukova would like to thank Prof. Dr. Sergei Pereverzyev, Johann Radon Institute for Computational and Applied Mathematics (RICAM) of the Austrian Academy of Sciences, for sharing his wisdom and support.

References

1. Bargi, A., Xu, R.Y.D., Piccardi, M.: AdOn HDP-HMM: an adaptive online model for segmentation and classification of sequential data. IEEE Trans. Neural Netw. Learn. Syst. **PP**(99), 1–16 (2018). https://doi.org/10.1109/TNNLS.2017.2742058
2. Bauer, F., Pereverzev, S., Rosasco, L.: On regularization algorithms in learning theory. J. Complex. **23**(1), 52–72 (2007). https://doi.org/10.1016/j.jco.2006.07.001
3. Fong, S., Fiaidhi, J., Mohammed, S., Moutinho, L.: Real-time decision rules for diabetes therapy management by data stream mining. IT Prof. **PP**(99), 1 (2017). https://doi.org/10.1109/MITP.2017.265104658
4. Goh, C.Y., Dauwels, J., Mitrovic, N., Asif, M.T., Oran, A., Jaillet, P.: Online map-matching based on hidden Markov model for real-time traffic sensing applications. In: 2012 15th International IEEE Conference on Intelligent Transportation Systems, pp. 776–781 (2012). https://doi.org/10.1109/ITSC.2012.6338627
5. Harchaoui, Z., Vallet, F., Lung-Yut-Fong, A., Cappe, O.: A regularized kernel-based approach to unsupervised audio segmentation. In: 2009 IEEE International Conference on Acoustics, Speech and Signal Processing, pp. 1665–1668 (2009). https://doi.org/10.1109/ICASSP.2009.4959921
6. Jackson, Z., Souza, C., Flaks, J., Nicolas, H.: Jakobovski/free-spoken-digit-dataset v1.0.7 (2018). https://doi.org/10.5281/zenodo.1136198
7. Kimeldorf, G., Wahba, G.: Some results on Tchebycheffian spline functions. J. Math. Anal. Appl. **33**(1), 82–95 (1971). https://doi.org/10.1016/0022-247X(71)90184-3
8. Kohlmorgen, J., Lemm, S.: A dynamic hmm for on–line segmentation of sequential data. In: Proceedings of the 14th International Conference on Neural Information Processing Systems: Natural and Synthetic, NIPS 2001, pp. 793–800. MIT Press, Cambridge (2001). http://dl.acm.org/citation.cfm?id=2980539.2980642
9. Krajzewicz, D., Erdmann, J., Behrisch, M., Bieker, L.: Recent development and applications of SUMO - simulation of urban MObility. Int. J. Adv. Syst. Meas. **5**(3&4), 128–138 (2012)
10. Kriukova, G., Panasiuk, O., Pereverzyev, S.V., Tkachenko, P.: A linear functional strategy for regularized ranking. Neural Netw. **73**, 26–35 (2016). https://doi.org/10.1016/j.neunet.2015.08.012
11. Kriukova, G., Pereverzyev, S., Tkachenko, P.: Nyström type subsampling analyzed as a regularized projection. Inverse Probl. **33**(7), 074001 (2017). http://stacks.iop.org/0266-5611/33/i=7/a=074001
12. Kriukova, G., Shvai, N., Pereverzyev, S.V.: Application of regularized ranking and collaborative filtering in predictive alarm algorithm for nocturnal hypoglycemia prevention. In: 2017 9th IEEE International Conference on Intelligent Data Acquisition and Advanced Computing Systems: Technology and Applications (IDAACS), vol. 2, pp. 634–638 (2017). https://doi.org/10.1109/IDAACS.2017.8095169
13. Lu, S., Pereverzev, S.V.: Regularization theory for ILL-posed problems: selected topics. Inverse and Ill-Posed Problems Series. De Gruyter, Berlin (2013). http://cds.cern.ch/record/1619978
14. Muandet, K., Fukumizu, K., Sriperumbudur, B., Schölkopf, B.: Kernel mean embedding of distributions: a review and beyond. Found. Trends Mach. Learn. **10**(1–2), 1–141 (2017). https://doi.org/10.1561/2200000060
15. Newson, P., Krumm, J.: Hidden Markov map matching through noise and sparseness. In: Proceedings of the 17th ACM SIGSPATIAL International Conference on Advances in Geographic Information Systems, GIS 2009, pp. 336–343. ACM, New York (2009). https://doi.org/10.1145/1653771.1653818

16. OpenStreetMap contributors: Planet dump retrieved from https://planet.osm.org (2017). https://www.openstreetmap.org
17. Smola, A., Gretton, A., Song, L., Scholkopf, B.: A hilbert space embedding for distributions. In: Hutter, M., Servedio, R.A., Takimoto, E. (eds.) Algorithmic Learning Theory, pp. 13–31. Springer, Heidelberg (2007)
18. Smola, A.J., Schlkopf, B.: Sparse greedy matrix approximation for machine learning, pp. 911–918. Morgan Kaufmann (2000)
19. Song, L., Boots, B., Siddiqi, S.M., Gordon, G., Smola, A.: Hilbert space embeddings of hidden Markov models. In: Proceedings of the 27th International Conference on International Conference on Machine Learning, ICML 2010, pp. 991–998. Omnipress, Madison (2010). http://dl.acm.org/citation.cfm?id=3104322.3104448
20. Song, L., Huang, J., Smola, A., Fukumizu, K.: Hilbert space embeddings of conditional distributions with applications to dynamical systems. In: Proceedings of the 26th Annual International Conference on Machine Learning, ICML 2009, pp. 961–968. ACM, New York (2009). https://doi.org/10.1145/1553374.1553497
21. Sudakov, O., Kriukova, G., Natarov, R., Gaidar, V., Maximyuk, O., Radchenko, S., Isaev, D.: Distributed system for sampling and analysis of electroencephalograms. In: 2017 9th IEEE International Conference on Intelligent Data Acquisition and Advanced Computing Systems: Technology and Applications (IDAACS), vol. 1, pp. 306–310 (2017). https://doi.org/10.1109/IDAACS.2017.8095095
22. Williams, C., Seeger, M.: Using the Nyström method to speed up kernel machines. In: Advances in Neural Information Processing Systems 13, pp. 682–688. MIT Press (2001)

Application of Probabilistic-Algebraic Simulation for the Processing and Analysis of Large Amounts of Data

Viktor Smorodin[✉] [iD] and Elena Sukach

Francisk Skorina Gomel State University, Gomel, Belarus
smorodin@gsu.by

Abstract. The paper considers a way of processing and analysis of large volumes of data characterizing network objects. We propose the application of probabilistic-algebraic simulation, which takes into account the probabilistic nature of the object and substantially reduces the computational complexity of the analysis at all stages of object research, namely: the processing stage, the analysis stage, the interpretation stage.

Keywords: Modeling · Simulation · Large data
Multiply connected network objects · Neural networks · Management
Reliability · Efficiency · Knowledge bases · Statistics processing

1 Introduction

Network objects, which include industrial, information, transport, and other complex systems with a clearly expressed graph structure, can include a number of components. The properties of the components (reliability, efficiency, cost, etc.) are of probabilistic nature and define the properties of the object as a whole.

The modern level of technological development, accompanying the operation of network structures, allows capturing and accumulating the data that reflect dynamics of changes in the properties of components of a selected class of objects. These data, obtained over a long period of time, have a large volume. They differ in variety and in some cases are poorly structured.

The growth in the number of the components and connections between them is accompanied by an exponential growth in calculations when trying to evaluate the probabilistic estimates of the properties.

The increase in the computational complexity of algorithms and the increase in the amount of heterogeneous data during the study of real-life network objects that tend to increase in size require the use of modern methods and automation tools that can cope with the tasks of collecting, systematizing, processing, and interpreting available information in order to provide an increase in efficiency and reliability of operation of the objects under study.

Simulating and analytical modeling methods are able to provide certain effectiveness when studying network objects and taking into account the probabilistic properties of their elements and the parameters of the external environment.

O. Chertov et al. (Eds.): ICDSIAI 2018, AISC 836, pp. 348–357, 2019.
https://doi.org/10.1007/978-3-319-97885-7_34

Logical-probabilistic methods provide a mathematical apparatus for the study of the probabilistic properties of the system taking into account the structural specifications of the systems and the definition of criteria of their functioning [1–3]. However, efficiency of those methods decreases with the increase in the number of parameters of components being taken into account and the alteration of the system structure during its operation. In [4], a list of examples of network structures is given, as well as a method of estimation of probabilistic features of reliability of systems. The method is intended for studying objects, which usually contain a set of components with a limited number of connections (degree of a vertex of the graph does not exceed 3) and are characterized by two or three states of components of the structures under study. Specified restrictions limit the possibility of the described method application greatly when solving the real-life problems of reliability estimation.

In order to estimate integral properties of network structures, which operate under conditions of random impacts, a number of simulating modeling systems were developed. Those systems implement the methods that form their basis and imply performing series of simulations and the subsequent averaging of the obtained results [5, 6]. Simulating modeling is the tool that often allows to study the dynamics of the interactions between the components of the object under study with high level of detail of its inner processes.

In the process of solving practical problems of estimating probabilistic properties of network objects, it was established that the method of probabilistic-algebraic simulation, which combines the advantages of both simulation and analytical calculation, can be effective when applied to a number of problems [7]. The method removes limitations of the logical-probabilistic methods on the number of structural components and allows to take into account the evolutionary dependencies between components when considering the object in dynamics. Unlike the simulating modeling, the probabilistic-algebraic simulation does not require a large amount of resources, is universal when studying different areas, and is implemented with the help of transparent and simple data processing algorithms from the ControlSyst software complex [8].

2   Essence of the Probabilistic-Algebraic Simulation of Network Structures

The probabilistic-algebraic simulation implements a process of constructing vectors of probabilities of network structure states from vectors of probabilities of the states of system components considering their connections.

It is assumed that the structure under study is described by a graph $G(N, K)$, where N is a finite number of vertices, K is a set of edges. There are two vertices in the set that define system entrance ($N1 \in N$) and system exit ($N2 \in N$). Edges of the graph match components $K = \{K_i\}$, $i = \overline{1, m}$, the coordinated interactions of which provide the operation of the object under study. The components can be in one of the incompatible states $S = \{S_j\}$, $j = \overline{0, n - 1}$, which characterize the property under study of the network structure.

As a result of sequential processing of the observational data, simulation parameters that define the probability of the states of the components are being formed:

$$P^{it} = (p_0^{it}, p_1^{it}, \ldots . p_n^{it}), \quad \sum_{j=0}^{n} p_j^{it} = 1, \quad i = \overline{1, m}, \quad t = \overline{1, T}. \tag{1}$$

It is assumed that components of the system are independent, and connections between them are described by a set of functions $F = \{F_z\}$.

A task is set to find the vector of probabilities of the network property under study states using probabilities of the states of its components:

$$P^{net\_t} = (p_0^{net\_t}, p_1^{net\_t}, \ldots, p_n^{net\_t}), \quad \sum_{j=0}^{n} p_j^{net\_t} = 1, \quad i = \overline{1, m}, \quad t = \overline{1, T}. \tag{2}$$

Probability vector elements of result vector P^3, obtained as a result of probabilistic-algebraic multiplication of P^1 and P^2, are calculated using the formula:

$$p_k^3 = \sum_{j=0}^{n} \sum_{i=0}^{n} a_{ij}^k p_i^1 p_j^2, \quad i, j, k = \overline{0, n}. \tag{3}$$

Coefficients a_{ij}^k are called coefficients of the probabilistic-algebraic simulation and satisfy the following requirements:

$$\forall i, j, k, a_{ijk} \geq 0 \text{ and } \sum_{k=0}^{n} a_{ijk} = 1. \tag{4}$$

In the case when the connections between the components of the object under study are of deterministic nature, coefficients of the probabilistic-algebraic simulation are determined by the formula:

$$\begin{cases} a_{ij}^k = 1, & \text{if } k = F(i, j) \\ a_{ij}^k = 0, & \text{if } k \neq F(i, j) \end{cases}. \tag{5}$$

The probabilistic-algebraic simulation is based on the algebraic apparatus, which is a useful tool when studying probabilistic properties of network structures. The apparatus allows to extend common properties of algebraic structures to the studied domain.

3   Theoretical Basis of the Probabilistic-Algebraic Simulation Method

Operation *, defined on a set of vectors (1), generates algebra A^*. This means that for each P^1 and P^2, $P^3 = P^1 * P^2$, and the distributive laws apply to the operation.

The algebra is defined by the structural coefficients (4), while the elements of the result vector P^3 are calculated using (3). The algebra, structural coefficients of which satisfy (4), is called a stochastic algebra because stochastic matrices $M = \|m_{jk}\|$ are elements of its representations. Elements of the matrices $M = \|m_{jk}\|$ are defined by the following formula:

$$m_{jk} = \sum_{i=0}^{n} a_{ij}^{k} p_{i}, \tag{6}$$

where a_{ij}^{k} are structural coefficients of the algebra, $p_i, i = \overline{0,n}$ are elements of the possibilities vector P.

The type of the stochastic matrices, which are the elements of the representations of stochastic algebras, is defined by the operations generating the algebras.

Property 1. Algebra $A*$ is associative when the function $F(i,j)$, which defines operation $*$ that generates algebra $A*$, is associative.

Property 2. For associative algebra $A*$, generated by operation $*$, and for any vectors P^1, P^2, and P^3, which have corresponding stochastic matrices M_1, M_2 and M_3, where $P^3 = P^1 * P^2$, the relation $M_3 = M_1 \cdot M_2$ is correct, and M_3 is a stochastic matrix of the same structure as matrices M_1 and M_2.

Therefore, we have a set of stochastic matrices for the specified operation, which form a semigroup.

4 Automated Information Processing During Initial Simulation Data Generation

Two stages of data processing and structuring are proposed: the network object structure data processing, and the field experiment data processing. The object formalization scheme and the parameterized model type are being selected based on the obtained information [8].

4.1 Network Object Structural Data Processing

Creation of the model structure is implemented in stages with the help of the intelligent graphical editor. The object decomposition is being performed taking into account the selected level of detail, the graph structure type (simple/complex) is being determined, the dimensionality of the model graph is being reduced, variants for splitting the network into n-pole substructures are being proposed, the object representation in the form of a generalized graph structure with complex vertices (n-pole substructures) is being created.

A detailed diagram of the network object with indicated structural components is used as initial data for this process. The diagram is used as a basis and provides efficiency when adding new information. Reduction in the dimensionality of the graph is being achieved through a step-by-step analysis of the graph of the model. At the first step, number of the structural elements, the influence of which can be neglected in the process of evaluating the property under study, is being reduced. At the second step, fragments of a simple graph structure are selected and an algorithm of probabilistic-algebraic modeling is being implemented in order to calculate the replacement probabilistic value of the particular property of the network object fragment. Then, possible variants of the model graph representation as a composition of n-pole substructures ($n = 2, 3, 4$) are being generated by special algorithms implemented in the graphical editor for the multiply connected network objects of large dimension. Substructures are used as elements of the generalized model. Probabilistic properties of the substructures are estimated using one of the methods designed to estimate properties of objects of limited dimensionality [7]. One of the variants is then used to estimate integral values of probabilistic properties of the object under study.

Efficiency of the process of construction of a graph model with an ability to vary the level of detail is provided by such properties of the intellectual graphical editor as its structural hierarchy, scalability, interactivity, flexibility, and transparency. The result of the first stage is selection of a parameterized model of a structure corresponding to the object under study.

4.2   Processing Observed Data of the Network Object Operation and Interpreting Simulation Results

During implementation of the second stage of data processing, parameters of the simulation are being generated. The ControlSyst software complex provides technical tools of interface with the database, which allow the generation of arrays of numerical values of physical parameters (such as failures in working with classification of causes, capacity, cost of performing a typical operation, etc.), which characterize the operation of the structural components of the object under study. This information is being systematized and saved to the database. Then, it is being processed using the tools of integration of heterogeneous data [8] and universal procedures of statistical packages. As a result of sequential processing of the observed data, simulation parameters (1) are being formed.

Implementation of the multi-step process of simulating the network object under study using the constructed model is accompanied by saving results of simulation (2) to the database.

Visualization tools in the ControlSyst complex provide a high level of automation for displaying calculated simulation responses graphically in the form of diagrams and graphs.

Described tools for processing large amounts of data, necessary for organizing management of network objects, are implemented within the framework of a single technological process that meets the principles of system organization, centralization, and integration of data processing.

5   Automated Analysis of Probabilistic Properties of Network Objects

5.1   Static Probabilistic-Algebraic Simulation

Static probability-algebraic simulating is carried out using one of selected parameterized models corresponding to the formalization scheme at the selected level of detail.

The graphical editor implements a graphical representation of model elements, relationships between them, and the nesting levels of functions describing the relationships. The representation allows to set levels of the hierarchy of functional relationships and to ensure replacement of functional relationships between the elements of the model by probabilistic calculations using coefficients of probabilistic-algebraic simulation (4).

For network objects of a simple structural organization, a model from the library of parameterized models of the ControlSyst complex is being selected. The model has no restrictions on the number of elements and their states, and implements the probability-algebraic multiplication of vectors (1) according to functions that describe their interaction. For networks of complex structure, a model is used that implements the technique of reducing models with an arbitrary number of states to a set of binary models. For multiply connected network objects of large dimension, a numerical method of probabilistic-algebraic simulation of multiplication of n-pole substructures is being implemented [7].

Effectiveness of analysis conducted within the probabilistic-algebraic apparatus for multiply connected network objects of different structural organization is determined by automation of static simulation using different levels of probability-algebraic multiplication (simple elements/n-pole substructures) and special techniques that reduce computational complexity of the algorithms.

5.2   Predictive Probabilistic-Algebraic Simulation of a Network Object

Predictive simulation is used in cases when it is necessary to adapt a (potentially dangerous) network object to the probabilistic changes in properties of its components in order to ensure its stable operation.

The method of predictive probabilistic-algebraic simulation of a network object, which includes generation of control actions based on preliminary calculation of probabilistic values of the properties of the network object, is being realized by the following sequence of steps (Table 1).

Stage 1.1 Initial Data Generation. During a field experiment in the following time interval, the information about the failures of components with the classification of their causes is being obtained from a set that covers the time interval T of the network object's operation.

Stage 1.2 Classification of the Component Failures. According to selected types of failures, the set of all states of components of the network object is divided into the following groups: states of reliable operation of components $\{S_r\}$, component states of simple failures $\{S_f\}$, and component states of dangerous failures $\{S_d\}$.

Table 1. Stages of predictive probabilistic-algebraic simulation

Stage name	Contents of the stage
1 Initial data generation	1.1 Obtaining statistical data for the l-th time interval $\{\Delta t_l, l = \overline{1, z}\}$
	1.2 $K_i \to S = \{S_j\} \to \{\{S_r\}, \{S_f\}, \{S_d\}\}$
	1.3 Generation of vectors P_i
	1.4 Construction of the model graph
2 The next lth iteration	2.1 Selecting and configuring a parameterized model
	2.2 Implementation of the l-th iteration
	2.3 Saving results of simulation into the database
3 Analysis of results by the control unit	3.1 Generation of features of reliability for K_i and the network components: $\alpha_{li/net} = 0$—reliable operation, $\alpha_{li/net} = 1$—simple failure, $\alpha_{li/net} = 2$—dangerous failure
	3.2 Generation of options for adjusting the network (updating parameters/backup scheme for $\alpha_{li/net} = 1$, altering the structure for $\alpha_{li/net} = 2$)
4 Predictive $(l + 1)$-th iteration	4.1 Predictive simulation for various network modification variants
	4.2 Selecting the modification of the network according to the generalized criterion
5 Modification of the organization of the network object	5.1 Updating parameters and structural organization of the network according to the chosen modification variant
	5.2 Operation of network objects at the next $(l + 1)$-th time interval

Stage 1.3 Calculation of Current Network Object Simulation Parameters. According to available statistical data, probabilities of reliability states of components of the object under study are calculated in the form of vectors (1) for the current time interval Δt_l.

Stage 2 Realization of the lth Iteration of Probabilistic-Algebraic Simulation. For the given parameters (1), the lth iteration of simulation is performed. The result of the iteration is a set of vectors of the form (2) that are saved to the database of the modeling complex and are analyzed by the simulation dynamics control unit.

Stage 3 Analysis of the Results of the lth Iteration of Probabilistic-Algebraic Simulation. Values of features of reliability $\alpha_{li}, l = \overline{1, z}, i = \overline{1, m}$ for all ith components and α_{net} for the network as a whole are being established as a result of the analysis of the probability vectors of the states of reliability of components and the probability vector of network states. The value of the feature $\alpha_{li} = 0$ determines the reliable operation of the ith component during the time interval Δt_l. If $\alpha_{li} = 1$, then the operation of the ith component is accompanied by simple failures and it is possible to restore the reliability of its operation by switching to one of the redundancy schemes or updating the equipment. If $\alpha_{li} = 2$, then it is considered that the probability of occurrence of a dangerous accident that has destructive consequences at the level of the object under study as a whole is high.

Based on the obtained values of the features of reliability α_{li} for the following time interval Δt_l, the feature of the failure of the entire network is determined. Its value determines the possibility of the network operation in the established mode during the following time interval Δt_{l+1} or shows the necessity of generation of control actions.

Considering control rules established by an expert in the ControlSyst complex simulation control unit for the object under study, a set of measures defining the control actions, which will ensure the subsequent reliable operation of the object and will allow to prevent the occurrence of accidents during the following time interval Δt_{l+1}, is being generated for the ith iteration of the simulation. Corrective measures include three types of actions: altering parameters of the components operation as a result of equipment upgrades, transitioning to one of the options of the reservation of components, or altering the structural organization of the object.

In the event that the value of the reliability feature $\alpha_{li} \neq 0$, a decision is made to modify the structural organization of the object in accordance with the modeling rules. In case of emergency during the operation of the components ($\alpha_{li} = 2$), the distribution function estimates the time (t_r) necessary to carry out the work to eliminate consequences of failures and accidents that occurred during the operation of the ith component and to restore its reliability/safety. The time determines the number of simulation iterations, during which the corresponding component is excluded from the structural composition, thus bringing the organization of the object into line with the received reliability estimates of the components and the network as a whole. In case when the reliability feature $\alpha_{li} = 1$, parameters of the model component are corrected and schemes for its possible reservation are generated. When $\alpha_{net} = 2$, possible variants of modification of the entire structure of the object are generated.

Stage 4 Implementation of Predictive Probabilistic-Algebraic Simulation of Variants of Modification of a Network Object. For each variant of the modification of the object, model experiments are performed, initial parameters (1) and results (2) are saved to the simulation database and analyzed in accordance with the specified general reliability/safety criterion, which allows selecting the most reliable modification of the object, which requires a minimum of material costs for its implementation:

$$\max(W_h = \delta_{1h} \cdot P_h + \delta_{2h} \cdot 1/Q_h^*), \sum_{i=1}^{2} \delta_{ih} = 1, h = \overline{1, n}, \tag{7}$$

where P_h is the weight coefficient of the importance of the probability of a reliable/safe hth variant of the organization of the network, Q_h is the normalized value of material costs for elimination of consequences of failures and maintenance of the hth variant of network organization.

Stage 5 Modification of the Variant of the Network Object Organization. Altering of the parameters and structural organization in the object itself is being carried out using synchronization devices. For the updated version of network organization, transition to the first stage is made.

Upon completion of predictive simulation in the selected time interval T, values of vectors (1) and (2), as well as variants of the network organization for each of the iteration sequences, are stored in the ControlSyst simulation database.

6 Conclusions

The described approach, when being applied to processing, analyzing and interpreting large amounts of data during the process of estimating the probabilistic properties of the selected class of objects, allows:

- to use a large amount of data, diverse and probabilistic in nature, that characterize values of physical indicators of the properties under study (reliability, bandwidth, cost) of network structure components, in order to obtain predicted probabilistic properties of the objects under study;
- to provide high speed for initial data processing and for the result obtaining by the use of transparent algorithms of probabilistic-algebraic simulating, which allows to remove limitations on the number of components for the simple network structures, to reduce the computational complexity from n^m in case of the exhaustive search to 2^m (where n is the number of connections, m is the number of components) when working with complex network structures. For the structurally-complex systems of large dimension, it makes it possible to solve the problem of obtaining the probabilistic estimate of the property under study for the variants of objects organization as a result of their decomposition and subsequent probabilistic-algebraic multiplication of selected n-pole substructures;
- to carry out the initial data processing adequately according to the rate of its obtaining and updating as a result of the organization of the predictive probabilistic-algebraic simulation, therefore providing stable operation of the network object, including the change (increase) in the number of the structural components.

The approach has shown most efficiency when studying the transport network structure [5], information networks, and production systems [9, 10].

References

1. Ryabinin, I.A.: Reliability and Safety of Structurally Complex Systems. St. Petersburg University, St. Petersburg (2007). (in Russian)
2. Mozhaev, A.S., Gromov, V.N.: Theoretical Foundations of the General Logical-Probabilistic Method of Automated System Simulation. VITU Publishing House, St. Petersburg (2000). (in Russian)
3. Solozhentsev, E.D.: Risk Management Technologies (With Logic and Probabilistic Models). Springer, Dordrecht (2012)
4. Sahinoglu, M., Benjamin, R.: Wiley interdisciplinary reviews. Comput. Stat. **2**, 189–211 (2010)
5. Maximey, I.V., Sukach, E.I., Galushko, V.M.: Design modeling of service technology for passenger flow by city transport network. J. Autom. Inf. Sci. **41**(4), 67–77 (2009)

6. Maximey, I.V., Smorodin, V.S., Demidenko, O.M.: Development of Simulation Models of Complex Technical Systems. F. Skorina State University, Gomel (2014)
7. Sukach, E.I.: Probabilistic-Algebraic Modeling of Complex Systems of Graph Structure. F. Skorina State University, Gomel (2012)
8. PTK ControlSyst: certificate of registration of the computer program No 773/E.I. Sukach, Yu.V. Zherdetsky. – Minsk: NCIS, 2015. – Application No C20150031. – Date of submission: 15/04/2015
9. Sukach, E.I.: Designing of functionally complex systems based on probabilistic-algebraic modeling. J. Autom. Inf. Sci. **3**, 45–58 (2011)
10. Maximey, I.V., Demidenko, O.M., Smorodin, V.S.: Problems of Theory and Practice of Modeling Complex Systems. F. Skorina State University, Gomel (2015)

Multiblock ADMM with Nesterov Acceleration

Vladyslav Hryhorenko$^{(\boxtimes)}$ ⓘ and Dmitry Klyushin ⓘ

Taras Shevchenko National University of Kyiv, Kyiv, Ukraine
grigorenkovlad1993@gmail.com, dokmed5@gmail.com

Abstract. ADMM (alternating direction methods of multipliers) is used for solving many optimization problems. This method is particularly important in machine learning, statistics, image, and signal processing. The goal of this research is to develop an improved version of ADMM with better performance. For this purpose, we use combination of two approaches, namely, decomposition of original optimization problem into N subproblems and calculating Nesterov acceleration step on each iteration. We implement proposed algorithm using Python programming language and apply it for solving basis pursuit problem with randomly generated distributed data. We compare efficiency of ADMM with Nesterov acceleration and existing multiblock ADMM and classic two-block ADMM.

Keywords: Alternating direction method of multipliers · Multiblock ADMM
Fast ADMM · Nesterov acceleration

1 State-of-the-Art

The classic ADMM is used for solving the following optimization problem:

$$\text{minimize} \quad f_1(x_1) + f_2(x_2)$$
$$\text{subject to} \quad A_1 x_1 + A_2 x_2 = b$$

where $x_i \in \mathbb{R}^{m_i}$, $A_i \in \mathbb{R}^{n \times m_i}$, $c \in \mathbb{R}^n$ and $f_i \in \mathbb{R}^{n_i}$ are closed proper convex functions, $i = 1, 2$. This problem naturally arises in a number of areas. Often $f_1(x)$ is smooth and $f_2(x)$ is not smooth function like lasso or bridge penalty, l_1, l_2 are regularization parameters. l_1-penalized least-squares problems are often solved for selection of variables when only few regression parameters are nonzero [1]. The main idea of ADMM is breaking original optimization problem into smaller subproblems, each of which is significantly easier to solve [2]. On each iteration, our basic algorithm minimizes augmented Lagrangian separately by each x_i:

$$L_\rho(x_1, x_2, \lambda) = \sum_{i=1}^{N} f_i(x_i) - \lambda^T \left(\sum_{i=1}^{N} A_i x_i - b \right) + \frac{\rho}{2} \left\| \sum_{i=1}^{N} A_i x_i - b \right\|_2^2.$$

The iterative scheme of ADMM can be written as follows:

© Springer Nature Switzerland AG 2019
O. Chertov et al. (Eds.): ICDSIAI 2018, AISC 836, pp. 358–364, 2019.
https://doi.org/10.1007/978-3-319-97885-7_35

$$\begin{cases} x_1^{k+1} = \arg\min_{x_1} L_\rho(x_1, x_2^k, \lambda^k), \\ x_2^{k+1} = \arg\min_{x_2} L_\rho(x_1^{k+1}, x_2, \lambda^k), \\ \lambda^{k+1} = \lambda^k - \rho(A_1 x_1^{k+1} + A_2 x_2^{k+1} - b). \end{cases}$$

The goal of the paper is to develop a variant of ADMM with better performance than performance of existing methods. We show how to combine multiblock distributed ADMM and Nesterov acceleration for achieving this. The Nesterov method is most commonly used to accelerate gradient descent type (first order) methods. The Nesterov first order minimization scheme has global converge rate $O(1/k^2)$ [3], which has optimal convergence rate for the class of Lipschitz differentiable functions [4]. Accelerating is achieved by calculating over relaxation step on each iteration. The iterative scheme of Nesterov acceleration for gradient descent method can be written as follows:

$$\begin{cases} x_k = y_k - \tau \nabla F(y_k) \\ \alpha_{k+1} = (1 + \sqrt{4\alpha_k^2 + 1}/2 \\ y_{k+1} = x_k + (\alpha_k - 1)(x_k - x_{k-1})/\alpha_{k+1} \end{cases}$$

This approach was first used for acceleration of proximal descent method in [5].

The proposed method achieves global convergence rate $O(1/k^2)$, as well as Nesterov gradient descent method, which is probably optimal. After this, Nesterov techniques were successfully applied for splitting schemes (the first example is FISTA) [6]. In [7], this approach was extended for ADMM method. The proposed method is simply ADMM with corrector-predictor acceleration step. It is stable when both objective functions are convex.

The iterative scheme for fast ADMM is proposed below:

$$\begin{cases} x_1^{k+1} = \arg\min_{x_1} L_\rho(x_1, x_2^k, \lambda^k), \\ x_2^{k+1} = \arg\min_{x_2} L_\rho(x_1^{k+1}, x_2, \lambda^k), \\ \lambda^{k+1} = \lambda^k - \rho(A_1 x_1^{k+1} + A_2 x_2^{k+1} - c), \\ \alpha_{k+1} = \dfrac{1 + \sqrt{1 + 4\alpha_k^2}}{2}, \\ \widehat{x}_{k+1}^{(i)} = x_k^{(i)} + \dfrac{\alpha_k - 1}{\alpha_{k+1}}\left(x_k^{(i)} - x_{k-1}^{(i)}\right), \\ \widehat{\lambda}_{k+1} = \lambda_k + \dfrac{\alpha_k - 1}{\alpha_{k+1}}(\lambda_k - \lambda_{k-1}). \end{cases}$$

The convergence rate $O(1/k^2)$ of this method was proven in [7], and it is more efficient than classic ADMM.

In paper [8], solving multiple block separable convex optimization problem using ADMM was considered, and modification of ADMM was proposed, which extends alternating direction method of multipliers, decomposes original problem into N

smaller subproblems, and solves them separately on each iteration. Modified method outperforms classic two-block splitting approach.

Consider optimization problem with N block variables where the objective function is the sum of $N > 2$ separable convex functions:

$$\text{minimize } \sum_{i=1}^{N} f_i(x_i)$$

$$\text{subject to } \sum_{i=1}^{N} A_i x_i = c \tag{1}$$

The multiblock convex optimization problem arises in many practical applications. For example, RPCA (Robust Principal Component Analysis) problem [9] can be considered, where we need to recover low-rank and sparse components of the matrix L_0 from highly corrupted measurements $M = L_0 + S_0$.

ADMM for solving problem can be described as follows [8]:

$$\begin{cases} x_i^{k+1} = \arg\min_{x_i \in X_i} f_i(x_i) + \frac{\beta}{2} \left\| \sum_{j=1}^{i-1} A_j x_j^{k+1} + A_i x_i + \sum_{j=i+1}^{m} A_j x_j^k - b - \frac{\lambda^k}{\beta} \right\|^2, \\ \lambda^{k+1} = \lambda^k - \beta \left(\sum_{i=1}^{m} A_i x_i^{k+1} - b \right). \end{cases}$$

The ADMM convergence was established and proven for the case of two variables. For the multiblock ADMM ($N > 2$), convergence was proven in [10–12] when three conditions are satisfied: (a) for each i, matrix A_i has full column rank and matrices A_i are mutually near orthogonal; (b) parameter β is sufficiently small; (c) proximal terms are added to each of N subproblems (in this case, we don't need any assumptions about matrices A_i). Also adding proximal term to xi-subproblem guarantees convergence and existing of unique solution even for not strictly convex functions and we can improve established converge rate from $O(1/k)$ to $o(1/k)$. These important improvements were considered in [12].

2   Nesterov Accelerated Multiblock ADMM Iterative Scheme

We propose a new modification for multiblock ADMM, based on adding Nesterov acceleration step, similar to modifications for the two-block ADMM, described in [7]. On each iteration, we calculate predictor-corrector Nesterov acceleration step

$$\alpha_{k+1} = \frac{1 + \sqrt{1 + 4\alpha_k^2}}{2}.$$

Then, corrections $\widehat{x}_i^{(k+1)}$ and $\widehat{\lambda}_{k+1}$ are calculated to generate new iteration value by:

$$\widehat{x}_i^{(k+1)} = x_i^{(k)} + \frac{\alpha_k - 1}{\alpha_{k+1}}\left(x_i^{(k)} - x_i^{(k-1)}\right),$$

$$\widehat{\lambda}_{k+1} = \lambda_k + \frac{\alpha_k - 1}{\alpha_{k+1}}(\lambda_k - \lambda_{k-1}).$$

This modification allows as increase converge rate. Proposed iterative scheme for fast multiblock ADMM can be written as follows:

$$
\begin{cases}
x_i^{(k)} = \arg\min_{x_i \in X_i} f_i(x_i) + \frac{\beta}{2}\left\| \sum_{j=1}^{i-1} A_j\widehat{x}_j^{(k)} + A_i x_i + \sum_{j=i+1}^{m} A_j x_j^{(k-1)} - b - \frac{\lambda_k}{\beta} \right\|^2, \\[3mm]
\lambda^k = \widehat{\lambda}^k - \beta\left(\sum_{i=1}^{m} A_i x_i^{(k)} - b\right), \\[3mm]
\alpha_{k+1} = \frac{1 + \sqrt{1 + 4\alpha_k^2}}{2}, \\[3mm]
\widehat{x}_i^{(k+1)} = x_i^{(k)} + \frac{\alpha_k - 1}{\alpha_{k+1}}\left(x_i^{(k)} - x_i^{(k-1)}\right), \\[3mm]
\widehat{\lambda}_{k+1} = \lambda_k + \frac{\alpha_k - 1}{\alpha_{k+1}}(\lambda_k - \lambda_{k-1}).
\end{cases}
$$

This algorithm converges globally if A_i are mutually orthogonal and have full column rank. When this condition is not satisfied, we can use addition proximal correction, similar to described in [13].

3 Numerical Experiments

In this section we compare the efficiency of improved multiblock ADMM with Nesterov acceleration with the ordinary multiblock ADMM and the classic-two block ADMM. For this purpose we consider basis pursuit (BP) problem

$$\text{minimize } \|x\|_1$$
$$\text{subject to } Ax = b$$

where $x \in \mathbb{R}^m$, $A \in \mathbb{R}^{n \times m}$, $b \in \mathbb{R}^n$. This problem is usually applied in cases when we have an undetermined system of linear equation, and vector x should be recovered using a small number of observations $b(n \ll m)$ [14], which widely arises in compressive sensing, signal and image processing, statistics, and machine learning.

We split data into m blocks as follows: $x = (x_1, x_2, \ldots, x_m)$ and $A = (A_1, A_2, \ldots, A_m)$. Then, this problem can be viewed as (1), and we can apply iterative scheme for multiblock ADMM. Each of primal iteration can be viewed as

$$x_i^{(k)} = \arg\min_{x_i \in X_i} |x_i|_1 + \frac{\beta}{2} \left\| \sum_{j=1}^{i-1} A_j \widehat{x}_j^{(k)} + A_i x_i + \sum_{j=i+1}^{m} A_j x_j^{(k-1)} - b - \frac{\lambda_k}{\beta} \right\|^2$$

$$= shrinkage\left(\frac{\beta A_i^T}{\sqrt{a_i^T a_i}} \left[b + \frac{\lambda^k}{\beta} - \sum_{j=1}^{i-1} A_j \widehat{x}_j^{(k)} - \sum_{j=i+1}^{m} A_j x_j^{(k-1)} \right], \frac{1}{\beta(a_i^T a_i)} \right).$$

On each iteration, we calculate proximity operator, which is easy to compute:

$$\mathrm{Prox}_f^\tau(b) = \arg\min_{x \in X} \tau f(x) + \frac{1}{2}\|x - b\|^2, \text{where } \tau > 0 \text{ and } c \text{ are given.}$$

When $f(x) = \|x\|_1$, then proximal operator can be calculated as

$$\left(\mathrm{Prox}_{\|\cdot\|_1}^\tau(b) \right)_i = (shrinkage(b, \tau))_i = \mathrm{sgn}(b_i) \cdot \max(|b_i| - \tau, 0).$$

We get the new iterative scheme for solving BP problem using the multiblock ADMM with the Nesterov acceleration:

$$\begin{cases} x_i^{(k)} = shrinkage\left(\frac{\beta A_i^T}{\sqrt{a_i^T a_i}} \left[b + \frac{\lambda^k}{\beta} - \sum_{j=1}^{i-1} A_j \widehat{x}_j^{(k)} - \sum_{j=i+1}^{m} A_j x_j^{(k-1)} \right], \frac{1}{\beta(a_i^T a_i)} \right), \\[2mm] \lambda^k = \widehat{\lambda}^k - \beta\left(\sum_{i=1}^{m} A_i x_i^{(k)} - b \right), \\[2mm] \alpha_{k+1} = \frac{1 + \sqrt{1 + 4\alpha_k^2}}{2}, \\[2mm] \widehat{x}_i^{(k+1)} = x_i^{(k)} + \frac{\alpha_k - 1}{\alpha_{k+1}}\left(x_i^{(k)} - x_i^{(k-1)} \right), \\[2mm] \widehat{\lambda}_{k+1} = \lambda_k + \frac{\alpha_k - 1}{\alpha_{k+1}}(\lambda_k - \lambda_{k-1}). \end{cases}$$

For testing, we randomly generate vector x^* with $k(k \ll n)$ nonzeros elements from the standard Gaussian distribution. Matrix A was also generated from the standard Gaussian distribution, and then split into m blocks. All tests were performed in Python on the workstation with Intel Core i3 CPUs (2.4 GHz) and RAM 4 GB. We generated a dataset with $n = 500$ rows and $m = n \times 2.5$ columns. We ran algorithm for 300 iterations and plotted averaged dependency between residual and number of iterations for 100 random trials (Fig. 1). For multiblock ADMM and Nesterov accelerated multiblock ADMM, we split objective function into 5 subfunctions.

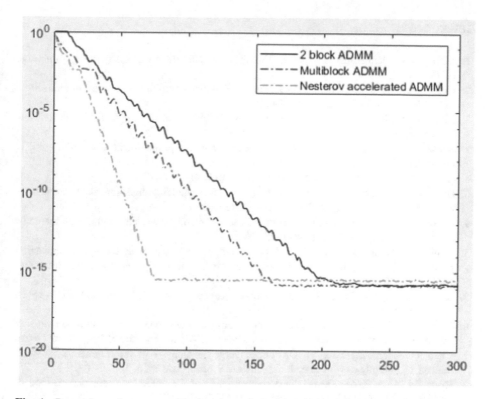

Fig. 1. Dependency between residual and number of iterations (basis pursuit problem where $n = 500$, $m = n * 2.5$, $N = 5$)

4   Conclusions

We have shown that multiblock ADMM is more efficient than the classic 2 block ADMM, and applying the Nesterov acceleration step on each iteration of multiblock ADMM allows us to significantly improve performance for the basis pursuit optimization problem.

ADMM is a very useful tool for solving a great variety of optimization problems in statistics, big data, machine learning, and signal processing, and it is well suited for distributed computation. In this paper, several existing techniques were considered that allow us to improve ADMM performance. We proposed a new approach, based on the combination of the multiblock ADMM and the Nesterov accelerated method. We used this algorithm for solving basis pursuit problem and demonstrated its efficiency in comparison with existing algorithms.

References

1. Boyd, S., Vandenberghe, L.: Convex Optimization. Cambridge University Press, Cambridge (2004)
2. Guler, O.: On the convergence of the proximal point algorithm for convex minimization. SIAM J. Control Optim. **29**, 403–419 (1991)
3. Nesterov, Y.: Introductory Lectures on Convex Optimization. Kluwer Academic Press, New York (2004)
4. Nesterov, Y.: A method of solving a convex programming problem with convergence rate O $(1/k^2)$. Sov. Math. Dokl. **27**, 372–376 (1983)
5. Glowinski, R., Marrocco, A.: Inf. Rech. Oper. **R-2**, 41–76 (1975)
6. Beck, A., Teboulle, M.: A fast iterative shrinkage-thresholding algorithm for linear inverse problems. SIAM J. Imaging Sci. **2**, 183–202 (2009)
7. Goldstein, T., O'Donoghue, B., Setzer, S.: Fast alternating direction optimization methods. SIAM J. Imaging Sci. **7**(3), 1588–1623 (2014)
8. Wang, X., Hong, M., Ma, S.: Solving multiple-block separable convex minimization problems using two-block alternating direction method of multipliers. arXiv preprint arXiv: 1308.5294 (2013)
9. Candès, E., Li, X., Ma, Y., Wright, J.: Robust principal component analysis. J. ACM **58**(1), 1–37 (2011)
10. Chen, C., He, B., Ye, Y.: The direct extension of ADMM for multi-block convex minimization problems is not necessarily convergent. Math. Program. **155**(1), 57–79 (2016)
11. Chen, C., Shen, Y., You, Y.: On the convergence analysis of the alternating direction method of multipliers with three blocks. Abstr. Appl. Anal. **2013**, 183961 (2013)
12. Hong, M., Luo, Z.-Q.: On the linear convergence of the alternating direction method of multipliers, March 2013. arXiv e-prints. arXiv:1208.3922v3
13. Deng, W., Lai, M., Peng, Z., Yin, W.: Parallel multi-block ADMM with $o(1/k)$ convergence. J. Sci. Comput. (2016). Springer, New York. https://doi.org/10.1007/s10915-016-0318-2
14. Goldstein, T., Setzer, S.: High-order methods for basis pursuit, under review (available as UCLA CAM report) (2010)

Input Information in the Approximate Calculation of Two-Dimensional Integral from Highly Oscillating Functions (Irregular Case)

Oleg M. Lytvyn[1], Olesia Nechuiviter[1(✉)], Iulia Pershyna[1], and Vitaliy Mezhuyev[2]

[1] Ukrainian Engineering and Pedagogical Academy, Kharkiv, Ukraine
olesia.nechuiviter@gmail.com
[2] University Malaysia Pahang, Gambang, Malaysia

Abstract. Nowadays, methods for digital signal and image processing are widely used in scientific and technical areas. Current stage of research in astronomy, radiology, computed tomography, holography, and radar is characterized by broad use of digital technologies, algorithms, and methods. Correspondingly, an issue of development of new or improvement of known mathematical models arose, especially for new types of input information. There are the cases when input information about function is given on the set of traces of the function on planes, the set of traces of the function on lines, and the set of values of the function in the points. The paper is dedicated to the improvement of mathematical models of digital signal processing and imaging by the example of constructing formulas of approximate calculation of integrals of highly oscillating functions of two variables (irregular case). The feature of the proposed methods is using the input information about function as a set of traces of function on lines. The estimation of proposed method has been done for the Lipschitz class and class of differentiable functions. The proposed formula is based on the algorithm, which is also effective for a class of discontinuous functions.

Keywords: Highly oscillating functions · Two-dimensional functions
Integrals of highly oscillating functions of two variables

1 Introduction

The current stage of development of many technical fields (astronomy, radiology, computed tomography, holography, radar etc.) is characterized by the rapid introduction of the new digital technologies, algorithms, and methods. Scientists have an issue of building novel or improving known mathematical models. Today, the problem should be solved in cases when the input information about function is a set of traces of function on planes, a set of traces of function on lines, or a set of values of the function in the points. Function recovery algorithms [1, 2] are very effective in this case. It has already proven to be able to provide new results in digital signal and image processing

© Springer Nature Switzerland AG 2019
O. Chertov et al. (Eds.): ICDSIAI 2018, AISC 836, pp. 365–373, 2019.
https://doi.org/10.1007/978-3-319-97885-7_36

[3], in space [4], computer and seismic imaging [5], and non-destructive testing. This paper demonstrates a method, which helps to create a theory of numerical integration of highly oscillating functions of many variables with using new information operators.

Oscillatory integrals have various applications, but the problem is that it is difficult to compute them by standard methods. There is a number of papers, where problems of highly oscillating functions are discussed for the regular case. Most dated paper was published by Filon in 1928 [6]. Let us also mention works [7–10]. A good review and analysis of these methods was given in [11, 12]. In [13], it was shown how the traditional Fourier method can be modified to give numerical methods of high order for calculating derivatives and integrals. Eckhoff's method for univariate and bivariate functions is described in detail in [14].

One and two-dimensional methods for computing integrals of highly oscillating functions in irregular case are also discussed in various papers. A collocation procedure for efficient integration of rapidly oscillatory functions is presented in [15]. This method is also extended to the two-dimensional integration, and numerical results show the efficiency for rapid irregular oscillations. In [16, 17], the methods for computation of highly oscillatory integrals (for one and two-dimensional irregular case) are explored. The outcomes are two families of methods, one is based on a truncation of the asymptotic series and the other one uses the idea of Filon. These papers come with numerical results that demonstrate the power of the proposed methods.

The problem of computing rapidly oscillating integrals of differentiable functions using various information operators is considered in [18]. In [3], we derived formulas of the evaluating of two dimensions of Fourier coefficients using splines. These formulas were constructed in two cases: input information about function is a set of traces of function on lines and a set of values of the function in the points. Main advantages of these methods are high precision of approximation and less amount of information about function.

Formulas of the calculation of 3D Fourier coefficients are presented in [19] by using new method on the class of differentiable functions. Information about a function is specified by the traces on the system of mutually perpendicular planes. In [20], the authors consider the formula for the approximate calculation of triple integrals of rapidly oscillating functions by using method with the optimal choice of the nodal planes for the approximation of the non-oscillating set. The error of the formula is estimated on the class of differentiable functions defined on a unit cube.

The aim of this paper is to show how we can effectively calculate two-dimensional integrals from highly oscillating functions in a more general case when the information about functions is a set of lines.

2   Formula for Calculating Two-Dimensional Integral of Irregular Highly Oscillating Functions on Various Classes

This paper considers $C^2_{2,L,\tilde{L}}$, i.e. the class of functions, which are defined in the domain $G = [0, 1]^2$, and for which the following holds:

$$\left|f(x_1,y) - f(x_2,y)\right| \leq L|x_1 - x_2|, \quad \left|f(x,y_1) - f(x,y_2)\right| \leq L|y_1 - y_2|,$$
$$\left|f(x_1,y_1) - f(x_2,y_1) - f(x_1,y_2) + f(x_2,y_2)\right| \leq \tilde{L}|x_1 - x_2||y_1 - y_2|,$$
$$\left|f(x,y)\right| \leq \tilde{M}.$$

Definition 2.1. By the traces of function $f(x,y)$ on the lines $x_k = k\Delta_1 - \Delta_1/2$, $y_j = j\Delta_1 - \Delta_1/2$, $k, j = \overline{1, \ell_1}$, $\Delta_1 = 1/\ell_1$, we understand a function of one variable $f(x_k, y)$, $0 \leq y \leq 1$ or $f(x, y_j)$, $0 \leq x \leq 1$.

Definition 2.2. By the traces of function $g(x,y)$ on the lines $x_p = p\Delta_2 - \Delta_2/2$, $y_s = s\Delta_2 - \Delta_2/2$, $p, s = \overline{1, \ell_2}$, $\Delta_2 = 1/\ell_2$ we understand a function of one variable $g(x_p, y)$, $0 \leq y \leq 1$ or $g(x, y_s)$, $0 \leq x \leq 1$.

A two-dimensional integral from highly oscillating functions of general view is defined as

$$I^2(\omega) = \int\limits_0^1 \int\limits_0^1 f(x,y) e^{i\omega g(x,y)} dx dy$$

for $f(x,y)$, $g(x,y) \in C^2_{2,L,\tilde{L}}$.

Let

$$h1_{0k}(x) = \begin{cases} 1, x \in X1_k, \\ 0, x \notin X1_k, \end{cases} \quad k = \overline{1, \ell_1}, \quad H1_{0j}(y) = \begin{cases} 1, y \in Y1_j, \\ 0, y \notin Y1_j, \end{cases} \quad j = \overline{1, \ell_1},$$
$$X1_k = \left[x_{k-1/2}, x_{k+1/2}\right], \quad Y1_j = \left[y_{j-1/2}, y_{j+1/2}\right],$$
$$x_k = k\Delta_1 - \Delta_1/2, \quad y_j = j\Delta_1 - \Delta_1/2, \quad k, j = \overline{1, \ell_1}, \quad \Delta_1 = 1/\ell_1,$$
$$h2_{0p}(x) = \begin{cases} 1, x \in X2_p, \\ 0, x \notin X2_p, \end{cases} \quad p = \overline{1, \ell_1}, \quad H2_{0j}(y) = \begin{cases} 1, y \in Y2_s, \\ 0, y \notin Y2_s, \end{cases} \quad s = \overline{1, \ell_1},$$
$$X1_p = \left[x_{p-1/2}, x_{p+1/2}\right], \quad Y1_s = \left[y_{s-1/2}, y_{s+1/2}\right],$$
$$x_p = p\Delta_2 - \Delta_2/2, \quad y_s = s\Delta_2 - \Delta_2/2, \quad p, s = \overline{1, \ell_2}, \quad \Delta_2 = 1/\ell_2.$$

Let us define two operators. The first one is as follows:

$$J_{\ell_1}(x,y) = \sum_{k=1}^{\ell_1} f(x_k, y) h1_{0k}(x) + \sum_{j=1}^{\ell_1} f(x, y_j) H1_{0j}(y)$$
$$- \sum_{k=1}^{\ell_1} \sum_{j=1}^{\ell_1} f(x_k, y_j) h1_{0k}(x) H1_{0j}(y),$$

and the second one is as follows:

$$O_{\ell_2}(x,y) = \sum_{p=1}^{\ell_2} g(x_p, y) h2_{0p}(x) + \sum_{s=1}^{\ell_2} g(x, y_s) H2_{0s}(y)$$

$$- \sum_{p=1}^{\ell_2} \sum_{s=1}^{\ell_2} g(x_p, y_s) h2_{0p}(x) H2_{0s}(y).$$

The following formula

$$\Phi^2(\omega) = \int_0^1 \int_0^1 J_{\ell_1}(x,y) e^{i\omega O_{\ell_2}(x,y)} dxdy$$

is proposed for numerical calculation of

$$I^2(\omega) = \int_0^1 \int_0^1 f(x,y) e^{i\omega g(x,y)} dxdy.$$

Theorem 2.1. Suppose that $f(x,y),\ g(x,y) \in C^2_{2,L,\tilde{L}}$. Let functions $f(x,y)$ and $g(x,y)$ be defined by $N = 2\ell_1 + 2\ell_2$ traces $f(x_k, y), k = \overline{1, \ell_1}$ $f(x, y_j), j = \overline{1, \ell_1}$ and $g(x_p, y), p = \overline{1, \ell_2}, g(x, y_s), s = \overline{1, \ell_2}$ on the systems of perpendicular lines in domain $G = [0,1]^2$. It is true that

$$\rho\big(I^2(\omega), \Phi^2(\omega)\big)$$
$$= \left| \int_0^1 \int_0^1 f(x,y) e^{i\omega g(x,y)} dxdy - \int_0^1 \int_0^1 J_{\ell_1}(x,y) e^{i\omega O_{\ell_2}(x,y)} dxdy \right|$$
$$\leq \tfrac{\tilde{L}}{16} \tfrac{1}{\ell_1^2} + \tilde{M} \min\left(2; \tfrac{\tilde{L}\omega}{16} \tfrac{1}{\ell_2^2}\right).$$

Proof. The integral $I^2(\omega)$ can be written in another form

$$I^2(\omega) = \int_0^1 \int_0^1 f(x,y) e^{i\omega g(x,y)} dxdy = \int_0^1 \int_0^1 J_{\ell_1}(x,y) e^{i\omega O_{\ell_2}(x,y)} dxdy$$
$$+ \int_0^1 \int_0^1 [f(x,y) - J_{\ell_1}(x,y)] e^{i\omega O_{\ell_2}(x,y)} dxdy + \int_0^1 \int_0^1 f(x,y) \left[e^{i\omega g(x,y)} - e^{i\omega O_{\ell_2}(x,y)} \right] dxdy.$$

Hence, it is sufficient to show that

$$\rho\big(I^2(\omega),\Phi^2(\omega)\big)$$

$$= \left| \int_0^1 \int_0^1 f(x,y)e^{i\omega g(x,y)}\,dxdy - \int_0^1 \int_0^1 J_{\ell_1}(x,y)e^{i\omega O_{\ell_2}(x,y)}\,dxdy \right|$$

$$\le \int_0^1 \int_0^1 |f(x,y) - J_{\ell_1}(x,y)|\,dxdy + \int_0^1 \int_0^1 |f(x,y)|\,\big|e^{i\omega g(x,y)} - e^{i\omega O_{\ell_2}(x,y)}\big|\,dxdy.$$

We use the fact that

$$e^{i\omega g(x,y)} - e^{i\omega O_{\ell_2}(x,y)} = \cos(\omega g(x,y)) + i\sin(\omega g(x,y)) - \cos(\omega O_{\ell_2}(x,y)) - i\sin(\omega O_{\ell_2}(x,y))$$

$$= -2\sin\frac{\omega g(x,y) + \omega O_{\ell_2}(x,y)}{2}\sin\frac{\omega g(x,y) - \omega O_{\ell_2}(x,y)}{2}$$

$$+ 2i\sin\frac{\omega g(x,y) - \omega O_{\ell_2}(x,y)}{2}\cos\frac{\omega g(x,y) + \omega O_{\ell_2}(x,y)}{2}$$

$$= 2i\sin\frac{\omega g(x,y) - \omega O_{\ell_2}(x,y)}{2}\left[i\sin\frac{\omega g(x,y) + \omega O_{\ell_2}(x,y)}{2} + \cos\frac{\omega g(x,y) + \omega O_{\ell_2}(x,y)}{2} \right]$$

$$= 2i\sin\frac{\omega g(x,y) - \omega O_{\ell_2}(x,y)}{2}\, e^{\frac{i\omega}{2}\left(g(x,y) + O_{\ell_2}(x,y)\right)}.$$

Thus,

$$\rho\big(I^2(\omega),\Phi^2(\omega)\big) \le \int_0^1 \int_0^1 |f(x,y) - J_{\ell_1}(x,y)|\,dxdy$$

$$+ \int_0^1 \int_0^1 |f(x,y)|\left| 2i\sin\frac{\omega g(x,y) - \omega O_{\ell_2}(x,y)}{2}\, e^{\frac{i\omega}{2}\left(g(x,y) + O_{\ell_2}(x,y)\right)} \right|dxdy$$

$$\le \sum_{k=1}^{\ell_1}\sum_{j=1}^{\ell_1} \int_{x_{k-\frac{1}{2}}}^{x_{k+\frac{1}{2}}}\int_{y_{j-\frac{1}{2}}}^{y_{j+\frac{1}{2}}} \widetilde{L}|x-x_k||y-y_j|\,dxdy$$

$$+ 2\widetilde{M}\sum_{p=1}^{\ell_2}\sum_{s=1}^{\ell_2} \int_{x_{p-\frac{1}{2}}}^{x_{p+\frac{1}{2}}}\int_{y_{s-\frac{1}{2}}}^{y_{s+\frac{1}{2}}} \left| \sin\frac{\omega(g(x,y) - O_{\ell_2}(x,y))}{2} \right|dxdy$$

$$\le \widetilde{L}\sum_{k=1}^{\ell_1}\sum_{j=1}^{\ell_1} \int_{x_{k-\frac{1}{2}}}^{x_{k+\frac{1}{2}}} |x-x_k|\,dx \int_{y_{j-\frac{1}{2}}}^{y_{j+\frac{1}{2}}} |y-y_j|\,dy$$

$$+ 2\widetilde{M}\sum_{p=1}^{\ell_2}\sum_{s=1}^{\ell_2} \int_{x_{p-\frac{1}{2}}}^{x_{p+\frac{1}{2}}}\int_{y_{s-\frac{1}{2}}}^{y_{s+\frac{1}{2}}} \min\left(1; \frac{\omega|g(x,y) - O_{\ell_2}(x,y)|}{2} \right)dxdy$$

$$\le \widetilde{L}\sum_{k=1}^{\ell_1}\sum_{j=1}^{\ell_1} \int_{x_{k-\frac{1}{2}}}^{x_{k+\frac{1}{2}}} |x-x_k|\,dx \int_{y_{j-\frac{1}{2}}}^{y_{j+\frac{1}{2}}} |y-y_j|\,dy$$

$$+2\widetilde{M}\sum_{p=1}^{\ell_2}\sum_{s=1}^{\ell_2}\int_{x_{p-\frac{1}{2}}}^{x_{p+\frac{1}{2}}}\int_{y_{s-\frac{1}{2}}}^{y_{s+\frac{1}{2}}}\min\left(1;\tfrac{\omega}{2}\widetilde{L}|x-x_p||y-y_s|\right)dxdy$$

$$\leq\widetilde{L}\ell_1^2\tfrac{\Delta_1^2}{4}\tfrac{\Delta_1^2}{4}$$

$$+2\widetilde{M}\min\left(\sum_{p=1}^{\ell_2}\sum_{s=1}^{\ell_2}\int_{x_{p-\frac{1}{2}}}^{x_{p+\frac{1}{2}}}\int_{y_{s-\frac{1}{2}}}^{y_{s+\frac{1}{2}}}dxdy,\tfrac{\widetilde{L}\omega}{2}\sum_{p=1}^{\ell_2}\sum_{s=1}^{\ell_2}\int_{x_{p-\frac{1}{2}}}^{x_{p+\frac{1}{2}}}\int_{y_{s-\frac{1}{2}}}^{y_{s+\frac{1}{2}}}|x-x_p||y-y_s|dxdy\right)$$

$$=\tfrac{\widetilde{L}}{16}\Delta_1^2+2\widetilde{M}\min\left(\ell_2^2\Delta_2^2,\tfrac{\widetilde{L}\omega}{2}\ell_2^2\tfrac{\Delta_2^2}{4}\tfrac{\Delta_2^2}{4}\right)$$

$$=\tfrac{\widetilde{L}}{16}\Delta_1^2+\widetilde{M}\min\left(2;\tfrac{\widetilde{L}\omega}{16}\Delta_2^2\right)=\tfrac{\widetilde{L}}{16}\tfrac{1}{\ell_1^2}+\widetilde{M}\min\left(2;\tfrac{\widetilde{L}\omega}{16}\tfrac{1}{\ell_2^2}\right).$$

Let $H^{2,r}\left(M,\widetilde{M}\right)$, $r\geq0$ be the class of functions, which are defined in the domain $G=[0,1]^2$, and

$$\left|f^{(r,0)}(x,y)\right|\leq M,\left|f^{(0,r)}(x,y)\right|\leq M,r\neq0,\left|f^{(r,r)}(x,y)\right|\leq\widetilde{M},\quad r\geq0.$$

Theorem 2.2. Suppose that $f(x,y)$, $g(x,y)\in H^{2,1}\left(M,\widetilde{M}\right)$. Let functions $f(x,y)$ and $g(x,y)$ be defined by $N=2\ell_1+2\ell_2$ traces $f(x_k,y),k=\overline{1,\ell_1}$, $f(x,y_j),j=\overline{1,\ell_1}$ and $g(x_p,y)$, $p=\overline{1,\ell_2}$, $g(x,y_s)$, $s=\overline{1,\ell_2}$ on the systems of perpendicular lines in domain $G=[0,1]^2$. It is true that

$$\rho\left(l^2(\omega),\Phi^2(\omega)\right)=\left|\int_0^1\int_0^1 f(x,y)e^{i\omega g(x,y)}dxdy-\int_0^1\int_0^1 J_{\ell_1}(x,y)e^{i\omega O_{\ell_2}(x,y)}dxdy\right|$$

$$\leq\frac{\widetilde{M}}{16}\frac{1}{\ell_1^2}+M\min\left(2;\frac{\widetilde{M}\omega}{16}\frac{1}{\ell_2^2}\right).$$

Proof. It is important to note that

$$f(x,y)-J_{\ell_1}(x,y)=\int_{x_k}^x\int_{y_j}^y f^{(1,1)}(\xi,\eta)d\xi d\eta,$$

and

$$g(x,y)-O_{\ell_2}(x,y)=\int_{x_s}^x\int_{y_p}^y g^{(1,1)}(\xi,\eta)d\xi d\eta.$$

Thus,

$$\rho\big(I^2(\omega), \Phi^2(\omega)\big) \le \int_0^1 \int_0^1 |f(x,y) - J_{\ell_1}(x,y)| dxdy$$

$$+ \int_0^1 \int_0^1 |f(x,y)| \left| 2i \sin\frac{\omega g(x,y) - \omega O_{\ell_2}(x,y)}{2} e^{\frac{i\omega}{2}\big(g(x,y) + O_{\ell_2}(x,y)\big)} \right| dxdy$$

$$\le \sum_{k=1}^{\ell_1} \sum_{j=1}^{\ell_1} \int_{x_{k-\frac{1}{2}}}^{x_{k+\frac{1}{2}}} \int_{y_{j-\frac{1}{2}}}^{y_{j+\frac{1}{2}}} \left| \int_{x_k}^x \int_{y_j}^y f^{(1,1)}(\xi,\eta) d\xi d\eta \right| dxdy$$

$$+ 2\tilde{M} \sum_{p=1}^{\ell_2} \sum_{s=1}^{\ell_2} \int_{x_{p-\frac{1}{2}}}^{x_{p+\frac{1}{2}}} \int_{y_{s-\frac{1}{2}}}^{y_{s+\frac{1}{2}}} \min\left(1; \frac{\omega}{2} \left| \int_{x_p}^x \int_{y_s}^y g^{(1,1)}(\xi,\eta) d\xi d\eta \right| \right) dxdy$$

$$\le \frac{\tilde{M}}{16}\Delta_1^2 + 2\tilde{M}\min\left(\ell_2^2\Delta_2^2, \frac{\tilde{M}\omega}{2}\ell_2^2 \frac{\Delta_2^2}{4}\frac{\Delta_2^2}{4}\right)$$

$$= \frac{\tilde{M}}{16}\Delta_1^2 + \tilde{M}\min\left(2; \frac{\tilde{M}\omega}{16}\Delta_2^2\right) = \frac{\tilde{M}}{16}\frac{1}{\ell_1^2} + \tilde{M}\min\left(2; \frac{\tilde{M}\omega}{16}\frac{1}{\ell_2^2}\right).$$

3 Numerical Results

Let us show validity of Theorem 2.2 on the example. We want to calculate

$$I_s^2(\omega) = \int_0^1 \int_0^1 f(x,y)\sin(\omega g(x,y))dxdy$$

by formula

$$\Phi_s^2(\omega) = \int_0^1 \int_0^1 J_{\ell_1}(x,y)\sin(\omega O_{\ell_2}(x,y))dxdy,$$

$$J_{\ell_1}(x,y) = \sum_{k=1}^{\ell_1} f(x_k,y)h1_{0k}(x) + \sum_{j=1}^{\ell_1} f(x,y_j)H1_{0j}(y)$$

$$- \sum_{k=1}^{\ell_1} \sum_{j=1}^{\ell_1} f(x_k,y_j)h1_{0k}(x)H1_{0j}(y),$$

$$O_{\ell_2}(x,y) = \sum_{p=1}^{\ell_2} g(x_p,y)h2_{0p}(x) + \sum_{s=1}^{\ell_2} g(x,y_s)H2_{0s}(y)$$

$$- \sum_{p=1}^{\ell_2} \sum_{s=1}^{\ell_2} g(x_p,y_s)h2_{0p}(x)H2_{0s}(y),$$

when $f(x,y) = \sin(x+y)$, $g(x,y) = \cos(x+y)$, $\omega = 2\pi$, and $\omega = 5\pi$ in MathCad 15.0. For $\omega = 2\pi$, $\omega = 5\pi$, we have exact values:

$$I_s^2(2\pi) = 0.062699216073162,$$
$$I_s^2(5\pi) = 0.022780463640219.$$

Let us denote the error of calculation by

$$\varepsilon_{ex} = \left| I_s^2(\omega) - \Phi_s^2(\omega) \right|.$$

It is evident that $\varepsilon_{ex} = \varepsilon_{ex}(\ell_1, \ell_2)$. We want to show that $\varepsilon_{ex} \leq \varepsilon_{th}$,

$$\varepsilon_{th} = \frac{\widetilde{M}}{16} \frac{1}{\ell_1^2} + \widetilde{M} \min\left(2; \frac{\widetilde{M}\omega}{16} \frac{1}{\ell_2^2}\right),$$

for various ℓ_1, ℓ_2. In our case, for $f(x,y) = \sin(x+y)$, $g(x,y) = \cos(x+y)$, we have $M = \widetilde{M} = 1$, and

$$\varepsilon_{th} = \frac{1}{16\ell_1^2} + \min\left(2; \frac{\omega}{16\ell_2^2}\right).$$

In Table 1, we present the results of calculation of $I_s^2(\omega)$ by formula $\Phi_s^2(\omega)$ for $\omega = 2\pi$, $\omega = 5\pi$, and the values of ε_{th} and ε_{ex} when ℓ_1, ℓ_2 are changed. Numerical results show that $\varepsilon_{ex} \leq \varepsilon_{th}$.

Table 1. Calculation of $I_s^2(\omega)$ by formula $\Phi_s^2(\omega)$.

ω	ℓ_1	ℓ_2	$\Phi_s^2(\omega)$	ε_{ex}	ε_{th}
2π	4	4	0.062432583948326	$2.6 \cdot 10^{-4}$	$2.8 \cdot 10^{-2}$
2π	7	7	0.062683978467995	$1.5 \cdot 10^{-5}$	$9.2 \cdot 10^{-3}$
5π	6	6	0.022786668787906	$6.2 \cdot 10^{-6}$	$2.9 \cdot 10^{-2}$
5π	10	4	0.022808425368659	$2.7 \cdot 10^{-5}$	$6.1 \cdot 10^{-2}$
5π	10	10	0.02277048162594	$9.9 \cdot 10^{-6}$	$1.04 \cdot 10^{-2}$

4   Conclusions

This paper is devoted to the development of a formula for approximate calculation of two-dimensional irregular highly oscillatory integrals. The feature of the proposed formula is using sets of traces of the functions on lines as the input information about functions. Estimation of numerical integration has been obtained on the Lipschitz class and class of differentiable functions. A numerical experiment demonstrates the high accuracy of the formula, and also confirms the theoretical results.

References

1. Lytvyn, O.M.: Interlineation of function and its application, 544 p. Osnova, Kharkiv (2002). (in Ukrainian)
2. Mezhuyev, V., Lytvyn, O.M., Zain, J.M.: Metamodel for mathematical modelling surfaces of celestial bodies on the base of radiolocation data. Indian J. Sci. Technol. **8**(13), 70630 (2015)
3. Lytvyn, O.M., Nechuyviter, O.P.: Methods in the multivariate digital signal processing with using spline-interlineation. In: Proceedings of the IASTED International Conferences on Automation, Control, and Information Technology (ASIT 2010), Novosibirsk, 15–18 June 2010, pp. 90–96 (2010)
4. Lytvyn, O.N., Matveeva, S.Yu.: Aerospace pictures processing by means of interstripation of functions of two variables. J. Autom. Inf. Sci. **45**(3), 53–67 (2013)
5. Sergienko, I.V., Dejneka, V.S., Lytvyn, O.N., Lytvyn, O.O.: The method of interlineation of vector functions $\bar{w}$ (x, y, z, t) on a system of vertical straight lines and its application in crosshole seismic tomograph. Cybern. Syst. Anal. **49**(3), 379–389 (2013)
6. Filon, L.N.G.: On a quadrature formula for trigonometric integrals. Proc. R. Soc. Edinb. **49**, 38–47 (1928)
7. Luke, Y.L.: On the computation of oscillatory integrals. Proc. Cambridge Phil. Soc. **50**, 269–277 (1954)
8. Flinn, E.A.: A modification of Filon's method of numerical integration. JACM **7**, 181–184 (1960)
9. Zamfirescu, I.: An extension of Gauss's method for the calculation of improper integrals. Acad. R.P Romune Stud. Cerc. Mat. **14**, 615–631 (1963). (in Romanian)
10. Bakhvalov, N.S., Vasileva, L.G.: Comput. Math. Phys. **8**, 241 (1968)
11. Iserles, A.: On the numerical quadrature of highly-oscillating integrals I: fourier transforms. IMA J. Numer. Anal. **24**, 365–391 (2004)
12. Milovanovic, G.V., Stanic M.P.: Numerical integration of highly oscillating functions. In: Analytic Number Theory, Approximation Theory, and Special Functions, pp. 613–649 (2014)
13. Eckhoff, K.S.: Accurate reconstructions of functions of finite regularity from truncated Fourier series expansions. Math. Comput. **64**(210), 671–690 (1995)
14. Adcock, B.: Convergence acceleration of Fourier-like series in one or more dimensions. Cambridge Numerical Analysis Reports (NA2008/11)/DAMPT. University of Cambridge, 30 p
15. Levin, D.: Procedures for computing one and two-dimensional integrals of functions with rapid irregular oscillations. Math. Comput. **38**(158), 531–538 (1982)
16. Iserles, A.: On the numerical quadrature of highly oscillating integrals II: irregular oscillators. IMA J. Numer. Anal. **25**, 25–44 (2005)
17. Iserles, A., Norsett, S.P.: Efficient quadrature of highly oscillatory integrals using derivatives. Proc. R. Soc. Lond. Ser. A Math. Phys. Eng. Sci. **461**, 1383–1399 (2005)
18. Sergienko, I.V., Zadiraka, V.K., Melnikova, S.S., Lytvyn, O.M., Nechuiviter, O.P.: Optimal algorithms of calculation of highly oscillatory integrals and their applications, p. 447. Monography, T.1. Algorithms, Kiev (2011) (Ukrainian)
19. Lytvyn, O.N., Nechuiviter, O.P.: 3D fourier coefficients on the class of differentiable functions and spline interflatation. J. Autom. Inf. Sci. **44**(3), 45–56 (2012)
20. Lytvyn, O.N., Nechuiviter, O.P.: Approximate calculation of triple integrals of rapidly oscillating functions with the use of lagrange polynomial interflatation. Cybern. Syst. Anal. **50**(3), 410–418 (2014)

A New Approach to Data Compression

Oleh M. Lytvyn[⊠] and Oleh O. Lytvyn

Ukrainian Engineering-Pedagogical Academy, Kharkiv, Ukraine
academ_mail@ukr.net

Abstract. The idea of the method of data processing in this paper is to replace the data with operators that use functions (traces of unknown functions on the specified geometric objects such as points, lines, surfaces, stripes, tubes, or layers). This approach allows: first, to carry out data processing with parallelization of calculations; second, if the data are time dependent, to construct forecast operators at the functional level (extrapolation operators); third, to compress information. The paper will analyze the methods of constructing interlineation operators of functions of two or more variables, interflatation of functions of three or more variables, interpolation of functions of two or more variables, intertubation of functions of three or more variables, interlayeration of functions of three or more variables, interlocation of functions of two or more variables. Then, we will compare them with the interpolation operators of functions of corresponding number of variables.

The possibility of using known additional information about the investigated object as well as examples of objects or processes that allow to test the specified method of data processing in practice. Examples are given of constructing interpolation operators of the function of many variables using interlineation and interflatation, which require less data about the approximate function than operators of classical spline-interpolation. In this case, the order of accuracy of the approximation is preserved.

Keywords: Information · Data compression · Interpolation · Interlineation
Interflatation · Interstripation · Intertubation · Interlayeration · Interlocation

1 Introduction

The main idea of mathematical modeling is to replace the phenomenon, process, or object with their description with the help of mathematical symbols. The most well-known mathematical models are based on the use of fundamental laws of nature, such as the law of universal attraction, laws of mechanics, the law of interaction of electric charges, Maxwell's differential equations describing the propagation of electromagnetic waves, etc. Variational principles play an important role in the formation of mathematical models. In all methods that lead to creation of mathematical models in one form or another, basic geometric concepts such as points, lines, and surfaces are used. Classical methods of constructing mathematical models of a phenomenon, process, or object use information about the object under study in the form of functions located using data in separate points at different times. But the last century is marked by a significant expansion of information operators used in the construction of

© Springer Nature Switzerland AG 2019
O. Chertov et al. (Eds.): ICDSIAI 2018, AIS 836, pp. 374–381, 2019.
https://doi.org/10.1007/978-3-319-97885-7_37

mathematical models. The work of Johann Radon [1], in which the author proves that each function of many variables can be uniquely represented using integrals along a system of lines or planes, started with this. This statement has found its real application in modern computer tomographs, both medical and technical.

2   Non-traditional Types of Geometric Objects for Storing Information About a Function

2.1   Classical and Non-classical Ways for Storing Information

Classical ways for storing information about a function include:

1. Values of a desired function of several variables in a given point system. Between these points, one needs to approximately restore an unknown function. This case of data processing is called *interpolation*.
2. Information about an unknown function is given by traces on the system of lines. This case is called the *interlineation* of a function of two or more variables. An unknown function must be approximately restored between these lines.
3. Information on the functions of three or more variables is given on the system of surfaces. This case is called *interflatation*. An unknown function must be approximately restored between these surfaces.
4. Information about functions of two or more variables is given by tracks on the system of bands (stripes). This case is called the *interstripation* of functions of two or more variables. An unknown function must be approximately restored between these bands (stripes).
5. Information about an unknown function of three or more variables is given by traces in the pipes (tubes) system. This case is called *intertubation* of functions of three or more variables. An unknown function must be approximately restored between these pipes (tubes).
6. Information about an unknown function of three or more variables is given on the system of layers. This case is called *interlayeration* of functions of three or more variables. An unknown function must be approximately restored between these layers.
7. Information about an unknown function is given on a system of disjoint bounded domains, i.e. generalizations of points. This case is called *interlocation* of functions of two or more variables. It is necessary to restore an approximately unknown function between these regions called loci.

For convenience, we reduce the types of approximation operators appearing in this information in Tables 1 and 2 [2].

In this paper, authors draw attention to the fact that when switching from operators using information about approximating a function on the lines, the surfaces, or the point information, there are new possibilities for compressing the amount of data with the appropriate choice of points of the task.

Table 1. Type of information about function $f(x, y)$.

Type of geometric object	Method of assignment	Information about the provided function	Approximation operator
Point	$(x - a)^2 + (y - b)^2 = 0$	$f^{(s,p)}(a, b)$, $0 \leq s + p \leq n$ – function and its derivatives at point (a, b)	Taylor polynomial
System of points (x_k, y_k), $k = \overline{1, M}$	$(x - x_k)^2 + (y - y_k)^2 = 0$	$f^{(p,s)}(x_k, y_k)$, $k = \overline{1, M}$, $0 \leq s + p \leq n$ – function and its derivatives at point (x_k, y_k)	Lagrange interpolation for $n = 0$ and Hermit interpolation for $n \geq 1$
Line	$y = y(x)$ explicit line assignment	$\frac{\partial^s f}{\partial y^s}(x, y(x))$, $s = \overline{0, n}$ – function and its derivatives traces on the specified line	A generalized Taylor formula in the vicinity of a line
	$w(x, y) = 0$ implicit line assignment	$\frac{\partial^s f}{\partial \nu^s}\big|_{w(x,y)=0}$, $s = \overline{0, n}$ – function and its normal derivatives traces on the specified line	
	$x = u(t)$, $y = v(t)$ parametric line assignment	$\frac{\partial^s f}{\partial \nu^s}(x(t), y(t))$, $s = \overline{0, n}$ – function and its normal derivatives traces on the specified line	
System of lines	$x = x_k(y)$ or $y = y_k(x)$, $k = \overline{1, M}$ explicit system of lines assignment	$f^{(0,p)}(x, y_k(x))$, $p = \overline{0, n} f^{(s,0)}$ $(x_k(y), y)$ $s = \overline{0, n}$, $k = \overline{1, M}$ – function and its derivatives or other differential operators traces on the specified system of lines	Lagrange interlineation for $n = 0$ and Hermit interlineation for $n \geq 1$
	$w_k(x, y) = 0$, $k = \overline{1, M}$ implicit system of line assignment	$\frac{\partial^s f}{\partial \nu_k^s}\big|_{w_k(x,y)=0}$, $k = \overline{1, M}$, $s = \overline{0, n}$ – function's traces on the specified system of lines	
	$x = u_k(t)$, $y = v_k(t)$, $k = \overline{1, M}$, parametric system of line assignment	$\frac{\partial^s f}{\partial \nu_k^s}(u_k(t), v_k(t))$, $s = \overline{0, n}$, $f(x_k(t), y_k(t))$, $k = \overline{1, M}$ – function's and its normal derivative's traces on the specified system of lines	
System of loci	$w_k^2(x, y) \leq R_k^2$, $k = \overline{1, M}$	$f^{(p,s)}(x, y)\big|_{w_k^2(x,y) \leq R_k^2}$, $0 \leq s + p \leq n$ – function and its derivatives on the sets	Lagrange interlocation for $n = 0$ and Hermit interlocation for $n \geq 1$ on the system of loci. If $R_k = 0$, we obtain interpolation
System of stripes	$\Pi_k : \alpha_k \leq w_k(x, y) \leq \beta_k$, $k = \overline{1, M}$	$f^{(p,s)}(x, y)\big|_{\Pi_k}$, $k = \overline{1, M}$, $0 \leq s + p \leq n$ – function's and its derivative's traces on the system of stripes	Lagrange interstripation for $n = 0$ and Hermit interstripation for $n \geq 1$

Table 2. Type of information about function $f(x, y, z)$

Type of geometric object	Method of assignment	Information about the provided function	Approximation operator $f(x, y)$
Point (a, b, c)	$(x - a)^2 + (y - b)^2$ $+ (z - c)^2 = 0$	$f^{(s,p)}(a, b, c)$, $0 \leq s + p \leq n$ – function and its derivatives at the point (a, b, c)	Taylor polynomial
System of points $(x_k, y_k, z_k) k = \overline{1, M}$	$(x - x_k)^2 + (y - y_k)^2$ $+ (z - z_k)^2 = 0$	$f^{(p,s)}(x_k, y_k, z_k)$, $k = \overline{1, M}$, $0 \leq s + p \leq n$ – function and its derivatives at the point (x_k, y_k, z_k)	Lagrange interpolation for $n = 0$ and Hermit interpolation for $n \geq 1$
System of lines	$x = u_k(t)$, $y = v_k(t)$ $z = w_k(t)$, $k = \overline{1, M} a \leq t \leq b$ parametric system of lines assignment	$f^{(p,r,s)}(u_k(t), v_k(t), w_k(t))$, $k = \overline{1, M}$, $0 \leq p + r + s \leq n$ – function and its derivative traces on the specified system of lines	Lagrange interlineation for $n = 0$ and Hermit interlineation for $n \geq 1$
System of loci	$w_k^2(x, y, z) \leq R_k^2$, $k = \overline{1, M}$	$f^{(p,r,s)}(x, y, z)\big|_{w_k^2(x,y,z) \leq R_k^2}$, $0 \leq p + r + s \leq n$ – function and its derivatives on the sets	Lagrange interlocation for $n = 0$ and Hermit interlocation for $n > 1$ on the system of loci
System of tubes	$\Pi_k : (x - x_k)^2 +$ $+ (y - y_k)^2 \leq R_k^2(z)$, $k = \overline{1, M}$, $z \in [a_k, b_k]$	$f^{(p,r,s)}(x, y, z)\big|_{\Pi_k}$, $k = \overline{1, M}$, $0 \leq p + r + s \leq n$ – function and its derivative traces on system of tubes Π_k	Lagrange intertubeation for $n = 0$ and Hermit intertubeation for $n \geq 1$
System of layers	$L_k : u_k(x, y) \leq z$, $z \leq v_k(x, y)$, $(x, y) \in R^2$, $k = \overline{1, M}$ $u_k(x, y) < v_k(x, y) <$ $< u_{k+1}(x, y) <$ $< v_{k+1}(x, y) < \ldots <$ $< u_M(x, y) <$ $< v_M(x, y)$	$f^{(p,r,s)}(x, y, z)\big|_{L_k}$, $k = \overline{1, M}$, $0 \leq p + r + s \leq n$ – function's and its derivative's traces on system of layers L_k	Lagrange interlayeration for $n = 0$ and Hermit Interlayeration for $n \geq 1$

2.2   Examples

Example 1 [3]. Using of the interlineation of function for data compression.

Let a function $f(x, y)$ be a mathematical model of some object, and information about this function is known only on the system of lines $x = X_i$, $y = Y_j$ $0 \leq i, j \leq N$, $X_i = i/N$, $Y_j = j/N$. Then, operator

$$O_N f(x,y) = (O_{1,N} + O_{2,N} - O_{1,N} O_{2,N}) f(x,y),$$

where $O_{1,N} f(x,y) = \sum_{i=0}^{N} f(X_i, y) h(Nx - i)$, $O_{2,N} f(x,y) = \sum_{j=0}^{N} f(x, Y_j) h(Ny - j)$, $h(t) = \frac{|t+1| - 2|t| + |t-1|}{2}$, has an error $\varepsilon = O(N^{-4})$, $N \gg 1$ by approximate $f(x_i, y) \in C^{2,2}[0,1]^2$.

The error of approximation $|f - \tilde{O}_N f| \leq C_1 \Delta^4$, $C_1 = \text{const}$. If one replaces the traces $f(X_i, y)$, $f(x, Y_j)$ by interpolation formulas $f(X_i, y) \approx \sum_{\ell=0}^{N^2} f(X_i, \frac{\ell}{N^2}) h(N^2 y - \ell)$,

$f(x, Y_j) \approx \sum_{k=0}^{N^2} f(\frac{k}{N^2}, Y_j) h(N^2 x - k)$, then the result is an operator

$$\tilde{O}_N f(x,y) = (\tilde{O}_{1,N} + \tilde{O}_{2,N} - O_{1,N} O_{2,N}) f(x,y),$$

where

$$\tilde{O}_{1,N} f(x,y) = \sum_{i=0}^{N} \sum_{\ell=0}^{N^2} f(X_i, \frac{\ell}{N^2}) h(N^2 y - \ell) h(Nx - i),$$

$$\tilde{O}_{2,N} f(x,y) = \sum_{j=0}^{N} \sum_{k=0}^{N^2} f(\frac{k}{N^2}, Y_j) h(N^2 x - k) h(Ny - j).$$

The error of approximation $|f - \tilde{O}_N f| \leq C_1 \Delta^4$, $C_1 = \text{const}$. For comparison, let us note that the classical formula is spline-interpolation $O_{1N^2} O_{2N^2} f(x,y) = \sum_{k=0}^{N^2} \sum_{\ell=0}^{N^2}$

$f(\frac{k}{N^2}, \frac{\ell}{N^2}) h(N^2 x - k) h(N^2 y - \ell)$. For error $|f - O_N f|$, inequality $|f - O_N f| \leq C_2 \Delta^4$, $\Delta \gg 1$, $C_2 = \text{const}$. Operators O_N and $\tilde{O}_N$ have the same accuracy in order of approximation Δ. The operator has the same accuracy of approximation $O_{1N^2} O_{2N^2}$, but formula

$$O_{1N^2} O_{2N^2} f(x,y) = \sum_{k=0}^{N^2} \sum_{\ell=0}^{N^2} f\left(\frac{k}{N^2}, \frac{\ell}{N^2}\right) h(N^2 x - k) h(N^2 y - \ell)$$

uses $Q_1 = (N^2 + 1)^2$ values of the function being approximated. Formula $\tilde{O}_N f(x,y)$ requires only $Q_2 = 2(N+1)(N^2 + 1)$ values of the function being approximated to achieve comparative accuracy. That is, to achieve the same order of accuracy, an order of magnitude smaller amount of data about the approximate function is required.

It is shown in [3] that operators of interpolation of functions of three variables are constructed on the basis of operators of the interflatation of differentiable functions of three variables, which give an opportunity to use the necessary accuracy to use a considerably smaller number of values of the approximate function compared to classical spline-interpolation with the same accuracy of approximation.

Example 2 [3]. Use of the interflatation function for compression of information.

Let $n = 3$. We construct for a function $f(x, y, z)$ an interflatation operator, operators of classical spline-interpolation, and spline-interpolation operators constructed using spline-interflatation operators. We believe that $f(x, y, z) \in C^{2,2,2}\left([0, 1]^3\right)$. Function $h(t)$ is as given in Example 1. Construct for the given N operators

$$O_{1,N}f(x, y, z) = \sum_{i=0}^{N} f(\frac{i}{N}, y, z)h(Nx - i),$$

$$O_{2,N}f(x, y, z) = \sum_{j=0}^{N} f(x, \frac{j}{N}, z)h(Ny - j),$$

$$O_{3,N}f(x, y, z) = \sum_{k=0}^{N} f(x, y, \frac{k}{N})h(Nz - k),$$

$$O_{1,N}O_{2,N}f(x, y, z) = \sum_{i=0}^{N}\sum_{j=0}^{N} f(\frac{i}{N}, \frac{j}{N}, z)h(Nx - i)h(Ny - j),$$

$$O_{1,N}O_{3,N}f(x, y, z) = \sum_{i=0}^{N}\sum_{k=0}^{N} f(\frac{i}{N}, y, \frac{k}{N})h(Nx - i)h(Nz - k),$$

$$O_{2,N}O_{3,N}f(x, y, z) = \sum_{j=0}^{N}\sum_{k=0}^{N} f(x, \frac{j}{N}, \frac{k}{N})h(Ny - j)h(Nz - k),$$

$$O_{1,N}O_{2,N}O_{3,N}f(x, y, z) = \sum_{i=0}^{N}\sum_{j=0}^{N}\sum_{k=0}^{N} f(\frac{i}{N}, \frac{j}{N}, \frac{k}{N})h(Nx - i)h(Ny - j)h(Nz - k).$$

Then, operator $O_N f(x, y, z) = (O_{1,N} + O_{2,N} + O_{3,N} - O_{1,N}O_{2,N} - O_{1,N}O_{3,N} - O_{2,N}O_{3,N} + O_{1,N}O_{2,N}O_{3,N})f(x, y, z)$ has error $\varepsilon_N = O(N^{-6})$.

The same error is given by the classical interpolation operator on the points system $\left(\frac{i}{N}, \frac{j}{N}, \frac{k}{N}\right)$, $0 \leq i, j, k \leq N$ and the spline-interpolation operator constructed on the basis of spline-interflatation

$$\tilde{O}_N f(x, y, z) =$$
$$(\tilde{O}_{1,N} + \tilde{O}_{2,N} + \tilde{O}_{3,N} - \tilde{O}_{1,N}\tilde{O}_{2,N} - \tilde{O}_{1,N}\tilde{O}_{3,N} - \tilde{O}_{2,N}\tilde{O}_{3,N} + O_{1,N}O_{2,N}O_{3,N})f(x, y, z),$$

Where

$$\tilde{O}_{1,N}f(x,y,z) = \sum_{i=0}^{N}\left[\sum_{j'=0}^{N^{\frac{3}{2}}}\sum_{k'=0}^{N^3}f\left(\frac{i}{N},\frac{j'}{N^{\frac{3}{2}}},\frac{k'}{N^3}\right)h(N^{\frac{3}{2}}y-j')h(N^3z-k')\right.$$

$$+\sum_{k'=0}^{N^{\frac{3}{2}}}\sum_{j'=0}^{N^3}f\left(\frac{i}{N},\frac{j'}{N^3},\frac{k'}{N^{\frac{3}{2}}}\right)h(N^3y-j')h(N^{\frac{3}{2}}z-k')$$

$$\left.-\sum_{j'=0}^{N^{\frac{3}{2}}}\sum_{k'=0}^{N^{\frac{3}{2}}}f\left(\frac{i}{N},\frac{j'}{N^{\frac{3}{2}}},\frac{k'}{N^{\frac{3}{2}}}\right)h(N^{\frac{3}{2}}y-j')h(N^{\frac{3}{2}}z-k')\right]h(Nx-i),$$

$$\tilde{O}_{2,N}f(x,y,z) = \sum_{j=0}^{N}\left[\sum_{i'=0}^{N^{\frac{3}{2}}}\sum_{k'=0}^{N^3}f\left(\frac{i'}{N^{\frac{3}{2}}},\frac{j}{N},\frac{k'}{N^3}\right)h(N^{\frac{3}{2}}x-i')h(N^3z-k')\right.$$

$$+\sum_{k'=0}^{N^{\frac{3}{2}}}\sum_{i'=0}^{N^3}f\left(\frac{i'}{N^3},\frac{j}{N},\frac{k'}{N^{\frac{3}{2}}}\right)h(N^3x-i')h(N^{\frac{3}{2}}z-k')$$

$$\left.-\sum_{i'=0}^{N^{\frac{3}{2}}}\sum_{k'=0}^{N^{\frac{3}{2}}}f\left(\frac{i'}{N^{\frac{3}{2}}},\frac{j}{N},\frac{k'}{N^{\frac{3}{2}}}\right)h(N^{\frac{3}{2}}x-i')h(N^{\frac{3}{2}}z-k')\right]h(Ny-j),$$

$$\tilde{O}_{3,N}f(x,y,z) = \sum_{k=0}^{N}\left[\sum_{i'=0}^{N^{\frac{3}{2}}}\sum_{j'=0}^{N^3}f\left(\frac{i'}{N^{\frac{3}{2}}},\frac{j'}{N^3},\frac{k}{N}\right)h(N^{\frac{3}{2}}x-i')h(N^3y-j')\right.$$

$$+\sum_{j'=0}^{N^{\frac{3}{2}}}\sum_{i'=0}^{N^3}f\left(\frac{i'}{N^3},\frac{j'}{N^{\frac{3}{2}}},\frac{k}{N}\right)h(N^{\frac{3}{2}}y-j')h(N^3y-i')$$

$$\left.-\sum_{i'=0}^{N^{\frac{3}{2}}}\sum_{j'=0}^{N^{\frac{3}{2}}}f\left(\frac{i'}{N^{\frac{3}{2}}},\frac{j'}{N^{\frac{3}{2}}},\frac{k}{N}\right)h(N^{\frac{3}{2}}x-i)h(N^{\frac{3}{2}}y-j)\right]h(Nz-k),$$

$$\tilde{O}_{1,N}\tilde{O}_{2,N}f(x,y,z) = \sum_{i=0}^{N}\sum_{j=0}^{N}\left[\sum_{k'=0}^{N^3}f\left(\frac{i}{N},\frac{j}{N},\frac{k'}{N^3}\right)h(N^3z-k')\right]h(Nx-i)h(Ny-j),$$

$$\tilde{O}_{1,N}\tilde{O}_{3,N}f(x,y,z) = \sum_{i=0}^{N}\sum_{k=0}^{N}\left[\sum_{j'=0}^{N^3}f\left(\frac{i}{N},\frac{j'}{N^3},\frac{k}{N}\right)h(N^3y-j')\right]h(Nx-i)h(Nz-k),$$

$$\tilde{O}_{2,N}\tilde{O}_{3,N}f(x,y,z) = \sum_{j=0}^{N}\sum_{k=0}^{N}\left[\sum_{i'=0}^{N^3}f\left(\frac{i'}{N^3},\frac{j}{N},\frac{k}{N}\right)h(N^3x-i')\right]h(Ny-j)h(Nz-k).$$

For $n = 3$, direct calculation shows that in order to achieve the accuracy of the approximation of a function $f(x, y, z)$ with continuous derivatives of the 6th order, the operator $\tilde{O}_N f(x,y,z)$ needs to use less than $Q_1 = O(N^{5.5})$ the values of the approximate function in the nodes of the grid $\left(\frac{i}{N},\frac{j'}{N^{\frac{3}{2}}},\frac{k'}{N^3}\right)$, or $\left(\frac{i'}{N^{\frac{3}{2}}},\frac{j'}{N^3},\frac{k}{N}\right)$, or $\left(\frac{i'}{N^3},\frac{j}{N},\frac{k'}{N^{\frac{3}{2}}}\right)$. For comparison, for the approximation of function $f(x,y,z) \in C^{2,2,2}\left([0,1]^3\right)$ with error $\varepsilon = O(N^{-6})$, $Q_2 = (N+1)^9$ values of the approximate function $f(x, y, z)$ are required.

3 Conclusions

A similar statement can be deduced for $n \geq 4$. This paper proposes one of the possible methods to combat a known curse of dimensionality [4]. Other examples include information compression using Fourier coefficients [5], and when constructing mathematical models of processes, phenomena or objects that are described by discontinuous functions [6] for discontinuous processes.

References

1. Radon, J.: Uber die Bestimmung von Functionen durch ihre Integralverte 1 ngs gewisser Manningfaltigkeiten. Berichte Sachsische Academieder Wissenschaften. Leipzig, Mathem. - Phys. Kl. **69**, 262–267 (1917)
2. Sergienko, I.V., Lytvyn, O.M.: New information operators in mathematical modeling (a review). Cybern. Syst. Anal. **54**(1), 21–30 (2018)
3. Lytvyn, O.M.: Interlineation functions and some of its applications, 545 p. Osnova, Kharkiv (2002). (in Ukrainian)
4. Powell, W.B.: Approximate Dynamic Programming: Solving the Curses of Dimensionality. Wiley, Hoboken (2007). ISBN 0-470-17155-3
5. Lytvyn, O.N., Nechuiviter, O.P.: 3D Fourier coefficients on the class of differentiable functions and spline interflatation. J. Autom. Inf. Sci. **44**(3), 45–56 (2012)
6. Lytvyn, O.N., Pershina, Y.I., Sergienko, I.V.: Estimation of discontinuous functions of two variables with unknown discontinuity lines (rectangular elements). Cybern. Syst. Anal. **50**(4), 594–602 (2014)

Author Index